Advances in Sustainability Science and Technology

Series Editors

Robert J. Howlett, Bournemouth University and KES International, Shoreham-by-Sea, UK

John Littlewood, School of Art & Design, Cardiff Metropolitan University, Cardiff, UK

Lakhmi C. Jain, KES International, Shoreham-by-Sea, UK

The book series aims at bringing together valuable and novel scientific contributions that address the critical issues of renewable energy, sustainable building, sustainable manufacturing, and other sustainability science and technology topics that have an impact in this diverse and fast-changing research community in academia and industry.

The areas to be covered are

- Climate change and mitigation, atmospheric carbon reduction, global warming
- Sustainability science, sustainability technologies
- Sustainable building technologies
- Intelligent buildings
- Sustainable energy generation
- Combined heat and power and district heating systems
- Control and optimization of renewable energy systems
- Smart grids and micro grids, local energy markets
- Smart cities, smart buildings, smart districts, smart countryside
- Energy and environmental assessment in buildings and cities
- Sustainable design, innovation and services
- Sustainable manufacturing processes and technology
- Sustainable manufacturing systems and enterprises
- Decision support for sustainability
- Micro/nanomachining, microelectromechanical machines (MEMS)
- Sustainable transport, smart vehicles and smart roads
- Information technology and artificial intelligence applied to sustainability
- Big data and data analytics applied to sustainability
- Sustainable food production, sustainable horticulture and agriculture
- Sustainability of air, water and other natural resources
- Sustainability policy, shaping the future, the triple bottom line, the circular economy

High quality content is an essential feature for all book proposals accepted for the series. It is expected that editors of all accepted volumes will ensure that contributions are subjected to an appropriate level of reviewing process and adhere to KES quality principles.

The series will include monographs, edited volumes, and selected proceedings.

More information about this series at https://link.springer.com/bookseries/16477

Robert J. Howlett · Lakhmi C. Jain ·
John R. Littlewood · Marius M. Balas
Editors

Smart and Sustainable Technology for Resilient Cities and Communities

Springer

Editors
Robert J. Howlett
KES International Research
Shoreham-by-sea, UK

Lakhmi C. Jain
KES International
Selby, UK

John R. Littlewood
Cardiff Metropolitan University
Wales, UK

Marius M. Balas
Aurel Vlaicu University
Arad, Romania

ISSN 2662-6829   ISSN 2662-6837 (electronic)
Advances in Sustainability Science and Technology
ISBN 978-981-16-9103-4   ISBN 978-981-16-9101-0 (eBook)
https://doi.org/10.1007/978-981-16-9101-0

This Springer imprint is published by the registered company Springer Nature Singapore Pte Ltd.
The registered company address is: 152 Beach Road, #21-01/04 Gateway East, Singapore 189721,
Singapore

Preface

There is a great awareness of the urgent need to eliminate carbon emissions and improve the operational energy efficiency of the built environment to reduce the harmful effects on the ecosystem of human economic development and mitigate climate change reality. This has led to a huge growth in research around the science and technology of sustainable and resilient development. The series 'Advances in Sustainability Science and Technology (ASST)' was created by Springer Nature and KES International to respond to the need for a publication channel for the latest high-quality research on a broad range of sustainability topics.

The rise of the COVID-19 pandemic led researchers around the world to apply themselves to devise measures to alleviate its harmful effects upon society. This included not just clinical researchers, but also those working in areas such as engineering, computer science, and the built environment. In March 2021 KES International, a professional organisation for researchers in high-technology subjects such as artificial intelligence and sustainability, held the 'COVID-19 Challenge International Virtual Summit'. This had the theme 'A Transition to a more Resilient World' and provided the opportunity for those undertaking innovative research on measures in response to the pandemic, to present their work.

After the summit, selected authors were invited to write chapters for a book in the ASST series entitled 'Smart and Sustainable Technology for Resilient Cities and Communities'. The aim was to create a volume of research applicable to the mitigation of the COVID-19 pandemic and furthermore to look beyond it to improve the resilience of society, to make it better able to respond and withstand future disruptive challenges, including both pandemics and increasing problems due to climate change.

The outcome was this book which contains an overview and reports of 21 research investigations from mainly non-medical researchers in universities around the world.

This book is directed to engineers, scientists, researchers, practitioners, academics, and all those who are interested in developing and using sustainability science and technology for the betterment of our planet and humankind.

Thanks are due to the authors and reviewers for their expertise and time. The assistance provided by the support team at KES International and Springer Nature during the development phase of this book is gratefully acknowledged.

Shoreham-by-Sea, UK   Robert J. Howlett
Wales, UK   John R. Littlewood
Selby, UK   Lakhmi C. Jain
Arad, Romania   Marius M. Balas

Contents

1 Smart and Sustainable Technology for Resilient Cities
   and Communities—An Overview 1
   Robert J. Howlett and John R. Littlewood

Part I Changes in Work Practices and Employability in Response
        to the Covid-19 Pandemic

2 Examining Pedagogical Approaches in Developing
   Employability Skills in the Wake of the COVID-19 Pandemic 11
   John Aliu and Clinton Aigbavboa

3 Relating Work-Integrated Learning to Employability Skills
   in the Post-COVID-19 Era 29
   John Aliu and Clinton Aigbavboa

4 Opportunities and Barriers of Digitization in the COVID-19
   Crisis for SMEs 47
   Ralf-Christian Härting, Anna-Lena Rösch, Gianluca Serafino,
   Felix Häfner, and Jörg Bueechl

5 New Urban Mobility Strategies After the COVID-19 Pandemic 61
   Domenico Suraci

6 Integration of Indoor Air Quality Concerns in Educational
   Community Through Collaborative Framework of Campus
   Bizia Laboratory of the University of the Basque Country 73
   Iñigo Rodriguez-Vidal, Xabat Oregi, Jorge Otaegi,
   Gaizka Vallespir-Etxebarria, José Antonio Millán-García,
   and Alexander Martín-Garín

**Part II   Smart Techniques for Monitoring a Pandemic and
            Forecasting its Course**

7   **An Overview of Methods for Control and Estimation
    of Capacity in COVID-19 Pandemic from Point Cloud
    and Imagery Data**   91
    Jesús Balado, Lucía Díaz-Vilariño, Elena González,
    and Antonio Fernández

8   **Modeling and Evaluating the Impact of Social Restrictions
    on the Spread of COVID-19 Using Machine Learning**   107
    Mostafa Naemi, Amin Naemi, Romina Zarrabi Ekbatani,
    Ali Ebrahimi, Thomas Schmidt, and Uffe Kock Wiil

9   **Forecasting the COVID-19 Spread in Iran, Italy, and Mexico
    Using Novel Nonlinear Autoregressive Neural Network
    and ARIMA-Based Hybrid Models**   119
    Amin Naemi, Mostafa Naemi, Romina Zarrabi Ekbatani,
    Thomas Schmidt, Ali Ebrahimi, Marjan Mansourvar,
    and Uffe Kock Wiil

10 **Spatial Statistics Models for COVID-19 Data Under
    Geostatistical Methods**   137
    S. Zimeras

11 **Intelligent Multi-Sensor System for Remote Detection
    of COVID-19**   149
    G. Zaz, M. Alami Marktani, A. Elboushaki, Y. Farhane,
    A. Mechaqrane, M. Jorio, H. Bekkay, S. Bennani Dosse,
    A. Mansouri, and A. Ahaitouf

12 **A Comparative Study of Deep Learning Models for COVID-19
    Diagnosis Based on X-Ray Images**   163
    Shah Siddiqui, Elias Hossain, Rezowan Ferdous,
    Murshedul Arifeen, Wahidur Rahman, Shamsul Masum,
    Adrian Hopgood, Alice Good, and Alexander Gegov

13 **Fuzzy Cognitive Maps Applied in Determining the Contagion
    Risk Level of SARS-COV-2 Based on Validated Knowledge
    in the Scientific Community**   175
    Márcio Mendonça, Rodrigo H. C. Palácios, Ivan R. Chrun,
    Acácio Fuziy, Douglas F. da Silva, and Augusto A. Foggiato

Part III Changes in Teaching and Learning Practices in Response to a Pandemic

14 Education After COVID-19 193
Manuel Mazzara, Petr Zhdanov, Mohammad Reza Bahrami,
Hamna Aslam, Iouri Kotorov, Muwaffaq Imam, Hamza Salem,
Joseph Alexander Brown, and Ruslan Pletnev

15 Equipping European Higher Education Teachers
for Successful and Sustainable e-Learning with Home Remote
Work 209
Inés López-Baldominos, Vera Pospelova, and Luis Fernández-Sanz

16 Fully Online Project-Based Learning of Software
Development During the COVID-19 Pandemic 223
Atsuo Hazeyama, Kiichi Furukawa, and Yuki Yamada

17 A Tale of Two Zones: Pandemic ERT Evaluation 233
Enamul Haque, Tanvir Mahmud, Shahana Shultana,
Iqbal H. Sarker, and Md Nour Hossain

Part IV Adapting for Improved Resilience

18 Anticipating and Preparing for Future Change
and Uncertainty: Building Adaptive Pathways 255
Jeremy Gibberd

19 A Health-Energy Nexus Perspective for Virtual Power
Plants: Power Systems Resiliency and Pandemic Uncertainty
Challenges 267
Sambeet Mishra and Chiara Bordin

20 A Sustainable Nutritional Behavior in the Era of Climate
Changes 285
Gavrilaş Simona

21 The Development of a Smart Tunable Full-Spectrum
LED Lighting Technology Which May Prevent and Treat
COVID-19 Infections, for Society's Resilience and Quality
of Life 297
U. Thurairajah, John R. Littlewood, and G. Karani

22 Energy-Efficient Technologies for Ultra-Low Temperature
Refrigeration 309
Cosmin Mihai Udroiu, Adrián Mota-Babiloni,
Carla Espinós-Estévez, and Joaquín Navarro-Esbrí

Editors and Contributors

About the Editors

Prof. Robert J. Howlett is the Executive Chair of KES International, a non-profit organization that facilitates knowledge transfer and the dissemination of research results in areas including intelligent systems, sustainability and knowledge transfer. He is a Visiting Professor at 'Aurel Vlaicu' University of Arad, Romania, and Bournemouth University in the UK. Professor Howlett's technical expertise is in the use of intelligent systems to solve industrial problems. He has been successful in applying artificial intelligence, machine learning and related technologies to sustainability and renewable energy systems; condition monitoring, diagnostic tools and systems; and automotive electronics and engine management systems. His current research work is focused on the use of smart microgrids to achieve reduced energy costs and lower carbon emissions in areas such as housing and protected horticulture.

Prof. Lakhmi C. Jain Ph.D., M.E., B.E. (Hons), Fellow (Engineers Australia), is with Liverpool Hope University, UK and was formerly with the University of Technology Sydney, Australia. Professor Jain founded the KES International for providing a professional community the opportunities for publications, knowledge exchange, cooperation and teaming. Involving around 5,000 researchers drawn from universities and companies worldwide, KES facilitates international cooperation and generates synergy in teaching and research. KES regularly provides networking opportunities for professional community through one of the largest conferences of its kind in the area of KES.

Dr. John R. Littlewood graduated in Building Surveying, holds a Ph.D. in Building Performance Assessment of Zero Heating Housing, and is a Chartered Building Engineer. He is Head of the Sustainable and Resilient Built Environment research group in Cardiff School of Art & Design at Cardiff Metropolitan University (UK). He coordinates three Professional Doctorates in Art & Design, Engineering and Sustainable Built Environment. John's research is industry focused, investigating methods to optimise the fire safety, production, and thermal performance for existing and new dwellings during the design, manufacture, construction, operation, or during and after retrofit stages. The outcomes of John's research enhance occupant quality of life and increase the environmental sustainability and resilience of the built environment. He has authored and co-authored 155 peer-reviewed publications.

Dr. Marius M. Balas IEEE Senior Member, is a Professor at the Engineering Faculty of Aurel Vlaicu University of Arad, Romania. His research topics are in Systems Engineering, Electronic Circuits, Intelligent and Fuzzy Systems, Adaptive Control, Greenhouses, Modeling and Simulation. He is author of 16 books and book chapters, 120 indexed papers and 7 invention patents. His main contributions: the fuzzy-interpolative systems, the passive greenhouses, the intelligent rooftop greenhouses, the constant time to collision traffic optimization, the rejection of the switching controllers' instability by phase trajectory analysis and the Fermat neuron.

Contributors

A. Ahaitouf LSIGER, FST-FES, Sidi Mohammed Ben Abdellah University, Fez, Morocco

Clinton Aigbavboa cidb Centre of Excellence, University of Johannesburg, Johannesburg, South Africa

John Aliu cidb Centre of Excellence, University of Johannesburg, Johannesburg, South Africa

Murshedul Arifeen Time research and innovation (Tri), Southampton, United Kingdom;
Khilgaon Dhaka, Bangladesh

Hamna Aslam Innopolis University, Innopolis, Russia

Mohammad Reza Bahrami Innopolis University, Innopolis, Russia

Jesús Balado Universidade de Vigo, CINTECX, GeoTech Group, Vigo, Spain

H. Bekkay LESETI, ENSA-OUJDA, Mohamed Premier University, Oujda, Morocco

Chiara Bordin UiT, The Arctic University of Norway, Tromsø, Norway

Joseph Alexander Brown Innopolis University, Innopolis, Russia

Jörg Bueechl Aalen University, Aalen, BW, Germany

Ivan R. Chrun Technical-Professional Innovation and Engineering University (FEITEP), Maringá, Brazil

Douglas F. da Silva State University of Northern Paraná (UENP), Jacarezinho, Brazil

Lucía Díaz-Vilariño Universidade de Vigo, CINTECX, GeoTech Group, Vigo, Spain

S. Bennani Dosse LSIGER, ENSA-FES, Sidi Mohammed Ben Abdellah University, Fez, Morocco

Ali Ebrahimi Center for Health Informatics and Technology, The Maersk Mc-Kinney Moeller Institute, University of Southern Denmark, Odense, Denmark

Romina Zarrabi Ekbatani Swinburne University of Technology, Melbourne, Australia

A. Elboushaki LSIGER, ENSA-FES, Sidi Mohammed Ben Abdellah University, Fez, Morocco

Carla Espinós-Estévez Centro Nacional de Investigaciones Cardiovasculares Carlos III (CNIC), Madrid, Spain

Y. Farhane LSIGER, ENSA-FES, Sidi Mohammed Ben Abdellah University, Fez, Morocco

Rezowan Ferdous Time research and innovation (Tri), Southampton, United Kingdom;
Khilgaon Dhaka, Bangladesh

Luis Fernández-Sanz Universidad de Alcalá, Alcalá de Henares, Spain

Antonio Fernández Universidade de Vigo, CINTECX, GeoTech Group, Vigo, Spain

Augusto A. Foggiato Foggiato Research Institute, Jacarezinho, Brazil

Kiichi Furukawa Graduate School of Education, Tokyo Gakugei University, Koganei, Tokyo, Japan

Acácio Fuziy Foggiato Research Institute, Jacarezinho, Brazil

Alexander Gegov The University of Portsmouth (UoP), School of Computing, Faculty of Technology, Portsmouth, UK

Jeremy Gibberd Council for Scientific and Industrial Research, Pretoria, South Africa

Elena González Universidade de Vigo, CINTECX, GeoTech Group, Vigo, Spain

Alice Good The University of Portsmouth (UoP), School of Computing, Faculty of Technology, Portsmouth, UK

Enamul Haque University of Waterloo, Waterloo, Canada

Atsuo Hazeyama Tokyo Gakugei University, Koganei, Tokyo, Japan

Adrian Hopgood The University of Portsmouth (UoP), School of Computing, Faculty of Technology, Portsmouth, UK

Elias Hossain Time research and innovation (Tri), Southampton, United Kingdom; Khilgaon Dhaka, Bangladesh

Md Nour Hossain Indiana University Kokomo, Kokomo, USA

Robert J. Howlett KES International Research, Shoreham-by-sea, UK

Felix Häfner Aalen University, Aalen, BW, Germany

Ralf-Christian Härting Aalen University, Aalen, BW, Germany

Muwaffaq Imam Innopolis University, Innopolis, Russia

M. Jorio LSIGER, FST-FES, Sidi Mohammed Ben Abdellah University, Fez, Morocco

G. Karani Cardiff School of Health Sciences, Environmental Public Health Group, Cardiff Metropolitan University, Cardiff, UK

Iouri Kotorov North Karelia University of Applied Sciences, Joensuu, Finland

John R. Littlewood The Sustainable & Resilient Built Environment Research Group, Cardiff School of Art & Design, Cardiff Metropolitan University, Cardiff, UK

Inés López-Baldominos Universidad de Alcalá, Alcalá de Henares, Spain

Tanvir Mahmud Harrisburg University of Science and Technology, Harrisburg, USA

A. Mansouri LSIGER, ENSA-FES, Sidi Mohammed Ben Abdellah University, Fez, Morocco

Marjan Mansourvar Center for Health Informatics and Technology, The Maersk Mc-Kinney Moeller Institute, University of Southern Denmark, Odense, Denmark

M. Alami Marktani LSIGER, ENSA-FES, Sidi Mohammed Ben Abdellah University, Fez, Morocco

Alexander Martín-Garín ENEDI Research Group, Department of Thermal Engineering, Faculty of Engineering of Gipuzkoa, University of the Basque Country UPV/EHU, Donostia-San Sebastián, Spain

Shamsul Masum The University of Portsmouth (UoP), School of Computing, Faculty of Technology, Portsmouth, UK

Manuel Mazzara Innopolis University, Innopolis, Russia

A. Mechaqrane LSIGER, FST-FES, Sidi Mohammed Ben Abdellah University, Fez, Morocco

Márcio Mendonça Federal University of Technology—Parana (UTFPR), Cornélio Procópio, Brazil

José Antonio Millán-García ENEDI Research Group, Department of Thermal Engineering, Faculty of Engineering of Gipuzkoa, University of the Basque Country UPV/EHU, Donostia-San Sebastián, Spain

Sambeet Mishra TalTech, Tallinn University of Technology, Tallinn, Estonia

Adrián Mota-Babiloni ISTENER Research Group, Department of Mechanical Engineering and Construction, Universitat Jaume I (UJI), Castelló de la Plana, Spain

Amin Naemi Center for Health Informatics and Technology, The Maersk Mc-Kinney Moeller Institute, University of Southern Denmark, Odense, Denmark

Mostafa Naemi The University of Melbourne, Melbourne, Australia

Joaquín Navarro-Esbrí ISTENER Research Group, Department of Mechanical Engineering and Construction, Universitat Jaume I (UJI), Castelló de la Plana, Spain

Xabat Oregi CAVIAR Research Group, Department of Architecture, University of the Basque Country UPV/EHU, Donostia-San Sebastián, Spain

Jorge Otaegi CAVIAR Research Group, Department of Architecture, University of the Basque Country UPV/EHU, Donostia-San Sebastián, Spain

Rodrigo H. C. Palácios Federal University of Technology—Parana (UTFPR), Cornélio Procópio, Brazil

Ruslan Pletnev Innopolis University, Innopolis, Russia

Vera Pospelova Universidad de Alcalá, Alcalá de Henares, Spain

Wahidur Rahman Time research and innovation (Tri), Southampton, United Kingdom;
Khilgaon Dhaka, Bangladesh

Iñigo Rodriguez-Vidal CAVIAR Research Group, Department of Architecture, University of the Basque Country UPV/EHU, Donostia-San Sebastián, Spain

Anna-Lena Rösch Aalen University, Aalen, BW, Germany

Hamza Salem Innopolis University, Innopolis, Russia

Iqbal H. Sarker Chittagong University of Engineering and Technology, Chittagong, Bangladesh

Thomas Schmidt Center for Health Informatics and Technology, The Maersk Mc-Kinney Moeller Institute, University of Southern Denmark, Odense, Denmark

Gianluca Serafino Aalen University, Aalen, BW, Germany

Shahana Shultana Prairie View A&M University, Prairie View, USA

Shah Siddiqui The University of Portsmouth (UoP), School of Computing, Faculty of Technology, Portsmouth, UK;
Time research and innovation (Tri), Southampton, United Kingdom;
Khilgaon Dhaka, Bangladesh

Gavrilaş Simona Faculty of Food Engineering, Tourism and Environmental Protection, "Aurel Vlaicu" University of Arad, Arad, Romania

Domenico Suraci Graduated at Politecnico di Milano in Civil Engineering and Transport Engineering, Politecnico di Milano, Milan, Italy

U. Thurairajah Cardiff School of Art and Design, Sustainable and Resilient Built Environment Group, Cardiff Metropolitan University, Cardiff, UK

Cosmin Mihai Udroiu ISTENER Research Group, Department of Mechanical Engineering and Construction, Universitat Jaume I (UJI), Castelló de la Plana, Spain

Gaizka Vallespir-Etxebarria ENEDI Research Group, Department of Thermal Engineering, Faculty of Engineering of Gipuzkoa, University of the Basque Country UPV/EHU, Donostia-San Sebastián, Spain

Uffe Kock Wiil Center for Health Informatics and Technology, The Maersk McKinney Moeller Institute, University of Southern Denmark, Odense, Denmark;
The University of Melbourne, Melbourne, Australia

Yuki Yamada Graduate School of Education, Tokyo Gakugei University, Koganei, Tokyo, Japan

G. Zaz LSIGER, FST-FES, Sidi Mohammed Ben Abdellah University, Fez, Morocco

Petr Zhdanov Innopolis University, Innopolis, Russia

S. Zimeras Department of Statistics and Actuarial-Financial Mathematics, University of the Aegean, Samos, Greece

Chapter 1
Smart and Sustainable Technology for Resilient Cities and Communities—An Overview

Robert J. Howlett and John R. Littlewood

The COVID-19 pandemic has led to enormous human suffering and tragedy with the World Health Organisation estimating the global death toll to have reached nearly five million by late 2021 [1]. It has also had disastrous economic consequences, with the contraction of economies and the likelihood of recession in many parts of the world, which is itself likely to lead to indirect additional loss of life.

However, we should not be too surprised when global disasters occur as they happen regularly and come in many forms. Coronavirus pandemics have occurred regularly before COVID-19, for example the 1918 flu pandemic, SARS, and MERS. Natural disasters also occur regularly, from the Indian Ocean Tsunami of 2004, the Haiti earthquake of 2010 to the Fukushima Daiichi power plant disaster in 2011, following a volcanic eruption under the Pacific Ocean, resulting in a Tsunami. Also, food and water shortages are continuing disasters in many parts of the world, as is the decimation of animal species and reduction in biodiversity. All of this is influenced by the huge challenge of our times: man-made climate change. The European heat wave of 2003, influenced by human activity induced climate change, is not often thought of as a major disaster, but researchers have put the death toll at 70,000 with the observation that "global warming constitutes a new health threat" [2]. Further, the "heat domes" over Northwest Canada and the USA in the summer of 2021 seen as the worst heat wave on record in those parts of the earth saw thousands of deaths where entire conurbations were wiped out by raging fires and external temperatures peaking at mid-50 °C [3, 4].

R. J. Howlett (✉)
KES International Research, Shoreham-by-sea, UK
e-mail: rjhowlett@kesinternational.org

J. R. Littlewood
The Sustainable & Resilient Built Environment Research Group, Cardiff School of Art & Design, Cardiff Metropolitan University, Cardiff, UK

There is a need for the development of innovative methodologies and behaviour to enable societies to be more resilient in the face of disasters that occur from time to time, but also the continuing disasters which continue around ourselves due to climate change reality.

When the COVID-19 pandemic struck in 2020, researchers around the world reacted quickly to apply their research to alleviate its harmful effects on humanity [5]. There is the opportunity to learn from this research to help build a better world for society, not just in the short term during the COVID-19 pandemic, but also afterwards if and when it is eradicated. We should be investing in low-to-zero-carbon technologies resilient to future global challenges and disasters, whether natural or man-made.

There needs to be thought leadership on the way that society might evolve in response to current events; for example the dramatic expansion in homeworking and the evolution of smart digital productivity tools to enable this to happen. Cities and rural communities need to evolve to become smarter, more resilient, and self-sufficient.

This book contains this introduction (Chapter 1) and 21 subsequent chapters, divided into four parts. Chapters are from a range of countries and are on a range of topics. Many of the chapters describe research aiming to combat the spread or effects of COVID-19 but containing lessons to be considered for future. In other chapters, the research is on topics that have the potential to improve the resilience of society.

1   Part I: "Changes in Work Practices and Employability in Response to the COVID-19 Pandemic"

This part of the book contains five chapters describing research into improved employability skills and work-integrated learning, digital transformation of small-to-medium enterprises (SMEs), urban mobility, and air quality in the workplace.

Chapter 2, "Examining Pedagogical Approaches in Developing Employability Skills in the Wake of the COVID-19 Pandemic", describes research investigating the various approaches to teaching and learning that can be employed by institutions of higher education to develop employability skills among built-environment students in response to the COVID-19 pandemic. The study found that multidisciplinary teaching approaches, work-integrated learning, and placement opportunities made the most significant contributions to developing graduates' employability. The authors concluded that adoption of these approaches would help make graduates more academically able and also equip them to more easily fit into new work paradigms evolving out of the COVID-19 pandemic and future disruptions.

Chapter 3, "Relating Work-Integrated Learning to Employability Skills in the Post-COVID-19 Era", considers the place of work-integrated learning as higher education institutions around the world reshape their curricula attempting to alleviate the disruption due to the COVID-19 pandemic. The authors make a case for universities adopting a more innovative approach to ensure that work-integrated learning is successfully employed because of the benefits to students, despite the challenges inherent in the approach.

Chapter 4, "Opportunities and Barriers of Digitization in the COVID-19 Crisis for SMEs", describes work on digital transformation in small companies, which has been a key driver of business model development for a number of years. The study examined the problems and opportunities of integrating digital technology into small-to-medium enterprises (SMEs) that are particularly affected by external factors, including the COVID-19 pandemic.

Chapter 5, "New Urban Mobility Strategies After the COVID-19 Pandemic", is on the subject of an investigation into the measures relating to urban mobility, conducted in various cities of the world, in response to the COVID-19 crisis. The project, which involved cities from Europe and North and South America, evaluated which measures had been the most effective, verifying their effects in both the short and long term.

Chapter 6, "Integration of Indoor Air Quality Concerns in Educational Community Through Collaborative Framework of Campus Bizia Lab of the University of the Basque Country", outlines a research project to monitor and analyse the indoor air quality of several educational facilities in the Gipuzkoa Campus of the University of the Basque Country (UPV/EHU). It is well recognised that good ventilation can reduce the spread of COVID-19 indoors, while poor quality air can increase it. This study examined indoor air quality in several types of classrooms and offices with different characteristics in order to assess the performance of each one of them.

2 Part II: "Smart Techniques for Monitoring the COVID-19 Pandemic and Forecasting Its Course"

This part of the book contains seven chapters. It describes research involving various intelligent, statistical, and advanced sensory techniques in modelling, forecasting, diagnosis, and risk prediction applicable to the COVID-19 pandemic, and beyond it.

Chapter 7, "An Overview of Methods for Control and Estimation of Capacity in COVID-19 Pandemic from Point Cloud and Imagery Data", looks at the way sensing technologies can be applied in monitoring and dealing with the COVID-19 pandemic. The chapter reviews sensing techniques from point cloud and imagery data related to population control and estimation of the capacity, people counting, biometric identification, monitoring of activities, distance measurement, and 3D modelling. The chapter presents current techniques and the algorithms most often used. The merits and disadvantages of point cloud data and imagery and current trends are reviewed.

Chapter 8, "Modeling and Evaluating the Impact of Social Restrictions on the Spread of COVID-19 Using Machine Learning", reports on an investigation into the effect of social restrictions imposed in response to the COVID-19 pandemic. Included are restrictions on schools, workplaces, public events, gatherings, internal and international flights, brought in to control the spread of the virus. Three machine learning models were applied to simulate the number of infected cases per day under different levels of restrictions. Different scenarios of social restrictions were simulated to study the impact of decisions on social restrictions and imposing more strict ones.

Chapter 9, "Forecasting the COVID-19 Spread in Iran, Italy and Mexico Using Novel Nonlinear Autoregressive Neural Network and ARIMA-Based Hybrid Models", presents an analysis of single- and two-wave COVID-19 outbreaks using novel hybrid machine learning and statistical models to simulate and forecast the spread of the infection. For this purpose, historical cumulative numbers of confirmed cases for three countries were used. The performance of the techniques under different conditions was analysed and results and conclusions presented.

Chapter 10, "Spatial Statistics Models for COVID-19 Data Under Geostatistical Methods", describes techniques that can be applied to model distributions of pandemic infection. Geostatistics techniques are explained as a way of quantifying spatial uncertainty, and statistics are applicable to allow probability densities to be employed. Illustration of the spatial modelling based on coordinates considering the epicentres of COVID-19 virus in a small region is considered. Hence, a range of techniques is presented that can be used in the numerical analysis of COVID-19 data.

Chapter 11, "Intelligent Multi-sensor System for Remote Detection of COVID-19", presents the development of a novel multi-sensor method of identifying COVID-19 at a distance. This system applies the principle of multi-sensor data fusion to provide a robust, precise, and complementary analysis of these symptoms to tell whether or not an individual is a carrier of COVID-19 disease. The authors report that the system, designed to be used in public venues, can also be adapted as a useful means of early detection of many other diseases.

Chapter 12 "A Comparative Study of Deep-Learning Models for COVID-19 Diagnosis Based on X-Ray Images", reports on a project to test and compare different deep learning algorithms on a dataset consisting of a large number of digital COVID-19 X-ray images. The motivation is to find an alternative to the reverse transcription polymerase chain reaction (RT-PCR) kits, widely regarded as the best available. The accuracy of RT-PCR is not 100%, it takes a few hours to deliver the test results, and there has been a world shortage of the kits. The authors report developments in their quest to develop an alternative to the RT-PCR test that is not subject to the same disadvantages.

Chapter 13, "Fuzzy Cognitive Maps Applied in Determining the Contagion Risk Level of SARS-COV-2 Based on Validated Knowledge in the Scientific Community", describes research that aims to devise a smart tool to estimate the risk of an individual becoming infected with COVID-19 based on their behaviour. The technique uses a fuzzy cognitive map intelligent paradigm to model data provided from pre-existing medical research and link behaviour to risk. The authors claim a number of benefits from the tool, including providing information to enable individuals to change their behaviour to mitigate their contagion risk.

3 Part III: "Changes in Teaching and Learning Practices in Response to COVID-19"

This part of the book contains four chapters on subjects related to pedagogic evolution made necessary by the pandemic, including experiences of moving to e-learning, new online training methods, and human factors associated with online learning.

Chapter 14, "Education After COVID-19", is based on the authors' experiences of the changes to pedagogic practice necessitated by the pandemic. Education has been very heavily affected by the COVID-19 pandemic, with delivery moving almost entirely to online form during lockdowns, and often for some time after. While experience has shown some features of pre-COVID-19 educational practice to be out-dated, person-to-person contact is essential for many aspects of the education process. The authors reflect on the past, present, and future of education, what they have learned from the new teaching methods adopted in response to the pandemic, and what beneficial changes can be carried forward into future teaching and learning.

Chapter 15, "Equipping European Higher Education Teachers for Successful and Sustainable E-Learning with Home Remote Work", reports on an analysis of the reaction of higher education to the COVID-19 pandemic. This will enable the lessons learned to be implemented during a possible future crisis.

The authors report on a combined study based on a literature review and a specific survey to higher education teachers in Europe. The study is reported to have provided consistent results to indicate how teachers should be trained and supported for the future.

Chapter 16, "Fully Online Project-Based Learning of Software Development During the COVID-19 Pandemic", contains the authors' reflections on their experiences of delivering software engineering training online during the pandemic. The authors state that they have many years' experience of using project-based learning to develop software education. The COVID-19 pandemic led to this being applied to develop fully online teaching during the 2020 academic year using the process and software engineering environment they developed. This is a combination of a software repository and an online meeting system. Having evaluated this environment through students feedback, they report it to be well suited to fully online remote project-based learning.

Chapter 17, "A Tale of Two Zones: Pandemic ERT Evaluation", describes an analysis of emergency remote teaching (ERT) during the COVID-19 pandemic. During the COVID-19 pandemic, many educational institutes switched from conventional teaching to all online classes. Few institutes were well prepared for this change, and some faced greater problems than others owing to deficiencies in technology infrastructure. Furthermore, differences were identified between teachers' and students' attitudes to online teaching. These issues formed the basis for the analysis described in this chapter, leading to the identification of opportunities for improvement.

4 Part IV: "Adapting for Improved Resilience in an Uncertain Future" Contains Six Chapters

This part of the volume contains six chapters on topics which can contribute to improved resilience in cities and communities. Topics include coping with change and uncertainty in the built environment, improved resilience in the power system, food supply security, improving resistance to infection, and sustainable low-temperature refrigeration.

Chapter 18, "Anticipating and Preparing for Future Change and Uncertainty: Building Adaptive Pathways", presents an investigation into the nature of future disruptive change and its implications for improving the resilience of buildings. A structured approach is proposed to prepare for, and respond to, change in a proactive, structured way. This methodology is called building adaptive pathways and is illustrated and tested through application to a case study. Findings indicate that methodology provides useful insights on how change and uncertainty can be addressed in built environments and recommends that further work on the approach be undertaken.

Chapter 19, "A Health-Energy Nexus Perspective for Virtual Power Plants: Power Systems Resiliency and Pandemic Uncertainty Challenges", introduces measures to change the power generation and distribution system with the aim of providing improved resilience. The chapter discusses the link between health and energy and the way in which the two are related. The opportunities and challenges presented by the interaction between health and energy are presented, and ways to address the changes in the power systems resiliency due to pandemic conditions are discussed.

Chapter 20, "A Sustainable Nutritional Behavior in the Era of Climate Changes", contains a discussion on the topic of food supply security during disruptions occurring due to climate change and other unforeseen events. The chapter presents the view that the earth's ecosystem is tightly coupled, and different elements are inherently linked. Food supply, the availability of water for drinking and crops, and disease vectors such as COVID-19 are all inter-related to energy availability and consumption and climate change.

The chapter concludes with a number of suggestions for future developments to help improve the resilience of the food supply, including that environmental considerations should be inherent in any future innovations.

Chapter 21, "The Development of a Smart Tunable Full-Spectrum LED Lighting Technology Which May Prevent and Treat COVID-19 Infections, for Society's Resilience and Quality of Life", introduces the concept that lighting with a spectrum that results in increased vitamin-D production in the human body may be beneficial in combatting COVID-19 and have other health advantages. The research aims to develop a sustainable full-spectrum lighting system using light emitting diode technology to provide a similar colour balance and intensity as daylight in buildings. This will result in increased vitamin-D production in the body, which has been shown to be beneficial for infection resistance in general, including resistance to COVID-19.

Chapter 22, "Energy Efficient Technologies for Ultra-Low Temperature Refrigeration", describes work on the ultra-low-temperature refrigeration technology that is required for the storage of some vaccines. New vaccines have been developed in response to the current COVID-19 pandemic, and some of these require ultra-low-temperature refrigeration (at $-80\,°C$). The technology is not mature, and the energy performance is often low because heat pump efficiency is poor with the large gap between source and sink temperature. There are also challenges associated with refrigerants and lubricating oils. This chapter presents the main characteristics of several applicable refrigeration technologies, ways of overcoming the challenges inherent in using them, and future directions for sustainable ultra-low-temperature refrigeration.

References

1. World Health Organisation. https://covid19.who.int/. Accessed 13 Oct 2021
2. Robine J-M, Cheung SLK, Le Roy S, Van Oyen H, Griffiths C, Michel J-P, Herrmann FR (2008) Death toll exceeded 70,000 in Europe during the summer of 2003. CR Biol 331(2):171–178
3. Reuters (2021) Death's surge in U.S. and Canada from worst heatwave on record. https://www.reuters.com/world/americas/dire-fire-warnings-issued-wake-record-heatwave-canada-us-2021-06-30/. Accessed 20 Oct 2021
4. Vaughan A (2021) The heat is on out west. New Scientist, 10 July 2021, vol 250, issue 3342, pp 10–11. https://www.sciencedirect.com/science/article/abs/pii/S0262407921011696. Accessed 20 Oct 2021
5. Chakrabortya I, Maity P (2020) COVID-19 outbreak: migration, effects on society, global environment and prevention. Sci Total Environ 728:138882 (1 August 2020). https://www.sciencedirect.com/science/article/abs/pii/S0048969720323998. Accessed 20 Oct 2021

Part I
Changes in Work Practices and Employability in Response to the Covid-19 Pandemic

Chapter 2
Examining Pedagogical Approaches in Developing Employability Skills in the Wake of the COVID-19 Pandemic

John Aliu and Clinton Aigbavboa

Abstract The construction industry today and its employers are gradually coming to terms with the fact that its activities and processes require fully equipped graduates who are furnished with the right skills to succeed after graduation. However, with the continuously evolving nature of the industry coupled with the COVID-19 pandemic, the pressure on higher education to review and revamp its existing curricula has intensified in recent times. The process of developing employability skills among students can be reflected in the pedagogical approaches employed by institutions of higher learning. This research aims to determine the various pedagogical approaches that can be employed by institutions of higher learning to develop employability skills among built-environment students as the world grapples with the after-effect of the COVID-19 pandemic. A qualitative Delphi approach was adopted to validate these approaches. Fourteen experts completed a two-stage iterative Delphi study process and reached a consensus on all 16 approaches identified. This study found that multidisciplinary teaching approaches, work-integrated learning, and placement opportunities are the most significant approaches in developing employability skills among students. It is recommended that universities across South Africa and beyond continue to ensure the inclusion of these approaches into their existing curricula to not only produce graduates who are academically sound but also produce graduates who will fit easily into the 'new normal' prompted by both the COVID-19 pandemic and the Fourth Industrial Revolution (4IR).

Keywords Active learning · Built environment · Construction industry · COVID-19 · Employability · Pedagogy · Skills

J. Aliu (✉) · C. Aigbavboa
cidb Centre of Excellence, University of Johannesburg, Johannesburg, South Africa

© The Author(s), under exclusive license to Springer Nature Singapore Pte Ltd. 2022   11
R. J. Howlett et al. (eds.), *Smart and Sustainable Technology for Resilient Cities and Communities*, Advances in Sustainability Science and Technology,
https://doi.org/10.1007/978-981-16-9101-0_2

1 Introduction

The present-day construction industry plays a pivotal role in every economy globally. Its role in the achievement of socio-economic development goals and provision of local amenities, employment, and infrastructural development is encouraged through its numerous activities. These activities which range from the construction of highways, dams, bridges, structures, canals among others are strong forces that act as a catalyst for the achievement of infrastructural development in our modern-day economy. One of the characteristics of the construction industry is the mobilization and utilization of both human and material resources in boosting economic efficiency in areas such as infrastructural development, job creation, and sustainability [14]. However, the construction industry is faced with its challenges which arise as a result of the influx of built-environment graduates who are not fully equipped with the right skills [16]. This further implies that the construction sector is heavily reliant on skilled professionals for its functions in infrastructural development, maintenance, and all construction-related tasks. The construction profession, like other professions, possesses high principles about professional ethics, service, and practice. These rules of ethics, therefore, increase the need for graduates with industry-ready traits and skills to function effectively.

Considering the above characteristics of the construction industry which highlights its dynamism and increasing demands, it, therefore, implies that present-day graduates are required to be well educated and fortified with a wide array of skills (academic and non-academic) to address arising and existing industry problems Aliu and Aigbavboa [10]. In fact, due to the advent of the Fourth Industrial Revolution (4IR) and the pandemic-induced changes to the educational system, employers have taken the search for competent and skilled graduates to a whole new level. Aside from been academically sound, the industry now requires graduates to be sufficiently furnished with the relevant skills, abilities, and competencies to fit into the world of work after graduation. Industry employers now place premium value on graduates who can exhibit confidence in communicating effectively; work as part of a team when necessary; possess good ethics; think critically; show a willingness to learn; display flexibility and adaptability; possess analytical skills; understand the dynamics of information and communication technology (ICT skills); possess problem-solving skills among many others [7, 11]. These required skills have placed significant pressure on universities across the world to review and revamp their existing curricula to meet industry needs. It is against this backdrop that this research determines the pedagogical approaches that can be employed by universities globally to develop employability skills among built-environment students as the world embraces 4IR technologies and grapple with the aftermaths of the pandemic. It is worthy to note that the educational landscape has been greatly impacted as a result of COVID-19 which has seen a sudden transition to online pedagogy [44]. Firstly, a large number of academic activities including conferences and teaching activities were canceled and postponed [43], then a few months later, universities around the world began the adoption of technological applications to execute their academic activities. Some of these tools includes Zoom, Skype, Microsoft Teams, WhatsApp among several others. This new technologically driven paradigm altered the existing ways of teaching and learning with a new dimension of pedagogy taking center stage.

This study was conducted in South Africa and its objectives align with one of the nation's long-term National Development Plan (NDP) 2030, which seeks to develop the skills, knowledge, and capabilities of its citizens through the provision of quality education across all levels. This policy document was drafted in 2012 by the National Planning Commission. This mandate also resonates with the Sustainable Development Goals (SDGs) proposed by the United Nations which plans to ensure quality and inclusive education for all by 2030. In achieving the objectives of this study, a qualitative Delphi approach was adopted to validate these pedagogical approaches as universities across South Africa and beyond embraces the dynamics of the 'new normal'.

2 Review of Literature

2.1 The Era of the 'New Normal'

As the world continues to grapple with the adverse impact of the COVID-19 pandemic, there have been increased uncertainties in several aspects of global economies, and the educational sector is one of them [44]. In a bid to curb the spread of the coronavirus disease, several educational institutions across the world stopped in-person lectures (contact meetings) over a year ago. According to the United Nations Educational, Scientific and Cultural Organization (UNESCO), educational institutions in more than 185 countries around the world were shut down at the end of April 2020 [4]. That meant roughly 74% of learners worldwide were affected by the shutdown, which prompted a new approach in the teaching and learning process. Currently, most educational institutions globally have adopted a remote learning (online learning or distance learning) approach to combat the spread of the virus, in what many now refer to as education in the 'new normal'. However, there are some dynamics with this new mode of learning [15]. Firstly, there are concerns that students may have to spend less time during the learning process when compared to physical classroom settings [38]. Secondly, there are also concerns that students are more prone to stress and anxiety during remote learning which may affect their ability to thrive in their academic work [38]. Thirdly, the lack of physical contact with their peers and educators (lecturers) may lead to decreased motivation to engage in learning activities [44]. Fourthly, students from disadvantaged backgrounds will struggle with accessing digital resources such as laptops, internet connections, and scanners among others [44]. More so, the switch from offline to online learning is likely to affect the occurrence of certain higher education activities such as supervised industrial work experience schemes for built-environment students, laboratory experiments [12]. These dynamics have changed the pedagogical landscape and universities across the world have been compelled to revamp and restructure their academic activities to align with the current climate of the pandemic-driven world (new normal). This also places significant pressure on higher education to develop innovative pedagogical approaches to develop employability skills among students as the world grapples with the effect of the pandemic.

2.2 Understanding the Concept of Pedagogy

While this research aims to determine pedagogical approaches that can develop employability skills in the era of the 'new normal', it is pertinent to understand the meaning of the overarching term—pedagogy. Several researchers have described the concept as an activity or group of activities that result in positive skill outcomes for learners. According to Watkins and Mortimore [55], pedagogy is described as 'any conscious activity by one person designed to enhance learning in another'. Similarly, pedagogy is a 'sustained process whereby someone acquires new forms or develops existing forms of conduct, knowledge, practice, and criteria from somebody or something deemed to be an appropriate provider and evaluator' [20]. The definition by Alexander [6] states that 'teaching is an act while pedagogy is both act and discourse'. This definition suggests that pedagogy is a broad spectrum that encompasses the following characteristics of the educator (teacher): ideas, knowledge, and attitudes, understanding of the curriculum; understanding of the learning outcomes required; and student's limitations during the teaching and learning process [6]. Based on the preceding definitions, the essential goal of pedagogy is to develop learning among students and ultimately their employability skills. Hence, this study adopts the definition by Westbrook et al. [56] which describes effective pedagogy as 'those teaching and learning activities which make some observable change in students, leading to greater engagement and understanding and/or a measurable impact on student learning' [56]. Furthermore, it is important to note that for the sake of this research, 'pedagogy' is different from 'teaching practices'. According to Alexander [6] and Thoonen et al. [54], teaching practices are specific and physical actions that occur during the training process to further stimulate the understanding of students. Teaching practices include some of the following: the communicative skills of the educator including giving explanations and instructions, citing examples, asking questions, elaborating on ideas, and accepting responses from students, the visual representation utilized during teaching including the adoption of diagrams, learning aids, boards, and experiments to improve the understanding of learners; actively engaging learners, so they can develop skills; encouraging social interactions by introducing team tasks; and monitoring students by making use of feedback and assessment mechanisms to track their progress [6].

2.3 Reviewing Pedagogical Approaches

Globally, the quest to develop the employability of built-environment students is dependent on the quality of teaching practices and pedagogical approaches employed by institutions of higher learning. These approaches have been discussed extensively across existing literature, and in recent times, they have become increasingly necessary due to the 'new normal'. Several researchers have discussed the various benefits of final-year research or semester projects and its contribution to graduate employability. These research projects provide an opportunity to test the individual's learning integrity and experience via an exploration of facts that culminates the knowledge gathered through various academic levels. Some of these research projects assist students in establishing connections between their chosen careers and the world of work. These research activities also promote holistic thinking among students, increase their self-confidence and motivation, enrich their academic understanding of their chosen discipline and development of key skills (problem-solving, critical thinking, organizational and interpersonal) [35]. In improving the employability of learners and their work profile, the role of career management and development cannot be over-emphasized. According to Kuijpers and Scheerens [36] and Aliu et al. [8], career advice and development help to prepare learners for the industry by developing their job-searching skills such as interview preparation, curriculum vitae (CV) design, self-reflection and self-assessment abilities, and networking competencies. Doyle [26] suggests that career development involves several activities that cater to the career needs of learners such as professional workshops and seminars. In improving the employability of learners, another key strategy is the encouragement of mentoring among students. Mentoring is a social learning and interactive opportunity that improves learners' transition from lecture-room setting to work setting via industry involvement. According to Aliu et al. [9], industry mentors provide students with the knowledge required to thrive in their chosen careers to succeed in the world of work. More so, the presence of industry mentorship encourages career development and outcomes among students; hence, employability improvement is guaranteed. Levesque et al. [40] also suggest that industry mentoring provides students with career information, increased commitment, political and material support, on-the-job training, motivation, workplace understanding and realities, facilitation of connections for students, and role modeling. It is based on this wealth of knowledge that various pedagogical approaches in developing employability skills as seen in Table 1 were assessed in this study.

Table 1 Review of pedagogical approaches

Pedagogical approaches	Literature sources
Final-year research projects	Ryder [50]; Shaw et al. [51]; Parker [45]
Multidisciplinary teaching approaches	Al Hassan [5]
Work-integrated learning and placement opportunities	Kinash et al. [34]
Part-time employment for students	Dustmann and Van Soest [27]; Creed et al. [24]
Mentoring	McIntyre and Hagger [42]
Career pathing of students	Cao and Thomas [22]; Landrum [37]
Project-based learning (PBL)	Bell [19]
Simulation and role-play in classroom	Qing [47]; Bhattacharjee and Ghosh [21]
Visits to industry events	Aliu et al. [9]
Field trips to construction sites	Aliu et al. [9]
Integration of technical competitions	Ahlgren and Verner [2]
Volunteering and community engagement	Hall et al. [29]; Parker et al. [46]
Extra-curricular activities	Bartkus et al. [18]
International exchange programs	Kinash et al. [34]
Student government participation	Alviento [13]
Engagement in sports activities	Telford et al. [53]; Bailey et al. [17]

Source Author's compilation

3 Research Methodology

3.1 The Delphi Process

To validate these pedagogical approaches that can be employed by institutions of higher learning to develop employability skills among built-environment students, this study adopted a qualitative Delphi approach. According to Green [28], one of the major strengths of the Delphi technique is that it cuts across both the quantitative and qualitative methods of data collection and analysis. Consequently, this allows the research results and conclusions to be generally represented to the wider population.

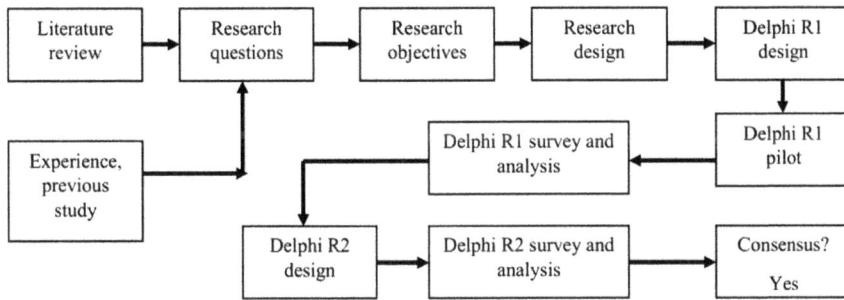

Fig. 1 Delphi design for this study (author's compilation)

It is critical to note that Delphi studies are considered reliable as professional opinions are obtained from subject matter experts [30]. More so, the various iterations and methodological rigor of the Delphi process further improves the dependability of research results [39]. Therefore, a Delphi approach was adopted to obtain professional opinions from employability experts on the various pedagogical approaches that were extracted from the review of the literature. Figure 1 provides an illustration of how the Delphi process was conducted.

According to Aliu et al. [9], some of the noteworthy features of the Delphi technique include the perseveration of the experts' anonymity, the obtained statistical responses and the repetitive process, which is known as iterations. These iterations or Delphi rounds allow experts to either maintain their opinions of the subsequent rounds or modify their responses which form the basis for a subsequent round. According to Hallowell and Gambatese [30], these rounds help to eradicate the variability of experts' responses and to reach a consensus on the subject matter. For this study, achieving consensus on the pedagogical approaches was one of the main objectives. More so, the choice of experts is another critical feature of the process as observed by Hasson et al. [31]. Moreover, adequate attention should be paid to the group responses rather than the individual ones. To measure the consistencies and central tendencies and ultimately consensus, this study considered parameters such as media and interquartile deviation (IQD). Another aspect of the Delphi study that has generated the opinions of several researchers is the identification of experts. To be called an expert, an individual should possess certain characteristics such as—full understanding of the subject matter, authors who publish articles relating to the subject matter, individuals who attend workshops and conferences relating to the subject matter, an individual who is willing to participate in several rounds if need be and an individual who communicates fluently in speaking and writing [49]. Other criteria include—years of working experience in the industry, employment with an accredited academic institution, serving as a member of academic committees, registered with a professional body, and possessing an academic degree related to the subject matter [1, 30]. Nevertheless, it has been widely agreed that a researcher has the final decision in determining who should be called an expert based on some of the above-named criteria depending on the study. For this current study, to be called an expert, three or more of the following criteria was required to be met. These include—possession of at least a Bachelor's degree in any of the disciplines within the built environment, working with the industry or academia, possession of at least five years of work experience, officially registered with a professional body, and an author of publications relating to disciplines within the built environment. As shown in Table 2, the fourteen experts who participated in the Delphi process satisfied three or more of the criteria. The fourteen experts considered for this research were from academia and the construction industry.

Before they were selected for the Delphi process, a comprehensive description detailing the requirement and guiding instructions was presented, to which the experts consented. After expressing their willingness to participate, the selected experts were requested to forward their curriculum vitae (CV) which helped to ascertain how many of the criteria were successfully met. While experts were required to possess three or more of the criteria to be eligible, all fourteen experts met the threshold. Five of them satisfied all the parameters, eight met four of them, while only one expert satisfied three of the criteria as shown in Table 2. Based on academic qualification, five experts possessed a PhD degree; seven had a Master's degree, while two had a Bachelor's degree. This shows that a majority of the experts considered for this study possessed postgraduate degrees, which improve the quality, credibility, and reliability of their responses as shown in Table 3. More so, all fourteen experts possessed built-environment backgrounds as shown in Table 4. Afterward, the experts were sent the first-round questionnaire survey, which contained close-ended and open-ended questions. This round also provided options for experts to rank the various factors as well as stating their opinions where necessary.

With regards to their academic qualifications, a majority (nine) of the experts possessed Engineering degrees (Civil, Electrical, and Mechanical), two of the experts possessed Construction Project Management degrees, another two possessed Architectural degrees, while only one expert possessed a Quantity Surveying qualification. The analysis further revealed that eight of the fourteen experts were from higher institutions based in South Africa, while six experts belonged to the construction industry. In terms of work experience, all fourteen experts possessed significant years of valuable work experience, which was pivotal to the realization of this study. This is further shown in Table 5.

On account of numbers of years of work experience, the experts had spent a considerable amount of time in their various positions. One expert had five years of work experience, while seven had between six and ten years of work experience. Table 5 further revealed that three experts possessed between eleven and twenty years of experience, two had between twenty-one and thirty years, while one expert possessed above thirty-one years of work experience. Other criteria that were considered were belonging to a recognized and accredited professional body. Therefore, five of the experts possessed Engineering Council of South Africa (ECSA) certifications; two were registered with the South African Council for the Project and Construction Management Professions (SACPCMP); a further two were registered with the South African Institution of Civil Engineering (SAICE), while one expert possessed Project Management South Africa (PMSA) certification. One of the main

Table 2 Assessment of Delphi expert qualifications

S/N	Eligibility criteria for experts	E1	E2	E3	E4	E5	E6	E7	E8	E9	E10	E11	E12	E13	E14
1.	Possess at least a Bachelor's degree	X	X	X	X	X	X	X	X	X	X	X	X	X	X
2.	Currently employed with a tertiary institution or professional in the construction industry	X	X		X			X	X	X	X	X		X	X
3.	At least five years of working experience with a tertiary institution or construction industry	X	X	X	X	X	X	X	X	X	X	X	X	X	X
4.	Affiliated with professional bodies	X	X	X	X	X	X	X		X	X	X	X	X	
5.	Author or co-author of a peer-reviewed publication	X	X	X	X		X	X	X		X		X		X
	Total	5	5	4	5	3	4	5	4	4	5	4	4	4	4

Source Author's compilation

Table 3 Panel of experts' qualifications

Highest qualification	Number of experts
Doctor of Philosophy (PhD)	5
Master's degree (MSc and ME)	7
Bachelor's degree (BE)	2
Total	14

Table 4 Panel of experts' field of specialization

Field of specialization	Number of experts
Architecture	2
Quantity Surveying	1
Construction Project Management	2
Engineering (Civil, Mechanical, Electrical)	9
Total	14

Table 5 Panel of experts' years of experience

Years of experience	Number of experts
5	1
6–10	7
11–20	3
21–30	2
Over 31 years	1
Total	14

aims of the iterations in the Delphi technique is to achieve consensus (convergence or agreement on opinions) among experts as noted by Holey et al. [32]. Consensus is attained by recording the media values and standard deviation (SD) values where a decrease in SD between iterations highlights higher levels of agreement among the experts [48]. More so, to reach consensus, Rayens and Hahn [48] suggested that the interquartile deviation (IQD) should be less than or equal to 1, which suggests that over 60% of experts were either largely positive or negative in their responses. This study draws from this assumption and states that consensus is achieved when IQD = 0.00 or ≤1 as further highlighted in Table 6.

Table 6 Consensus scales for this study

S/N	Consensus strength	Median	Mean	Interquartile deviation (IQD)
1.	Strong	9–10	8–10	≤ 1 and $\geq 80\%$ (8–10)
2.	Good	7–8.99	6–7.99	$\geq 1.1 \leq 2$ and $\geq 60\% \leq 79\%$ (6–7.99)
3.	Weak	≤6.99	≤5.99	$\geq 2.1 \leq 3$ and $\leq 59\%$ (5.99)

The Delphi method also addressed the reliability and validity questions surrounding its adoption for this study. According to Yousuf [57] and Creswell [25], both the reliability and validity considerations deal with the thoroughness of the entire process with regards to various components of the study such as determination of the expert panel, determination of the panel size, and determination of consensus. The reliability of the Delphi process was boosted by adequately explaining the process to all experts involved in this study. This was made possible by including detailed instructions in both rounds of the Delphi questionnaire to eradicate any doubts and ambiguity surrounding the questions (factors and sub-factors). The reliability of this study was also ensured by selecting experts from a similar background, which in this case is the built environment. The experts also belonged to both academia and the construction industry, which made certain a balance of opinions on the employability discussion. This study also stuck with the qualification criteria which also contributed to the reliability of the Delphi process. Furthermore, the validity of the Delphi process was ensured by maintaining the anonymity of experts all through the rounds to eliminate the 'bandwagon' effect. According to Aigbavboa [3], the 'bandwagon' effect is the tendency of experts to make decisions and offer opinions based on the choices of other experts. Internal validity of the study was also encouraged by providing experts with an option of willing participation. Furthermore, by ensuring multiple rounds, experts were able to make modifications to their responses where necessary and providing reasons for their opinions, ultimately improving the internal validity of the research.

4 Presentation of Findings

4.1 Delphi Round 1

The main objective of this article was to obtain respondents' opinions on the various pedagogical approaches through which employability skills can be developed among built-environment students. The responses were obtained using a 10-point Likert scale of 'no significance', 'low significance', 'medium significance', 'high significance', and 'very high significance'. While 'very high significance' had the highest weighting (9 and 10), 'no significance' was assigned the lowest weighting (1 and 2). Sixteen (16) pedagogical approaches were identified as shown in Table 1. More so, three of the approaches—work-integrated learning (placement opportunities), multidisciplinary teaching approaches, and final-year research projects—were highly selected by the experts based on the median score of 9.0 as shown in Fig. 2. From these three approaches, both work-integrated learning (placement opportunities) and final-year research projects achieved consensus based on IQD scores of 1.0 each. More so, volunteering and community engagement, integration of technical competitions, simulation and role-play in the classroom, field trips to construction sites, project-based learning (PBL), part-time employment for students, visits to industry

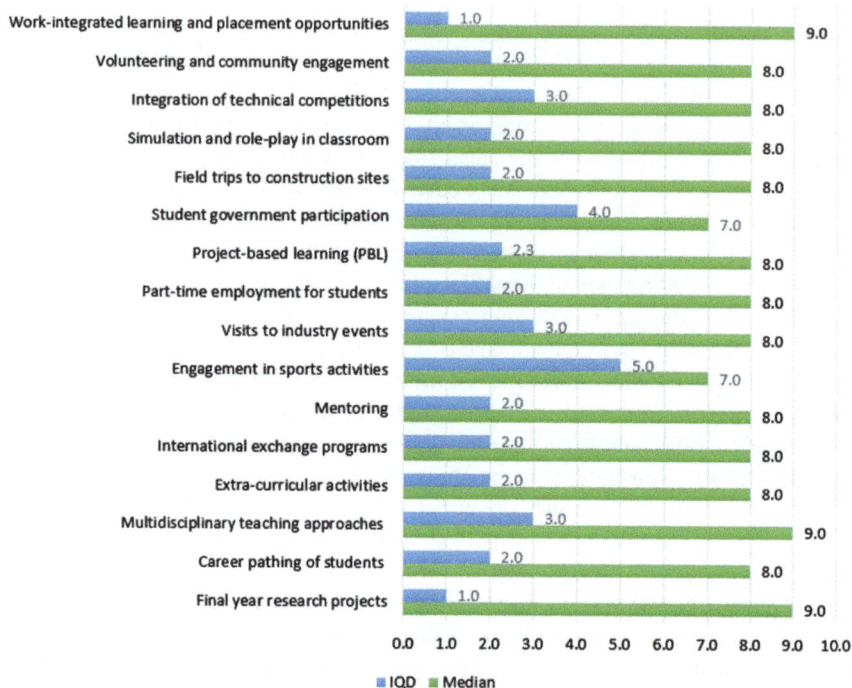

Fig. 2 List of pedagogical approaches

events, mentoring, international exchange programs, extra-curricular activities, and
career pathing of students recorded median scores of 8.0 each. Both student government participation and engagement in sports activities recorded median scores of 8.0
each. More so, no new approaches were introduced by the experts during the Delphi
first round.

4.2 Delphi Round 2

After the successful completion of the first round of the Delphi process, data was
analyzed and the second-round questionnaire was developed and distributed to the
experts. During the second round, experts were requested to accept the group media
value by simply returning them back to the researcher without making any changes.
The experts were also requested to also maintain their responses or to choose new
responses while providing justifications for the new additions. A summary of the
second-round responses showing the median, mean, standard deviation, and IQD are
presented in Table 7.

Table 7 Pedagogical approaches

Pedagogical approaches	Median	Mean $\bar{x}$	SD (σX)	IQD	Rank
Final-year research projects	9.0	8.36	1.65	0.00	1
Multidisciplinary teaching approaches	9.0	8.43	1.79	0.00	2
Work-integrated learning and placement opportunities	9.0	8.64	1.39	0.00	3
Part-time employment for students	8.0	8.00	0.88	0.00	4
Mentoring	8.0	7.93	1.00	0.00	5
Career pathing of students	8.0	7.79	1.25	0.00	6
Project-based learning (PBL)	8.0	7.64	1.50	0.00	7
Simulation and role-play in classroom	8.0	7.64	0.84	0.00	7
Visits to industry events	8.0	7.57	1.83	0.00	8
Field trips to construction sites	8.0	7.57	1.40	0.00	8
Integration of technical competitions	8.0	7.57	1.22	0.00	8
Volunteering and community engagement	8.0	7.50	1.34	0.00	9
Extra-curricular activities	8.0	7.43	1.99	0.00	10
International exchange programs	8.0	7.36	2.13	0.00	11
Student government participation	7.0	6.86	1.29	0.00	12
Engagement in sports activities	7.0	6.50	2.07	0.00	13

5 Discussion of Findings

After the successful completion of the second-round Delphi questionnaire, it was found that all 16 pedagogical approaches achieved good consensus. Hence, there was no need for a third- or fourth-round iteration. Findings emanating from the Delphi study suggest that the approaches which underwent both rounds resonate with what has been proposed by several researchers [41, 52]. As observed from Table 7, all 16 pedagogical approaches had median scores between 7.0 and 9.0, indicating very high significance rakings by the experts. Final-year research projects which had a median value of 9.0 and a mean value of 8.36 were identified as one of the major approaches by which employability skills can be developed. This resonates with the studies by Ryder [50] and Shaw et al. [51] who suggest that final-year research projects provide students with the opportunities to integrate knowledge learned in solving an academic task. Final-year research projects have also been found to improve the following— student's holistic understanding of their chosen discipline, abilities to work in teams or independently, ability to solve problems, and ability to handle responsibilities [45]. More so, with a median value of 9.0 and a mean value of 8.43, multidisci-plinary teaching approaches were highly ranked by experts and understandably so. Multidisciplinary teaching approaches are the inclusion of several disciplines (related or not) to bring diverse perspectives to elucidate a topic or lecture. Through these approaches, a specific title can be studied from the perspectives of several disciplines, thereby improving the overall understanding of students. These approaches further stimulate the interest of students in the learning process and provide answers to thought-provoking questions that can provide a memorable learning experience for built-environment students. More so, through these approaches, skills such as teamwork, critical thinking, innovativeness, and analytical reasoning can be developed among students [5]. With a median value of 9.0 and a mean value of 8.64, work-integrated learning (placement opportunities) was also highly ranked by the experts. According to several kinds of research, work opportunities provide students with the opportunity to actively apply what has been learned during conventional lectures. Apart from easing the transition from the classroom to the world of work, placement opportunities also improve the following—learning process of students, industry-readiness, problem-solving abilities, critical thinking skills, communication skills, self-confidence, and a holistic understanding of the workplace after graduation [34].

There also exists a strong consensus among experts on part-time employment for students who possessed a median value of 8.0 and a mean value of 8.00. This aligns with the works of Dustmann and Van Soest [27] and Creed et al. [24] who both suggest that part-time employments provide the following—improved transferable skills, self-confidence, interpersonal skills, financial management, and building professional networks. Mentoring with a median value of 8.0 and a mean value of 7.93 was also ranked highly by experts. By working with a mentor, students can benefit from the following—increased social and academic confidence, enhanced communication and study skills, gaining a sense of academic and career direction, building professional networks, improved decision-making abilities, exposure to refreshing and innovative ideas, and development of leadership skills [42]. These benefits of mentoring on students' performance are some of the reasons why institutions of higher learning across South Africa encourage a mentor–mentee relationship to improve the overall development of students. With a median value of 8.0 and a mean value of 7.79, career pathing of students was also ranked highly by experts, and this resonates with the studies of Cao and Thomas [22]. Career pathing has been found to improve the career choices of students and making them aware of their strengths and weaknesses. These can be achieved through career development and advancement learning and career development training [23, 37]. There also exists a strong consensus among experts on project-based learning (PBL) as it possessed a median value of 8.0 and a mean value of 7.64. This resonates with the study of Bell [19]. Also, ranked highly were simulation and role-play in the classroom with a median value of 8.0 and a mean value of 7.64. This aligns with the works of Qing [47] and Bhattacharjee and Ghosh [21] who suggest that simulation and role-play require students to portray real-life characters in classrooms to obtain a deeper understanding of the subject matter. Furthermore, visits to industry events were ranked highly with a strong median value of 8.0 and a mean value of 7.57. There also exists a strong consensus among experts on field trips to construction sites which possessed a median value of 8.0 and a mean value of 7.57. Furthermore, experts also ranked integration of technical competitions highly with a median value of 8.0 and a mean value also of 7.57. Through the inclusion of technical competitions into educational curricula, students are presented with the opportunity to experience deeper learning opportunities due to the healthy competitions such competitions present [2]. Both student government participation and engagement in sports activities were also ranked highly with median values of 7.0 and mean values of 6.86 and 6.50, respectively.

6 Conclusion

Given the various dynamics facing the present-day construction industry in the era of the 'new normal', it is essential for institutions of higher learning to revisit their existing undergraduate curricula and make modifications, and revamps were necessary to produce skilled graduates who will meet the expectations of employers. Aside from been academically sound, the industry requires graduates who are sufficiently furnished with the relevant skills, abilities, and competencies to fit into the world of work after graduation. From relevant works of literature, industry employers place premium value on graduates who can exhibit confidence in communicating effectively; work as part of a team when necessary; possess good ethics; think critically; show a willingness to learn; flexible and adaptable; possess analytical skills; understand the dynamics of information and communication technology (ICT skills); possess problem-solving skills among many others. It is against this backdrop that this study examines the pedagogical approaches that can be employed by institutions of higher learning to develop employability skills among built-environment students as online learning becomes the order of the day. In achieving the objectives of this study, a qualitative Delphi approach was adopted to validate these pedagogical approaches as universities across South Africa and beyond take measures in adapting to the dynamics of remote learning. For this study, the experts from the built environment who met the criteria for participation hailed from both industry practice and academia. As shown from the study after two rounds of Delphi analyses, all 16 approaches reached consensus with IQD values of 0.00. The study, therefore, achieved its stated objectives of validating these approaches among experts after two Delphi iterations.

As discussed throughout this research, the role of higher education in developing employability skills cannot be overstated. Hence, the implication of this study for higher education is to respond to the increased pressures it faces by enforcing meaningful and relevant teaching activities to ensure that their curricula develop graduates who are well equipped to handle industry positions in both design and supervisory roles after graduation. Universities are also expected to invest heavily in equipping educators with the required digital tools necessary for executing online lectures. The 'new normal' further requires universities to be more innovative in designing their pedagogical approaches as they demonstrate their reliability in producing quality-learning outcomes for students. This can be done by ensuring their curricula that are aligned to the 4IR technologies as the world embraces digitalization. More so, universities must adopt effective virtual learning environments to provide students with adequate educational resources that are essential for the learning process. Most of the pedagogical approaches discussed in this research are underpinned by the concept of active learning approaches in which students are actively engaged in the learning process. Unlike the passive approaches like traditional lectures, active-based learning promotes critical thinking skills, increases students' engagement, increases retention, and fosters problem-solving initiatives. The findings from this study are beneficial to higher education institutions, educators, students, policymakers, educational boards, Ministries of Education (MOE), researchers, industry professionals, and every stakeholder in the employability skills discussion.

References

1. Adler M, Ziglio E (1996) Gazing into the oracle: The Delphi method and its application to social policy and public health. Jessica Kingsley Publishers
2. Ahlgren DJ, Verner IM (2013) Socially responsible engineering education through assistive robotics projects: the robowaiter competition. Int J Soc Robot 5(1):127–138
3. Aigbavboa CO (2014) An integrated beneficiary centred satisfaction model for publicly funded housing schemes in South Africa. Doctoral dissertation, University of Johannesburg
4. Aigbavboa CO, Aghimien DO, Thwala WD, Ngozwana MN (2021) Unprepared industry meet pandemic: COVID-19 and the South Africa construction industry. J Eng Des Technol
5. Al Hassan IBM (2012) Multidisciplinary curriculum to teaching English language in Sudanese institutions (a case study). Theory Practice Lang Stud 2(2):402
6. Alexander RJ (2001) Culture and pedagogy: international comparisons in primary education. Blackwell, Oxford, pp 391–528
7. Aliu J, Aigbavboa C (2021) Reviewing problem-solving as a key employability skill for built environment graduates. Collaboration and integration in construction, engineering, management and technology. Springer, Cham, pp 399–403
8. Aliu J, Aigbavboa C, Oke A (2021) Examining undergraduate courses relevant to the built environment in the 4IR era: a Delphi study approach. Emerging research in sustainable energy and buildings for a low-carbon future. Springer, Singapore, pp 195–215
9. Aliu J, Aigbavboa C, Thwala W (2021) A 21st century employability skills improvement framework for the construction industry. Routledge
10. Aliu J, Aigbavboa C (2021a) Key generic skills for employability of built environment graduates. Int J Constr Manag 1–19
11. Aliu J, Aigbavboa CO (2019) Employers' perception of employability skills among built-environment graduates. J Eng Des Technol
12. Aliu J, Aigbavboa CO (2020) Structural determinants of graduate employability: impact of university and industry collaborations. J Eng Des Technol
13. Alviento S (2018) Effectiveness of the performance of the student government of North Luzon Philippines State College. Res Pedagogy 8(1):1–16
14. Anaman KA, Osei-Amponsah C (2007) Analysis of the causality links between the growth of the construction industry and the growth of the macro-economy in Ghana. Constr Manag Econ 25(9):951–961
15. Arowoiya VA, Oke AE, Aigbavboa CO, Aliu J (2020) An appraisal of the adoption internet of things (IoT) elements for sustainable construction. J Eng Des Technol
16. Ayarkwa J, Adinyira E, Osei-Asibey D (2012) Industrial training of construction students: perceptions of training organizations in Ghana. Education+ Training
17. Bailey R, Hillman C, Arent S, Petitpas A (2013) Physical activity: an underestimated investment in human capital? J Phys Act Health 10(3):289–308

18. Bartkus KR, Nemelka B, Nemelka M, Gardner P (2012) Clarifying the meaning of extra-curricular activity: a literature review of definitions. Am J Bus Educ 5(6):693–704
19. Bell S (2010) Project-based learning for the 21st century: skills for the future. Clearing House 83(2):39–43
20. Bernstein B (2000) Pedagogy, symbolic control, and identity: theory, research, critique, vol 5. Rowman & Littlefield
21. Bhattacharjee S, Ghosh S (2013) Usefulness of role-playing teaching in construction education: a systematic review. In: 49th ASC annual international conference, San Luis Obispo, CA
22. Cao J, Thomas D (2013) When developing a career path, what are the key elements to include?
23. Clanchy K, Sabapathy S, Reddan G, Reeves N, Bialocerkowski A (2019) Integrating a career development learning framework into work-integrated learning practicum debrief sessions. Augmenting health and social care students' clinical learning experiences. Springer, Cham, pp 307–330
24. Creed PA, French J, Hood M (2015) Working while studying at university: the relationship between work benefits and demands and engagement and well-being. J Vocat Behav 86:48–57
25. Creswell JW, Creswell JD (2017) Research design: qualitative, quantitative, and mixed methods approaches. Sage Publications
26. Doyle DE (2012) Developing occupational programs: a case study of community colleges. J Technol Stud 38(1):53–62
27. Dustmann C, Van Soest A (2008) Part-time work, school success and school leaving. In: The economics of education and training, pp 23–45. Physica-Verlag HD
28. Green RA (2014) The Delphi technique in educational research. SAGE Open 4(2):2158244014529773
29. Hall M, Lasby D, Ayer S, Davis WG (2009) Caring Canadians, involved Canadians: highlights from the 2007 Canada survey of giving, volunteering and participating. Minister of Industry, Ottawa
30. Hallowell MR, Gambatese JA (2010) Qualitative research: application of the Delphi method to CEM research. J Constr Eng Manag 136(1):99–107
31. Hasson F, Keeney S, McKenna H (2000) Research guidelines for the Delphi survey technique. J Adv Nurs 32(4):1008–1015
32. Holey EA, Feeley JL, Dixon J, Whittaker VJ (2007) An exploration of the use of simple statistics to measure consensus and stability in Delphi studies. BMC Med Res Methodol 7(52):1–10
33. Impey C (2020) Coronavirus: social distancing is delaying vital scientific research. The Conversation [Internet]
34. Kinash S, Crane L, Judd MM, Knight C (2016) Discrepant stakeholder perspectives on graduate employability strategies. High Educ Res Dev 35(5):951–967
35. Kinash S, Crane L, Knight C, Dowling D, Mitchell K, McLean M, Schulz M (2014) Global graduate employability research: a report to the Business20 human capital taskforce (draft)
36. Kuijpers MACT, Scheerens J (2006) Career competencies for the modern career. J Career Dev 32(4):303–319
37. Landrum RE (2015) Career development courses: preparing psychology majors for the workplace
38. Lavy S, Naama-Ghanayim E (2020) Why care about caring? Linking teachers' caring and sense of meaning at work with students' self-esteem, well-being, and school engagement. Teach Teach Educ 91:103046
39. Leung L (2016) Validity, reliability, and generalizability in qualitative research. J Fam Med Prim Care 4(3):324 (2015)
40. Levesque LL, O'Neill RM, Nelson T, Dumas C (2005) Sex differences in the perceived importance of mentoring functions. Career Dev Int 10(6/7):429–443
41. MacCallum J, Casey SC (2017) Enhancing skills development and reflective practise in students during their programme of study. New Dir Teach Phys Sci (12)
42. McIntyre D, Hagger H (2018) Mentoring: challenges for the future. In: Mentors in Schools (1996), pp 144–164. Routledge

43. Osunsanmi TO, Aigbavboa CO, Thwala WDD, Molusiwa R (2021) Modelling construction 4.0 as a vaccine for ensuring construction supply chain resilience amid COVID-19 pandemic. J Eng Des Technol
44. Oyedotun TD (2020) Sudden change of pedagogy in education driven by COVID-19: perspectives and evaluation from a developing country. Res Globalization 2:100029
45. Parker J (2018) Undergraduate research, learning gain and equity: the impact of final year research projects. High Educ Pedagogies 3(1):145–157
46. Parker EA, Myers N, Higgins HC, Oddsson T, Price M, Gould T (2009) More than experiential learning or volunteering: a case study of community service learning within the Australian context. High Educ Res Dev 28(6):585–596
47. Qing XU (2011) Role-play-an effective approach to developing overall communicative competence. Cross-Cultural Commun 7(4):36–39
48. Rayens MK, Hahn EJ (2000) Building consensus using the policy Delphi method. Policy Polit Nurs Pract 1(2):308–315
49. Rogers MR, Lopez EC (2002) Identifying critical cross-cultural school psychology competencies. J Sch Psychol 40(2):115–141
50. Ryder J (2004) What can students learn from final year research projects? Biosci Educ 4(1):1–8
51. Shaw K, Holbrook A, Bourke S (2013) Student experience of final-year undergraduate research projects: an exploration of 'research preparedness.' Stud High Educ 38(5):711–727
52. Talley NL (2018) Soft skill development for generation Z: through strengths based approach to internships. Doctoral dissertation, University of Phoenix)
53. Telford RD, Cunningham RB, Fitzgerald R, Olive LS, Prosser L, Jiang X, Telford RM (2012) Physical education, obesity, and academic achievement: a 2-year longitudinal investigation of Australian elementary school children. Am J Public Health 102(2):368–374
54. Thoonen EE, Sleegers PJ, Oort FJ, Peetsma TT, Geijsel FP (2011) How to improve teaching practices: the role of teacher motivation, organizational factors, and leadership practices. Educ Adm Q 47(3):496–536
55. Watkins C, Mortimore P (1999) Pedagogy: What do we know. In: Understanding pedagogy and its impact on learning, pp 1–19
56. Westbrook J, Durrani N, Brown R, Orr D, Pryor J, Boddy J (2014) Pedagogy, curriculum, teaching practice & teacher education in developing countries. A rigorous literature review
57. Yousuf MI (2007) Using experts' opinions through Delphi technique. Pract Assess Res Eval 12(1):4

Chapter 3
Relating Work-Integrated Learning to Employability Skills in the Post-COVID-19 Era

John Aliu and Clinton Aigbavboa

Abstract Due to COVID-19, there have been several disruptions to the status quo, which has triggered several changes within the workspace as well as the higher education sector. As the latter seeks to readjust its teaching and learning approach by embracing online and distance learning models, one of the most critical components of the built environment, work-integrated learning (WIL), needs to be given greater attention in the era of the 'new normal'. This is because, as the world embraces this new wave of digitalization, future graduates are required to possess innovative skills and attributes to transit easily into the world of work ('new economy'). This article seeks to re-evaluate the importance of WIL as higher education institutions (HEIs) around the world reshape their curriculum in a bid to adjust to the significant changes caused by COVID-19. For this study, a quantitative research approach was adopted with close-ended questionnaires developed and administered to built environment professionals based in the Gauteng province of South Africa. Data obtained were analyzed using several statistical tools such as descriptive statistics (DS), mean item score (MIS), one-sample T-test (OST), and exploratory factor analysis (EFA). Findings revealed four clusters highlighting the key WIL attributes that are critical to graduate success. These include understanding of job responsibilities, enhanced learning, exposure to multi-disciplinary teams, and developing professional identity. While it may be easy for present-day universities to overlook WIL due to the various intricacies in evaluating and monitoring students' participation due to online learning, this article makes a case for universities to adopt a more innovative approach to ensure that WIL is successfully executed due to its employability benefits for students. The outcomes of this study will be beneficial to university educators, higher education officials, policymakers, and even students.

Keywords Built environment · Construction industry · Covid-19 · Employability · Pedagogy · Skills · Work-integrated learning

J. Aliu (✉) · C. Aigbavboa
cidb Centre of Excellence, University of Johannesburg, Johannesburg, South Africa

© The Author(s), under exclusive license to Springer Nature Singapore Pte Ltd. 2022
R. J. Howlett et al. (eds.), *Smart and Sustainable Technology for Resilient Cities and Communities*, Advances in Sustainability Science and Technology,
https://doi.org/10.1007/978-981-16-9101-0_3

1 Introduction

As the world gradually aligns with the concept of the 'new normal', which is propelled by the COVID-19 pandemic, it remains to be seen how the educational sector will successfully adjust to meet the learning needs of students. In a bid to control the spread of the coronavirus disease, several educational institutions globally have adopted a remote learning (online learning or virtual learning) approach [12]. For built environment disciplines (areas that deals with the development of human-made environment), the impact of the COVID-19 pandemic has been huge because work-integrated learning (WIL) opportunities have been hampered and in some cases have been overlooked due to the ongoing disruption to global economies. It is widely known that the global pandemic resulted in the closure of several businesses and industrial firms, which resulted in reduced work placement opportunities for students. In addition, with the switch from on-site to online learning, WIL activities have either been shifted online or even cancelled altogether. This switch prompted by the COVID-19 pandemic has been regarded as 'panic-gogy' as noted by Dean and Campbell [13]. However, the adoption of online platforms for WIL is not new as observed by Grace and O'Neil [21] and Larkin and Beatson [26]. Zegwaard [42] likewise noted that the viability of this approach requires more research as well as the need to investigate the experience of students who participate in WIL activities online [35]. Nevertheless, some challenges of remote WIL include the absence of real-time opportunities for students to directly learn, limited exposure to the communication processes which occur within the workplace and limited productivity [6, 9]. Despite these challenges [30], opined that new learning opportunities are generated through online WIL as students can develop self-management skills and improved digital skills.

Many have predicted that the 'new normal' has come to stay and understandably so. The impact of the COVID-19 pandemic has caused significant disruptions that would take the world some years to recover from. According to Zegwaard et al. [43], the global economic repercussions of the COVID-19 pandemic will be massive and will linger for a while. With global GDP's of economies shrinking, deep national recessions have been predicted which will result in a further spike in unemployment rates as economic activities and investments struggle to break even. The implications of these are reduced work placement opportunities for students, which would require universities to provide alternatives and innovative solutions. Such alternatives could include providing an alternative subject or module for students to complete their qualification, redesigning the conventional placement-based WIL approach by integrating a virtual or simulated approach or scraping the program off the academic calendar. The main goal of this article is to re-evaluate the importance of WIL as HEIs around the world reshape their curriculum in a bid to adjust to the significant changes caused by COVID-19.

2 Review of Literature

2.1 Integrating Technology into the Work-Integrated Learning Concept

As the world embraces the technologies accompanying the fourth industrial revolution, it has become imperative that universities align with this 'technological revolution' to ensure its effectiveness in producing adequately skilled graduates [2]. Presently, the impact of technology across the world can be felt across various sectors, and it is germane to examine how work-integrated learning (WIL) activities can leverage on 4IR technologies to ensure its employability impact on students. While the WIL discussion in the COVID-19 era continues to gain the attention of the research community, it is important to ensure that WIL activities must meet the needs of the three key stakeholders: academia, industry, and students [40]. This study adopts the definition of WIL according to the International Journal of Work-Integrated Learning (IJWIL), which describes it as 'an educational approach that uses relevant work-based experiences to allow students to integrate theory with the meaningful practice of work as an intentional component of the curriculum. Defining elements of this educational approach require that students engage in authentic and meaningful work-related tasks, and must involve three stakeholders: the student, the university, and the workplace/community.' The IJWIL also clarified that 'such practice includes work placements, work-terms, internships, practicum, cooperative education (co-op), fieldwork, work-related projects/competitions, service learning, entrepreneurship, student-led enterprise, applied projects, simulations' (IJWIL [24], para. 3). Therefore, leaning towards the definition offered by the IJWIL, this study defines WIL as an educational approach that requires students to undergo work-based experiences which provides them with an opportunity to apply the principles and theory learnt in classroom in an actual work environment.

As the world currently grapples with the effects of the COVID-19 pandemic, several industries have developed innovative ways of working remotely. This implies that as the world embraces this new paradigm shift termed the 'new normal,' terms such as 'virtual WIL and online WIL' will become more prevalent. This has prompted studies like Wood et al. [40] and Chernikova et al. [8] to highlight terminologies such as remote WIL (working remotely through online platforms) and simulated WIL (an immersive WIL experience through virtual reality and simulations). While WIL may be virtual or simulated, Trede and Flowers [38] earlier stated that caution must be taken, as online connectedness does not automatically lead to effective learning, as there exist several dynamics regarding the use of technology in the WIL process. Nevertheless, this study makes a case for universities across the globe to align themselves with 4IR technologies in the wake of the COVID-19 pandemic, by revamping their curricula to ensure the survival of WIL activities, as their employability benefits are key to producing the future workforce of the construction industry.

2.2 Employability Implications of WIL Activities

While universities and industries globally have directly or indirectly experienced the effects of the COVID-19 pandemic, students may have been hit the hardest. With most universities and industries shifting from on-site to online activities, there exist fears that WIL activities may become deemed surplus to requirements if not adequately advocated for. With limited opportunities for students across industries partly because of the economic strains caused by the COVID-19 pandemic, universities will be required to be proactive in their approach to ensure WIL activities are not scraped off the academic calendar. Therefore, this study makes a case for WIL activities to be maintained in the academic setting due to the several employability benefits for students [5].

The positive impacts of WIL on the employability of built environment graduates have been discussed extensively across the globe. By engaging in WIL activities, students develop practical knowledge and acquire professional skills. The work opportunities also help students to realize their strengths, weaknesses, and expectations of their chosen professions [1]. More so, graduates who possess adequate work experience are often considered more desirables hires as observed by Wasonga and Murphy [39]. Through WIL activities, students are presented with opportunities to apply theoretical knowledge garnered from the classroom and to fulfill their academic program requirements. Aliu and Aigbavboa [5] also stated that apart from improving academic performance, work activities could also improve student curriculum vitae (CV), overall skills, and abilities as well as softening the transitioning from lecture room experience to the world of work. Table 1 provides a summary of the various employability impacts of WIL.

3 Research Methodology

This study adopted a quantitative approach due to its tendencies to deliver an unbiased analysis of mathematical, numerical, and statistical data. As noted by Aliu and Aigbavboa [3], several approaches and instruments can be employed to collect data such as case studies, focus groups, group discussions, oral histories, survey studies, and well-structured questionnaires. In this study which centers around WIL activities, secondary data was obtained by critically reviewing peer-reviewed publications such as research books, book chapters, journals, and conference articles. Meanwhile, primary data was obtained through a well-structured questionnaire that was designed based on several work-integrated learning (WIL) variables which were retrieved from existing employability studies. For this study, due to the ease of data collection and time-saving propensities, a close-ended set of questions were adopted using a 5-point Likert scale in obtaining responses. The target population for this study involved relevant professionals from the Councils for the Built Environment Professions (CBEP) from academia, construction industry, and government

Table 1 Impacts of WIL on employability

Code	Pedagogical approaches	Literature sources
WIL1	Acquisition of professional skills	Omar et al. [32]; Sattler et al. [36]
WIL2	Awareness of workplace culture	Sattler et al. [36]; Aliu and Aigbavboa [5]
WIL3	Boosted employment chances	Lowden et al. [27]
WIL4	Developed interpersonal values	Aliu and Aigbavboa [4]
WIL5	Developed practical knowledge	Lam and Ching (2006)
WIL6	Developing professional identity	Gill and Lashine [20]; Mihail [29]
WIL7	Enhanced learning	Callanan and Benzing [7]; Sattler et al. [36]
WIL8	Exposure to career opportunities	Lowden et al. [27]; Aliu and Aigbavboa [3]
WIL9	Exposure to multi-disciplinary teams	Aliu and Aigbavboa [3]
WIL10	Financial management skills	Aliu and Aigbavboa [3]
WIL11	Improved academic performance	Gault et al. [17]; Mihail [29]
WIL12	Knowledge of engineering designs	Aliu and Aigbavboa [3]
WIL13	Knowledge of ongoing issues	Omar et al. [32]
WIL14	Knowledge of quality control	Aliu and Aigbavboa [3]
WIL15	Mentorship opportunities	Sattler et al. [36]; Gault et al. [17]
WIL16	Networking with other interns	Aliu and Aigbavboa [3]
WIL17	Positive work attitude	Aliu and Aigbavboa [3]
WIL18	Site-supervision knowledge	Aliu and Aigbavboa [3]
WIL19	Technical report writing	Aliu and Aigbavboa [3]
WIL20	Understanding job responsibilities	Mihail [29]
WIL21	Understanding professional ethics	Wasonga and Murphy [39]; Sattler et al. [36]

Source Author's compilation

(including architects, construction managers, construction project managers, engineers, and quantity surveyors) in South Africa. In achieving this study's aims, the total number of registered and candidate members of the various relevant built environment professions was obtained from the annual reports as provided by the CBEP Web site.

Two categories of non-probability sampling techniques were considered for the research, namely purposive (judgment) and snowball techniques. The purposive sampling technique was considered because it relies on the judgment of the researcher when it comes to selecting the population that are of interest to the study. Through the purposive technique, industry professionals and academicians who were in attendance during South African conferences at the time were given the opportunity to participate in the study. This is because conferences provided a perfect opportunity for the built environment professionals to converge and share ideas on several issues relating to both academia and industry. Apart from being a time-effective sampling method, purposive sampling is also cost effective. In addition to the purposive technique, snowball sampling (chain sampling) was also adopted. Participants were requested to identify other professionals (referrals) who could contribute to the realization of the study. Like the purposive technique, the snowball sampling technique was also time and cost efficient. Apart from its proximity to the researcher, the Gauteng province was selected because the city contributes enormously to various sectors in South Africa such as construction, manufacturing, technology, and finance, among others, making it the nation's economic hub. Several mathematical formulas were considered to achieve the required sample size. The formula proposed by Yamane [41] was adopted for this study which yielded a target population size of 220 respondents. However, 204 responses were retrieved, indicating a 92% response rate.

Data cleaning and screening were carried out by the Statistical Consultation Service experts (STATKON) of the University of Johannesburg before the commencement of data analysis. Subsequently, the cleaned data were analyzed using several statistical tools such as descriptive statistics (DS), mean item score (MIS), one-sample T-test (OST), and exploratory factor analysis (EFA). Also, the reliability and validity of the research instrument were determined. The Cronbach's alpha value of 0.909 was recorded, indicating high reliability of the questionnaire survey. This was supported by Hair et al. [22] who suggested that the higher the values, the higher the reliability. On the other hand, the validity was achieved by pilot testing the questionnaires on a small sample as recommended by Ticehurst and Veal [37]. A pilot study helps to eliminate any flaws and weaknesses of the survey instrument before the main study. However, results from the pilot study were not analyzed and integrated into this current research. Instead, the pilot study helped to fine-tune the instructions given to respondents and ultimately improving the overall structure and grammatical patterns of the questionnaire. Subsequently, the normality of data was conducted using the Shapiro–Wilk normality test. According to Ghasemi and Zahediasl [19], the Shapiro–Wilk normality test is usually conducted to check the normality of data when a sample size of less than 2000 is considered such as in this study. The test conducted revealed that the variables were not normally distributed as p-values of all assessed skills had values above the 0.05 threshold for normality, allowing for a parametric test to be conducted. Thus, one-sample T-test, a parametric test used to ascertain the significant difference in opinions of different categories of respondents was adopted for this study to check the consistency of opinions of sampled professionals.

4 Findings and Discussions

4.1 Respondents' Background Information

This study conducted frequency distributions of the data of the participants (professionals from the built environment from academia, government establishments, and the construction industry) to obtain full profile of respondents. Background data obtained were respondents' level of education, original professional qualifications, years of experience, and type of institution to which their organizations. These data were required to ascertain the experience and knowledgeability of the respondents to enhance the credibility of the data provided. Findings from the descriptive statistics revealed that 29.4% of the total population possessed a master's degree, followed by 28.4% with a bachelor's degree. In addition, 27.0% possess an honors degree, while the respondents with doctorates and diplomas constituted 7.4% and 7.8% of the total population, respectively. The high percentage of respondents with a master's degree indicates the high quality of responses obtained during the field survey. The majority of the respondents were engineers, representing 36.3% of the total population, followed by quantity surveyors representing 23.0% and construction project managers representing 14.2%, while architects and construction managers each accounted for 13.2% of the population. Also, 48% of the respondents had between 1 and 5 years of work experience, 34.8% had 6–10 years, 10.3% had 11–15 years, 2.5% had 16–20 years, while 4.4% of the respondents had more than 20 years of work experience. This implies that 52% of the respondents had more than five years of work experience which suggests that the respondents who contributed to this research were well knowledgeable in their respective disciplines within the built environment. This improves the validity of the outcomes generated from this study. Furthermore, 56.4% of the respondents' belonged to the private sector, while 43.6% of the respondents' belonged to the public sector and organizations. Finally, most of the respondents were from the construction industry with 41.7% of the total population, while 36.3% were from higher education institutions and 22.1% were consultants from government establishments. This even distribution of the respondents provided a holistic perspective as employers, academicians, and consultants are relevant to the graduate employability discussion. Hence, their opinions on the subject matter could be used to draw conclusions and make inferences.

4.2 Mean Rankings of Work-Integrated Learning Factors

In determining the consistency of respondents' opinions, several parameters were used in this regard such as the mean values and Kruskal–Walis H-test. For each of the constructs, the perspectives of professionals were obtained. Table 2 shows the importance of WIL for graduate employability as ranked by the various groups of respondents. From the table, all WIL factors were ranked as important with a significant mean value of 3.0 and above which was adequate. From the table, 'developed practical knowledge,' 'technical report writing,' 'enhanced learning,' 'acquisition of professional skills,' and 'understanding professional ethics' were the highest ranked WIL factors with mean values of 4.39, 4.36, 4.32, 4.31, and 4.31, respectively. Across the three categories of respondents, these five factors were ranked highly. However, across the three categories of respondents, there were some interesting views. While the overall highest-ranked factor is 'providing practical knowledge,' consultants from the government ranked 'knowledge of engineering designs' as the most important

Table 2 Mean ranking of work-integrated learning factors

Work-integrated learning factors	University		Industry		Govt.		Overall	
	Mean	Rank	Mean	Rank	Mean	Rank	Mean	Rank
Developed practical knowledge	4.46	2	4.31	3	4.42	3	4.39	1
Technical report writing	4.36	4	4.39	1	4.31	10	4.36	2
Enhanced learning	4.39	3	4.24	6	4.38	7	4.32	3
Acquisition of professional skills	4.26	6	4.27	4	4.49	2	4.31	4
Understanding professional ethics	4.23	9	4.33	2	4.40	5	4.31	5
Mentorship opportunities	4.47	1	4.11	12	4.33	8	4.29	6
Knowledge of engineering designs	4.32	5	4.13	11	4.51	1	4.28	7
Improved academic performance	4.22	10	4.26	5	4.40	5	4.27	8
Understanding job responsibilities	4.24	7	4.21	7	4.18	14	4.22	9
Knowledge of quality control	4.20	11	4.15	10	4.27	12	4.20	10
Positive work attitude	4.09	17	4.20	8	4.33	8	4.19	11
Exposure to multi-disciplinary teams	4.24	7	4.08	15	4.29	11	4.19	12
Knowledge of ongoing issues	4.14	15	4.11	12	4.42	3	4.19	13
Developing professional identity	4.18	12	4.16	9	4.24	13	4.19	14
Awareness of work place culture	4.15	14	4.11	12	4.11	18	4.12	15
Developed interpersonal values	4.12	16	4.06	17	4.18	15	4.11	16
Exposure to career opportunities	4.16	13	3.98	18	4.18	15	4.09	17
Site-supervision knowledge	4.00	18	4.08	15	4.18	15	4.07	18
Networking with other interns	3.97	19	3.91	19	4.00	19	3.95	19
Boosted employment chances	3.47	20	3.55	21	3.60	20	3.53	20

(continued)

Table 2 (continued)

Work-integrated learning factors	University		Industry		Govt.		Overall	
	Mean	Rank	Mean	Rank	Mean	Rank	Mean	Rank
Financial management skills	3.24	21	3.79	20	3.27	21	3.48	21

WIL factor. Considering the job of consultants is to advise and share their built environment expertise in both design and supervisory capacities, it is no surprise that they ranked the learning of engineering designs highly. On the other hand, industry professionals selected 'technical report writing' as the most significant factor of WIL. As industry professionals, employers are constantly seeking graduates who can convey technical information on construction projects effectively in a clear and comprehensible manner. This is one of the main reasons why these sets of respondents' value technical report writing higher than any other WIL factor. Furthermore, from the table, it appears that respondents from academia ranked 'mentorship opportunities' as the most significant factor of WIL. This is not surprising because mentorship forums are highly revered by lecturers and academics. This is one of the main reasons why higher education includes WIL into their curricula to ensure students experience work opportunities before graduation. Through these work opportunities, students are assigned to industry professionals to improve their learning experience and hence, mentorship opportunities. Also from the table, the least ranked factor was financial management skill with a mean value of 3.48. This is possible because the main reason for WIL is to present work opportunities for students to ease their transition from the classroom to the world of work. In most cases, these opportunities are not monetized; hence, the issue of managing finances was not considered overly important to the employability discussion.

Subsequently, a one-sample t-test was conducted to further ascertain the level of importance of these WIL. Gleaning toward the study of Elliott and Woodward [15], the null hypothesis for each WIL factor was set at unimportant when (H_0: $U = U_0$). On the other hand, the alternative hypotheses state that the WIL factor was deemed important when (H_a: $U > U_0$). For both assumptions, U_0 represents the population mean, and in this study, it was pegged at 3.0, while the significance level was pegged at 95% confidence interval. Based on this, a WIL factor was considered important when it possesses a mean of 3.0 and above. In situations where two WIL factors have equal mean values, then the factor with the lowest standard deviation (SD) is allocated the highest ranking of importance as proposed by Othman et al. [33]. Therefore, in the case of 'acquisition of professional skills' and 'understanding of professional ethics' which had equal mean values, the SD of 'acquisition of professional skills' was 0.680, while 'understanding of professional ethics' was 0.657. Hence, 'understanding of professional ethics' was allocated the highest ranking of importance. As noted by Hassani et al. [23], the standard error (SE) is the standard deviation of the sample mean which is a measure of how a sample represents the population under question. This means, if the study was repeated several times, the SE represents the variability of the mean values. Therefore, a large SE indicates several differences among various sample means, while a small SE indicates similarities between most sample means and population mean [31].

From Table 3, the SE associated with respective means is close to zero, indicating that the sample truly reflects the population. A more critical look at Table 3 shows that the SD of 20 out of the 21 factors are less than 1.0, indicating little data variability. Hence, there is consistency in the agreement among respondents regarding these 20 factors. The SD of financial management skill was above 1.0, indicating that there might be some divergent opinions by the respondents on the ranking of that factor. Table 3 also shows the *p-value* that highlights the significance of each WIL factor. More so, the one-sample *t*-test significance values at two-tailed all fall below the 0.05 threshold, indicating no statistically significant differences in respondents' opinions. This further indicates consistency in the ranking of the factors based on the opinions of the respondents.

4.3   Exploratory Factor Analysis (EFA)

The 21 WIL factors were further subjected to exploratory factor analysis (EFA) using Statistical Package for the Social Sciences (SPSS) version 26. Fabrigar and Wegener [16] opined that EFA is often conducted to determine the possible correlation patterns that exist in a given dataset, which is then used to extract variables into various factor clusters. SPSS checked the data suitability for EFA through the correlation matrix. During the check, satisfactory coefficients were observed from the communalities extraction table which indicates suitability for factor analysis. From the communalities table given in Table 4, values between 0.40 and 0.70 were recorded which is in accordance with the studies of Costello and Osborne [11]. These values indicate the appropriateness of the variables measuring WIL.

All 21 WIL factors were subjected to principal component analysis (PCA). The Kaiser–Meyer–Olkin (KMO), which is the measure of sampling adequacy yielded 0.900, was above the 0.6 thresholds [22]. The result of Bartlett's test of sphericity yielded a high chi-squared value of 1797.587, with an associated significance level (Sig.) of 0.000, which is less than 0.050. Thus, the variables are factorable and are suitable for EFA [18]. As given in Table 4, a Cronbach's alpha of 0.909 was recorded which indicates a high internal consistency and reliability. Table 4 shows the total variance of the variables indicated by the eigenvalues using Kaiser's criterion. In this case, variables with eigenvalues >1 will be retained which means four WIL factors with eigenvalues above 1 resulted in 7.765, 1.955, 1.162, and 1.144 which explains 36.974%, 9.308%, 5.535%, and 5.448% of the variance, respectively. These four clusters of WIL have a cumulative 57.265% of the total importance of the 21WIL factors as shown in Table 5.

Table 3 Statistics of one-sample *t*-test

Work-integrated learning factors	Std. deviation	Test value = 3.0		95% confidence interval of the difference	
		Std. error mean	Sig. (2-tailed)	Lower	Upper
Developed practical knowledge	0.710	0.050	0.000	1.29	1.49
Technical report writing	0.640	0.045	0.000	1.27	1.45
Enhanced learning	0.661	0.046	0.000	1.23	1.41
Understanding of professional ethics	0.657	0.046	0.000	1.22	1.40
Acquisition of professional skills	0.680	0.048	0.000	1.22	1.41
Mentorship opportunities	0.788	0.055	0.000	1.18	1.40
Knowledge of engineering designs	0.734	0.051	0.000	1.18	1.39
Improved academic performance	0.745	0.052	0.000	1.17	1.38
Understanding job responsibilities	0.675	0.047	0.000	1.12	1.31
Knowledge of quality control measures	0.688	0.048	0.000	1.10	1.29
Positive work attitude	0.714	0.050	0.000	1.09	1.29
Exposure to multi-disciplinary teams	0.719	0.050	0.000	1.09	1.29
Knowledge of ongoing issues	0.733	0.051	0.000	1.09	1.29
Developing professional identity	0.739	0.052	0.000	1.08	1.29
Awareness of work place culture	0.722	0.051	0.000	1.02	1.22
Developed interpersonal values	0.679	0.048	0.000	1.01	1.20
Exposure to career opportunities	0.751	0.053	0.000	0.98	1.19
Site-supervision knowledge	0.702	0.049	0.000	0.98	1.17
Networking with other interns on-site	0.734	0.051	0.000	0.85	1.05
Boosted employment chances	0.912	0.064	0.000	0.41	0.66

(continued)

Table 3 (continued)

Work-integrated learning factors	Std. deviation	Test value = 3.0		95% confidence interval of the difference	
		Std. error mean	Sig. (2-tailed)	Lower	Upper
Financial management skill	1.205	0.084	0.000	0.31	0.64

Table 4 Communalities and reliabilities for WIL factors

	Initial	Extraction
Communalities		
Developed practical knowledge	1.000	0.684
Understanding of job responsibilities	1.000	0.523
Positive work attitude	1.000	0.645
Developed interpersonal values	1.000	0.524
Boosted employment chances	1.000	0.579
Awareness of work place culture	1.000	0.533
Understanding of professional ethics	1.000	0.680
Developing professional identity	1.000	0.677
Knowledge of ongoing issues	1.000	0.585
Improved academic performance	1.000	0.614
Exposure to multi-disciplinary teams	1.000	0.562
Knowledge of engineering designs	1.000	0.594
Knowledge of quality control measures	1.000	0.700
Site-supervision knowledge	1.000	0.542
Financial management skill	1.000	0.514
Networking with other interns on-site	1.000	0.528
Exposure to career opportunities	1.000	0.605
Enhanced learning	1.000	0.570
Acquisition of professional skills	1.000	0.605
Technical report writing	1.000	0.586
Mentorship opportunities	1.000	0.575
KMO and Bartlett's Test		
Kaiser–Meyer–Olkin measure of sampling adequacy		0.900
Bartlett's test of sphericity	Approx. chi-square	1797.587
	Df.	210
	Sig.	0.000
Cronbach's alpha		0.909

Table 5 Total variance explained and pattern matrix for WIL factors

Component	Initial eigenvalues			Extraction sums of squared loadings		
	Total	% of variance	Cumulative %	Total	% of variance	Cumulative %
1.	7.765	36.974	36.974	7.765	36.974	36.974
2.	1.955	9.308	46.282	1.955	9.308	46.282
3.	1.162	5.535	51.818	1.162	5.535	51.818
4.	1.144	5.448	57.265	1.144	5.448	57.265

Component

	1	2	3	4
Understanding of job responsibilities	0.704			
Awareness of workplace culture	0.690			
Boosted employment chances	0.651			
Positive work attitude	0.640			
Developed interpersonal values	0.571			
Financial management skill	0.518			
Enhanced learning		0.705		
Acquisition of professional skills		0.672		
Technical report writing		0.654		
Mentorship opportunities		0.605		
Exposure to career opportunities		0.565		
Developed practical knowledge		0.520		
Exposure to multi-disciplinary teams			0.773	
Knowledge of engineering designs			0.688	
Knowledge of quality control measures			0.678	
Site-supervision knowledge			0.624	
Networking with other interns on-site			0.584	
Developing professional identity				0.717
Knowledge of ongoing issues				0.651
Understanding of professional ethics				0.631
Improved academic performance				0.610

Cluster 1—Understanding of Job Responsibilities: A total of six factors loaded onto this cluster and they are 'Understanding of job responsibilities' (70.4%), 'Awareness of workplace culture' (69%), 'Boosted employment chances' (65.1%), 'Positive work attitude' (64%), 'Developed interpersonal values' (57.1%), and 'Financial management skill' (51.8%). These WIL factors all relate to the competencies for graduates to understand the various requirements of their chosen professions, and they explain a cumulative percentage variance of 36.974% of the total variance. By

engaging in WIL opportunities, students are presented with the necessary knowledge of the roles and responsibilities of a typical job position to function effectively in a given task [39]. By understanding the requirements needed to successfully execute their tasks, students gain a holistic understanding of the workplace culture as they interact with several professionals both on-site and off-site. As observed from the cluster, WIL also develops interpersonal values among students. This overarching factor encapsulates several other skills such as communication skills, time-management skills, and teamwork skills. This resonates with several studies such as those of Divine et al. [14] and McLennan and Keating [28]. Graduates who possess adequate work experience are often considered to be more desirable's hires [25].

Cluster 2—Enhanced Learning: A total of six factors loaded onto this cluster, and they are 'Enhanced learning' (70.5%), 'Acquisition of professional skills' (67.2%), 'Technical report writing' (65.4%), 'Mentorship opportunities' (60.5%), 'Exposure to career opportunities' (56.5%), and 'Developed practical knowledge' (52%). These WIL factors all relate to the various learning opportunities that students are exposed to by virtue of undergoing WIL, and they explain a cumulative percentage variance of 9.308% of the total variance. By engaging in work activities, students are presented with enhanced knowledge about their career choices and related occupations. Consequently, students are presented with an opportunity to learn about the strengths, weaknesses, and expectations of their chosen fields and how they can explore other possible options. These confirm the works of Gault et al. [17], Gill and Lashine [20], Mihail [29], and Wasonga and Murphy [39]. According to Omar et al. [32], exposure to the world of work helps students to establish rapport with industry professionals as they seek to bolster their careers. Through exposure to career opportunities, students can also develop practical knowledge, understudy professionals (mentors), and acquire technical skills [10, 27, 36]. Professional skills that can be acquired by virtue of undergoing WIL include adaptability skills, problem-solving ability, technical report writing, time-management skills, and site-supervision knowledge [27, 34]. According to Wasonga and Murphy [39] and Sattler et al. [36], work activities help students to apply information garnered from lecture room activities which further improve their learning and thought processes as they commence their professional careers.

Cluster 3—Exposure to Multi-disciplinary Teams: A total of five factors are loaded onto this cluster, and they are 'Exposure to multi-disciplinary teams' (77.3%), 'Knowledge of engineering designs' (68.8%), 'Knowledge of quality control measures' (67.8%), 'Site-supervision knowledge' (62.4%), and 'Networking with other interns on-site' (58.4%). These WIL factors all relate to students' exposure to a range of different disciplines or fields of expertise while undergoing work experience and they explain a cumulative percentage variance of 5.535% of the total variance. As students interact with different professionals from different backgrounds during work experience, their learning process is enhanced and ultimately well rounded. For example, a civil engineering student in the construction site will be exposed to other professionals outside the structural dimension during the construction process such as architects, surveyors, mechanical engineers, and electrical engineers among several others. Such interactions provide students with adequate knowledge of the job process, thereby enhancing the learning process. Students are also afforded the opportunities to network with other interns which occurs as a result of introducing work experience into the undergraduate curriculum.

Cluster 4—Developing Professional Identity: A total of four factors are loaded onto this cluster, and they are 'Developing professional identity' (71.7%), 'Knowledge of ongoing issues' (65.1%), 'Understanding of professional ethics' (63.1%), and 'Improved academic performance' (61%). These WIL factors all relate to the development of a professional outlook by virtue of undergoing work experience, and they explain a cumulative percentage variance of 5.448% of the total variance. By working in a professional setting, students develop a plethora of both technical and soft skills that are critical to their profession. By interacting with professionals, students' attitudes, motives, and experiences are clearly defined in relation to their chosen profession. More so, students who are exposed to WIL are exposed to the various principles that govern the workplace such as values, ethics, and codes of conduct. Similarly, Gault et al. [17] and Wasonga and Murphy [39] suggest that work programs can lead to improved academic performance among students because students' interest and enthusiasm in the classroom are enhanced when work activities are integrated. Furthermore, as students engage in work activities, they become more aware of ongoing industry issues and global challenges facing their profession and the world at large.

5 Conclusions

This study contributed to the employability body of knowledge by discussing the relevance of work-integrated learning (WIL) to the employability discussion in the COVID-19 era. Data was obtained from professionals within the built environment from academia, government establishments, and the construction industry. Background data obtained were respondents' level of education, original professional qualifications, years of experience, and type of institution to which their organizations belong. Subsequently, the cleaned data was analyzed using several statistical tools such as descriptive statistics, mean item score, one-sample T-test, and exploratory factor analysis. Results from the EFA revealed four clusters of WIL—understanding of job responsibilities, enhanced learning, exposure to multi-disciplinary teams, and developing professional identity. As noticed from the study, these four WIL clusters are broad and encompass other related competencies that can be acquired, developed, and enhanced. This places significant pressure on universities to reshape and revamp its curricula to align with the dynamics of the COVID-19 era to prepare graduates who will fit seamlessly into the world of work after graduation. As stated earlier, the 'new normal' has come to stay and understandably so. As the world continues to recover from the economic setbacks caused by the COVID-19 pandemic, businesses and industries will struggle to break even which means a decrease in placement of opportunities for students. This possibility places immense pressure on universities to provide alternatives and innovative solutions to ensure WIL activities are not chalked off the academic calendar. Therefore, this study makes a case for the revamp of WIL activities to align with the 'new normal' by revisiting the various employability implications of work opportunities for students.

References

1. Aliu J, Aigbavboa C, Oke A (2021) Examining undergraduate courses relevant to the built environment in the 4IR era: a Delphi study approach. Emerging research in sustainable energy and buildings for a low-carbon future. Springer, Singapore, pp 195–215
2. Aliu J, Aigbavboa C, Thwala W (2021) A 21st century employability skills improvement framework for the construction industry. Routledge
3. Aliu J, Aigbavboa C (2021c) Key generic skills for employability of built environment graduates. Int J Constr Manag 1–19
4. Aliu J, Aigbavboa CO (2019) Employers' perception of employability skills among built-environment graduates. J Eng Des Technol
5. Aliu J, Aigbavboa CO (2020) Structural determinants of graduate employability: impact of university and industry collaborations. J Eng Des Technol
6. Bowen T, Pennaforte A (2017) The impact of digital communication technologies and new remote-working cultures on the socialization and work-readiness of individuals in WIL programs. In: Work-integrated learning in the 21st century. Emerald Publishing Limited
7. Callanan G, Benzing C (2004) Assessing the role of internships in the career-oriented employment of graduating college students. Educ Training 46(2):82–89. https://doi.org/10.1108/00400910410525261
8. Chernikova O, Heitzmann N, Stadler M, Holzberger D, Seidel T, Fischer F (2020) Simulation-based learning in higher education: a meta-analysis. Rev Educ Res 90(4):499–541. https://doi.org/10.3102/0034654320933544
9. Cho E (2020) Examining boundaries to understand the impact of COVID-19 on vocational behaviors
10. Cooper L, Orrell J, Bowden M (2010) Work integrated learning: a guide to effective practice. Routledge, New York
11. Costello AB, Osborne JW (2005) Best practices in exploratory factor analysis: four recommendations for getting the most from your analysis. Pract Assess Res Eval 10(7):1–9
12. Crawford J, Butler-Henderson K, Rudolph J, Malkawi B, Glowatz M, Burton R, Magni P, Lam S (2020) COVID-19: 20 countries' higher education intra-period digital pedagogy responses. J Appl Learn Teach 3(1):1–20
13. Dean BA, Campbell M (2020) Reshaping work-integrated learning in a post-COVID-19 world of work. Int J Work Integr Learn 21(4):355–364
14. Divine RL, Linrud JK, Miller RH, Wilson JH (2007) Required internship programs in marketing: benefits, challenges and determinants of fit. Mark Educ Rev 17(2):45–52. https://doi.org/10.1080/10528008.2007.11489003
15. Elliott AC, Woodward WA (2007) Comparing one or two means using the t-test. Statistical analysis and quick reference guide book. SAGE Publications Inc., Thousand Oaks, pp 47–76
16. Fabrigar LR, Wegener DT (2011) Exploratory factor analysis. Oxford University Press, University of Oxford, United Kingdom
17. Gault J, Redington J, Schlager T (2000) Undergraduate business internships and career success: are they related? J Mark Educ 22(1):45–53

18. George D, Mallery P (2003) SPSS for windows step by step: a simple guide and reference, 11.0 update, 4th edn. Allyn & Bacon, Boston
19. Ghasemi A, Zahediasl S (2012) Normality tests for statistical analysis: a guide for non-statisticians. Int J Endocrinol Metab 10(2):486–489
20. Gill A, Lashine S (2003) Business education: a strategic market-oriented focus. Int J Educ Manag 175:188–194. https://doi.org/10.1108/09513540310484904
21. Grace S, O'Neil R (2014) Better prepared, better placement: an online resource for health students. Asia Pac J Coop Educ 15(4):291–304
22. Hair JF, Black WC, Babin BJ, Anderson RE (2010) Multivariate data analysis. Pearson Education Inc., Pearson, NJ
23. Hassani H, Ghodsi M, Howell G (2010) A note on standard deviation and standard error. Teach Math Appl 29(2):108–112
24. IJWIL (n.d.) Int J Work Integrated Learn—Defining WIL. https://www.ijwil.org/defining-wil
25. Lam T, Ching L (2007) An exploratory study of an internship program: the case of Hong Kong students. Int J Hospitality Manag 26(2):336–351. https://doi.org/10.1016/j.ijhm.2006.01.001
26. Larkin I, Beatson A (2014) Blended delivery and online assessment: scaffolding student reflections in work-integrated learning. Mark Educ Rev 24(1):9–14
27. Lowden K, Hall S, Ellio DD, Lewin J (2011) Employers' perceptions of the employability skills of new graduates. Edge Foundation, London
28. McLennan B, Keating S (2008, June) Work-integrated learning (WIL) in Australian universities: the challenges of mainstreaming WIL. In: Proceedings of the ALTC NAGCAS national symposium, pp 2–14
29. Mihail DM (2006) Internships at Greek universities: an exploratory study. J Workplace Learn 18(1):28–41. https://doi.org/10.1108/13665620610641292
30. Ogunsanya OA, Aigbavboa CO, Thwala DW, Edwards DJ (2019) Barriers to sustainable procurement in the Nigerian construction industry: an exploratory factor analysis. Int J Constr Manag 1–12
31. Oke AE, Aghimien DO (2018) Drivers of value management in the Nigerian construction industry. JEDT 16(2):270–284
32. Omar MZ, Rahman MNA, Kofli NT, Mat K, Darus MZ, Osman SA (2008) Assessment of engineering students' perception after industrial training placement. In: Proceedings of 4th WSEAS/IASME international conference on educational technologies (EDUTE'08)
33. Othman AR, Yin TS, Sulaiman S, Ibrahim MIM, Rashid MR (2011) Application of mean and standard deviation in questionnaire surveys. Discov Math 33(1):11–22
34. Pillai S, Khan MH, Ibrahim IS, Raphael S (2012) Enhancing employability through industrial training in the Malaysian context. High Educ 63(2):187–204. https://doi.org/10.1007/s10734-011-9430-2
35. Pretti TJ, Parrott PAW, Hoskyn K, Fannon AM, Church D, Arsenault C (2020) The role of work-integrated learning in the development of entrepreneurs. Int J Work Integr Learn 21(4):451–466
36. Sattler P, Wiggers RD, Arnold C (2011) Combining workplace training with postsecondary education: the spectrum of work-integrated earning (WIL) opportunities from apprenticeship to experiential learning. Can Apprenticeship J 5. Retrieved from: http://www.adapt.it/fareap prendistato/docs/combining_workplace_training.pdf
37. Ticehurst GW, Veal AJ (2000) Business research methods. Longman, Frenchs Forest, Australia
38. Trede F, Flowers R (2020) Preparing students for workplace learning: short films, narrative pedagogy, and community arts to teach agency. Int J Work Integr Learn
39. Wasonga TA, Murphy JF (2006) Learning from tacit knowledge: the impact of the internship. Int J Educ Manag 20(2):153–163
40. Wood YI, Zegwaard KE, Fox-Turnbull WH (2020) Conventional, remote, virtual, and simulated work-integrated learning: a meta-analysis of existing practice
41. Yamane T (1967) Statistics, an introductory analysis, 2nd edn. Harper and Row, New York

42. Zegwaard KE (2015) Building an excellent foundation for research: challenges and current research needs. Asia Pac J Coop Educ 16(2):89–99
43. Zegwaard KE, Pretti TJ, Rowe AD (2020) Responding to an international crisis: the adaptability of the practice of work-integrated learning

Chapter 4
Opportunities and Barriers
of Digitization in the COVID-19 Crisis
for SMEs

Ralf-Christian Härting, Anna-Lena Rösch, Gianluca Serafino, Felix Häfner, and Jörg Bueechl

Abstract Digital transformation has been a key driver of business model development for years. The COVID-19 pandemic affects organizations and society alike and makes it imperative to leverage emerging exogenous shocks to remain functioning and competitive. In this vein, COVID-19 has the revelatory power for organizations to re-invent themselves by re-adjusting their strategies including corresponding business models. Along with the need to engage in physical distancing, the call for digital solutions becomes even more critical. Small- and medium-sized enterprises (SMEs) differ quantitatively and qualitatively from large corporations. Therefore, SMEs can navigate themselves more flexibly through disruptive environments. They are often regarded as the backbone of economies and deserve the attention of researchers for these reasons. The aim of our study is to examine potentials of digitization for small and medium-sized enterprises that are particularly impacted by exogenous determinants. On the one hand, the opportunities of digitization must be considered. On the other hand, the challenges and restrictions of the COVID-19 crisis are particularly relevant.

Keywords Company culture · COVID-19 crisis · Digitization · Efficiency · IT architecture · SMEs

1 Introduction

In today's world, major changes in technology have consequences on the whole society. Digitization often goes hand in hand with the development and introduction of new technologies. Everything becomes faster and smarter, and the access to goods and services is much easier. People use their smartphones in nearly every situation.

R.-C. Härting (✉) · A.-L. Rösch · G. Serafino · F. Häfner · J. Bueechl
Aalen University, Beethovenstr. 1, 73430 Aalen, BW, Germany
e-mail: ralf.haerting@hs-aalen.de; kmu@hs-aalen.de

J. Bueechl
e-mail: joerg.bueechl@hsaalen.de

© The Author(s), under exclusive license to Springer Nature Singapore Pte Ltd. 2022   47
R. J. Howlett et al. (eds.), *Smart and Sustainable Technology for Resilient Cities and Communities*, Advances in Sustainability Science and Technology,
https://doi.org/10.1007/978-981-16-9101-0_4

Personal computers are in every workspace regardless of profession. Digitization still has got a lot of potential not only for society, but also for small- and medium-sized enterprises. Especially during the COVID-19 crisis when people work in their home office, remote desks, and other places, digital technologies play an essential role for every enterprise and organization [1]. To advance the research in this topic area, a unified definition of digitization is essential. There are several meanings of the term, while a general definition cannot be found in the literature. A current definition of digitization comprises "an intelligent business and value creation process under the usage of information as well as communication technologies, such as Big Data, Cloud and Mobile Computing, Internet of Things, and Social Software." [2, 3]. The term digitization is also associated with the digital revolution, digital transformation, and Industry 4.0 [4]. Furthermore, the degree of digitization in small- and medium-sized enterprises compared to large corporations differs. It can be noted that the pandemic affects SMEs much more negatively than large corporations [5]. A shutdown or a revenue decline hits SMEs harder due to their lower resources in form of monetary reserves or personnel. In addition, a significant share of the enterprises decides to adopt a new business model to be able to remain in the market at all [6].

In times of the pandemic, companies must be innovative and adapt to the new circumstances. In the current situation, enterprises therefore prefer a "start-up mentality," which is characterized by rapid changes of direction, experiments, or even collaborations [7]. Since the beginning of the pandemic, many companies are bankrupt because the virus affects the global economic framework. Companies, especially small- and medium-sized enterprises can reopen their offices if they are observing health protection guidelines.

This survey is based on a qualitative study which has been conducted in July 2020 as a pre-study. The aim of this quantitative study is to find out which positive and negative consequences the digitization revealed throughout the COVID-19 crisis, especially for SMEs. The central aspects queried were Efficiency, IT Architectures, and Changed Company Culture. Figure 1 shows the hypotheses model as a result of the qualitative survey which was considered.

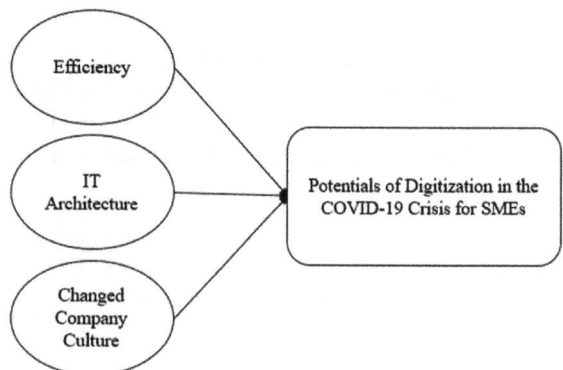

Fig. 1 Hypotheses model of the qualitative survey

New business models can be developed through the best possible application of digitization. Especially for small- and medium-sized enterprises, this insight plays a crucial role regarding the conduct of their daily business during the Corona pandemic. Digitization allows companies to generate new products and services, and, moreover, it leads to an improved value chain adapted to the crisis [8].

Generally, Efficiency describes the extend of operational excellence or productivity. Hence, Efficiency is concerned for example with minimizing costs and improving operational margins [9]. Regarding the COVID-19 crisis, faster decision-making processes and optimization can lead to more efficient working methods. Therefore, digitization enhances the skills of the people working with it because it improves the problem-solving and complex communication activities of employees [10]. New digital technologies offer opportunities to capture key information and make it transferable across actors and firm boundaries. Thus, processes can be simplified and made more comprehensible [11]. Digitization also brings an increased amount of data. Using it provides Efficiency by increasing performance volume and savings [12].

The term IT Architecture generally comprises hardware, software, and digital services [13]. All technologies for processing, using, and storing information can be called information technology [14]. IT Architectures s can enhance communication channels, which the COVID-19 crisis forced companies to explore further. The new communication channels that companies had to find allow a simpler and faster exchange, e.g., virtual communication. It allows a fast data exchange and a shorter time frame between question and respond [15]. Data security is an important challenge of digitization, especially for business models and technologies that depend on a secure use of the customers' data [16]. Possible risk factors for example could be hacker attacks or the abuse of private data.

All around the world, people had to adapt to the contact limitations and curfews due to COVID-19 crisis. It has not only affected social and economic relationships. Companies also had to adapt to these new circumstances to stay competitive [17]. People are now more acquainted with online environments, so time, money, and resources can be saved by holding online conferences [18].

2 Methodology

The following section explains the research methodology in this study to collect and analyze the relevant survey data. The potential use of digitization can be defined as the individual perceived capability of the implementation of digitization technologies [19].

2.1 Literature Research

The starting point for addressing the issue of possible opportunities for SMEs during the COVID-19 crisis regarding digitization was a systematic literature research (SLR) [20]. As a limitation, the authors decided to consider only articles from at least B-rated journals according to the VHB-JOURQUAL3 rating from 2013 onward. The authors preferred articles from A-rated journals which are, according to VHB, outstanding and leading or world-leading scientific journals in business and information management science. Because the study focused on small- and medium-sized companies and the findings in the SLR have been very limited, also B-rated journals and conference proceedings have been considered [21]. The following quantitative empirical design is derived from the SLR and a prior research project, a qualitative study [20]. Ten experts were interviewed using semi-structured interviews and grounded theory. Table 1 shows information about the experts and the companies. Based on this study, an initial hypotheses model was conducted (see [20] and Fig. 2).

Table 1 Experts in advanced qualitative research

Pseudonym	Industry	Size (employees)	Position	Age	Gender	Duration of employment
AMBA GmbH	Marketing Agency	55	Junior marketing manager	26	Female	1,5 years
Bentel Hotel	Hotel Industry	80	CEO	53	Female	27 years
DaRu-GmbH	Automotive Supplier	2050	Supplier auditor	30	Male	1 year
Energy BW GmbH	Energy Consultant	2700	Marketing manager (E-mobility)	26	Female	1 year
MediTec GbR	Medical Technology	2900	Technical support	25	Male	1 year
PeTe-GmbH	Interior Designer	130	CEO	55	Male	22 years
Plug-In GmbH	Mechanical Engineering	330	Product manager	27	Male	8 years
Staussinger GmbH	Steel Construction	85	Technical sales	24	Male	2,5 years
Wire GmbH	Automotive Supplier	40	CEO	58	Male	30 years
Witze KG	Metalworking Industry	3000	Strategic purchaser	27	Male	2,5 years

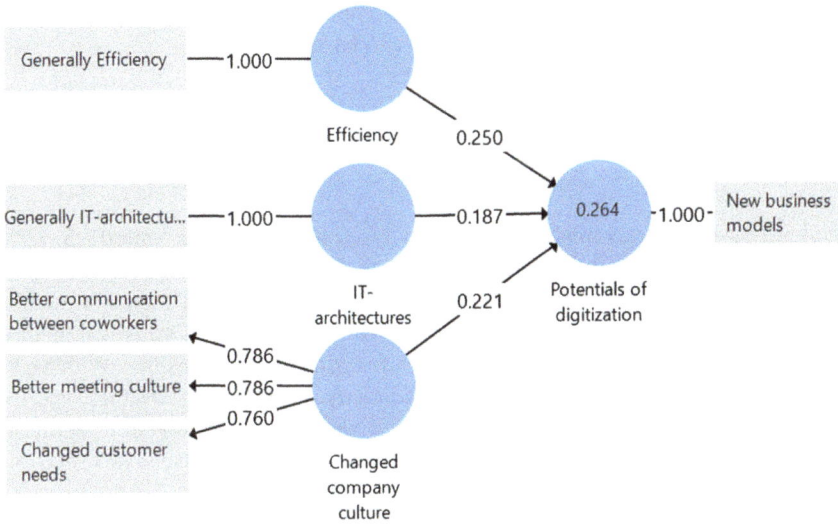

Fig. 2 Conceptual model with outer loadings, path coefficient, and R^2

2.2   Research Setting, Design, and Data Collection

In a further step, this study progressed our triangulation concept by employing a quantitative research approach. The use of a questionnaire to validate the generated hypotheses sheds light on emerging insights with specific cause–effect production. To analyze the collected data, structural equation modeling (SEM) was implemented. Influences of determinants can be evaluated with the analysis tool SmartPLS [22].

In order to obtain clean, scientific results, the target group for data generation includes only executives of the upper management level and IT experts, who can provide important input regarding the digital transformation. For this study, 114 experts of various medium-sized companies were surveyed with an online questionnaire about the opportunities and barriers for SMEs in times of the COVID-19 crisis. The conducted survey was compliant to the highest levels of data protection. It took about five minutes on average and was sent to the experts via various channels. The authors deliberately selected companies from different industries to ensure ecological validity.

An important aspect of a goal-oriented quantitative survey is the creation of a stringent questionnaire. It is divided up into three determinants: IT Architecture, and Changed Company Culture. These determinants emerged from the qualitative study and the literature review built on it. Closing the survey, experts were asked to answer socio-demographic questions regarding work experience and the company they work for. While analyzing datasets with the method of structural equation modeling, influences from different constructs on the research question can be discovered. Before a confirmation or a refutation of the hypotheses is possible, typical quality criteria must be considered. Appropriate criteria are composite reliability (CR), Cronbach's alpha (CA), and average variance extracted (AVE) [22].

2.3 Data Analysis

The built-up hypothesis model has been analyzed by SEM using SmartPLS. SEM allows statements about the relationships among the measured indicators by empirically testing the hypothesis model [22]. In SmartPLS, partial least squares structural equation modeling (PLS-SEM) is used. It is a second-generation statistical method based on multivariate data analysis approach for complex relationships. With SEM, unobservable variables can be measured indirectly by indicator variables, and the accounting for measurement error in observed variables is facilitated. In addition, the influence of the individual constructs on the research question can be tested and evaluated [23].

Three constructs are used in the calculation, and their influence on the potentials of digitization during the COVID-19 crisis is evaluated. In this context, the intermediate determinants in the hypothesis model are called latent exogenous variables. The structural model, also called the inner model, represents the latent variables (constructs) that are measured by the indicators. It also displays the relationships between the constructs and the endogenous latent variable. The measurement model displays the relationships between the latent variables and their indicator variables.

Evaluation of the Measurement Model

A reflective model was chosen, which means the indicator variables reflect the latent variable in its entirety. Therefore, it is implied that the latent variable will still have the same meaning after one indicator is dropped [23]. Goodness of fit measures appropriate for reflective measurement models are composite reliability (CR), Cronbach's alpha (CA), or average variance extracted (AVE). They are given in Table 2.

The single-item constructs always have a value of 1.0 for CR, CA, and AVE. The reason to use the single items for the constructs Efficiency and IT Architectures is the quality of the indicators used. After elaborating the results, different versions for each construct were tested, and the ones of the highest quality were selected. The indicators of Efficiency and IT Architecture describe the construct weaker than their respective general single items. Only for the construct "Changed Company Culture," the indicators were used in the model.

Table 2 Quality criteria of the measurement model

	CA	CR	AVE
Efficiency	1	1	1
IT architectures	1	1	1
Changed company culture	0.679	0.821	0.605

The required threshold of >0.7 for CA is not reached for Changed Company Culture. However, Cronbach's alpha may over- or underestimate scale reliability. In addition to that, CR may also estimate the true reliability better than CA. Therefore, CR may be a preferred alternative to CA. The CR value of Changed Company Culture of 0.821 even indicates a high reliability. The AVE target level for Changed Company Culture with >0.5 is also achieved, as given in Table 2.

Commonly, the size of the outer loading is also known as indicator reliability. The required minimum level of the outer loadings of all indicators to be statistically significant is 0.708 or higher. As Fig. 2 shows, the outer loadings on the arrows directed from the Changed Company Culture to its indicators are above that minimum level [22].

Evaluation of the Structural Model

In the inner model, there are standardized path coefficients pictured on the arrows between the latent variables in Fig. 2. The path coefficients are used to assess the strength of the effect the constructs have on the endogenous variable. Effects are called "significant" when the standardized path coefficients are greater than 0.2 [24]. According to the literature, the path coefficients of the constructs are subject to a hard threshold. The applicability of a hard threshold values for constructed models seems questionable in this case. All values of the SEM must be considered individually and analyzed in combination. The result is a comprehensive interpretation of the effects. Figure 2 shows that construct IT Architectures s does not have a significant influence in the structural model due to its path coefficient of 0.187. The path coefficients of 0.250 and 0.221 indicate that Efficiency and Changed Company Culture have a significant influence. Considering the path coefficients in combination with further statistical methods, the t values, as well as p values, a clearly improved situation emerges. Thus, the first observed value may be below the threshold. After performing the bootstrapping method (Table 3), for the construct IT Architectures, a t value of 1.961 is available. This is marginally above the threshold, indicating significance is present. The associated p value of $p \leq 0.05$ ($p = 0.050$) indeed corresponds to a low significance. In the literature, on the other hand, a clear significance is assumed [20]. Underlying error frequencies and mismatches in datasets could be the reason for a bias.

Table 3 shows the results of the bootstrapping method. A p value of $p < 0.1$ indicates a significant influence of the constructs on the latent variable, while the t values can be described as significant for $t > 1.96$ [22]. Efficiency and Changed Company Culture display p values even lower than 0.05 which indicates their high significance. IT Architectures are on the threshold of 0.05 and indicates a lower

Table 3 Bootstrapping results		t values	p values
	Efficiency	2.273	0.023
	IT architectures	1.961	0.050
	Changed company culture	2.229	0.026

significance. The *t* value of the first construct, Efficiency, is 2.273, and the *p* value is 0.023. The second construct IT Architectures shows a low significant influence with a t value of 1.961 and a *p* value of 0.05. The authors thus derive potentials of the determinant in terms of digitization. The reasons for the weak value presumably lie in the datasets and the resulting bias. Lastly, Changed Company Culture is significant as well, because of its *t* value of 2.229 and *p* value of 0.026.

The coefficient of determination R^2 indicates the percentage of variance of a latent endogenous variable that is explained by the independent variables assigned to it. Chin considers R^2 as "weak" at a value of 0.19, as "moderate" at 0.33, and as "substantial" at 0.66, although his specification can be generalized to a limited extent [24]. The coefficient of determination R^2 of this model is given as 0.264 and is therefore sufficient, although it is weak. The predictive relevance of the reflectively measured latent endogenous variable can be assessed by Q^2. For this model, the Q^2 value is above zero (0.231), and therefore, it has predictive relevance.

Since there is not one global quality criteria for PLS models, the above-discussed quality criteria must be considered together in their entirety [25]. Thus, the structural model of this study can be confirmed with intermediate quality, and all three latent variables show a significant influence on the potentials of digitization. Nevertheless, it should be noted that IT Architecture has got a weaker influence in the model than the other constructs.

3   Results

In the following section, the evaluations of the structural data are presented, and their results are explained. It is important to analyze the surveyed group of experts for providing background information and to interpret the data analyses in an optimized way. Figure 3 shows the gender distribution of the interviewed experts.

Fig. 3 Gender distribution of the experts

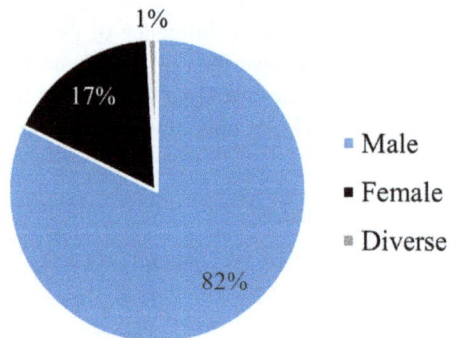

The absolut majority of the experts is male

1%
17%
82%

- Male
- Female
- Diverse

Fig. 4 Overview of industries in which the companies operate

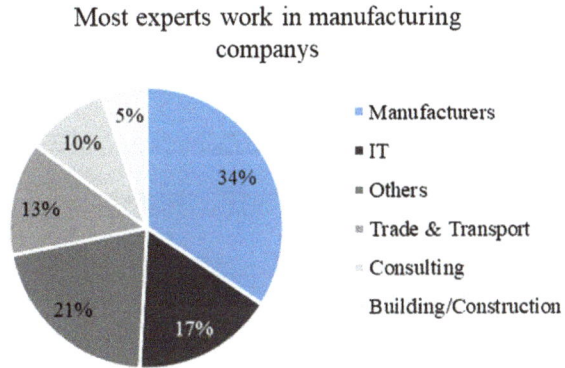

Most experts work in manufacturing companys

- Manufacturers
- IT
- Others
- Trade & Transport
- Consulting
- Building/Construction

5%
10%
13%
21%
17%
34%

The survey participants were predominantly male with a share of 82%. Only a small percentage of 17% indicates female. There was one participant who stated diverse. The unequal distribution between genders could be a limitation in data collection.

When analyzing the data, it is important to know in which industries the experts work. It was possible to cover many different companies. The largest proportion of experts work for companies in the manufacturing sector. Figure 4 shows that 34% of the experts can be assigned here. A further 19 experts come from information technology companies, which accounts for a share of 17%. These companies are very heavily dependent on digitization, which also affects their operating business. A total of 15 experts (13%) come from trade and transport sector. This sector also benefits from digitization approaches. The consulting industry presents 10% of the experts. A small proportion of 5% took part from the construction industry. Other sectors that make up a very small proportion of the sample, such as energy, finance, and healthcare, are grouped together in the section—Others.

A key aspect of the study's quality is demonstrated by the professional experience of the experts who participated in the study. A total of 48 participants reported having professional experience of more than 20 years which is shown in Fig. 5. Over two decades, the participants have been able to observe developments in digitization and thus offer high-quality research input. A further 33 experts have professional experience covering at least 9 years. With a share of more than 20%, 24 respondents still claim to have more than three years of experience in their industry. Only nine respondents have less than three years of professional experience. The sample shows that most experts are experienced participants.

Another insightful aspect is the activity of the experts in the companies. Figure 6 shows a quite balanced distribution, with experts from top-level management being represented twice as often as other areas. A total of 48% of the experts surveyed come from top management. A further 24% stated that they worked in the management area of the commercial or technical departments. The IT area is represented by experts with a share of 20%. The remaining 14% comprises all other management positions stated by the experts.

Surveyed Experts have a lot of work
experience

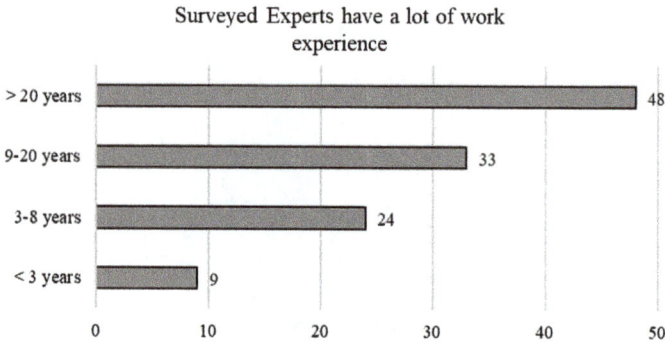

Fig. 5 Overview of the working experience of the interviewees

Fig. 6 Overview of the
function of the interviewees

Nearly half of experts are in top
management positions

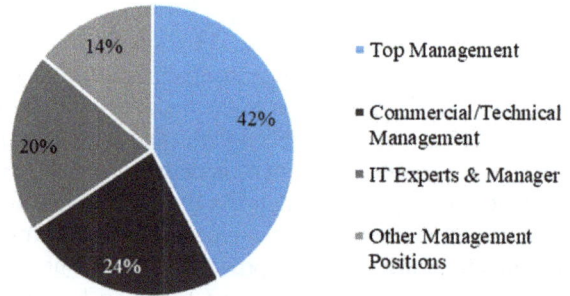

In summary, the respondents came from various industries, most of which are companies with fewer than 50 employees and more than 250 employees. Most of the interviewees were male and between 35 and 54 years old. They work in a position as a board member, in management, in commercial/technical, division management, as an IT expert or as IT department manager. In addition, 48 people noted that they had more than 20 years of professional experience, while half of the group (24) also has between three and eight years of professional experience.

4 Conclusion

The results show that all three determinants have an influence on the opportunities and barriers of digitization. However, the determinants Efficiency and Changed Company Culture have a more significant influence than IT Architectures. For the IT Architecture, a weak cause-and-effect relationship was established. In further research, the datasets have to be analyzed in detail, and reasons for the weak influences have to be identified. Based on digitization, efficiency can be increased, and companies can adapt more quickly and dynamically to the new circumstances. Overall, the interviewed experts agreed that opportunities of digitization lie in faster decision-making processes, faster work, and process optimization. Around the world, almost everyone had to adapt to new circumstances due to COVID-19. Contact limitations and curfews affected both social and business relationships which is why it is not surprising that Changed Company Culture can be seen as a potential of digitization.

The study also shows that IT Architectures are accompanied by certain risks and data security could be confirmed as a major risk factor of digitization. The second barrier that was identified in the qualitative study could not be confirmed. According to most of the interviewed experts, it is not true that digital communication has a negative impact on the relationship between employees.

Based on the findings, the opportunities must be seized, and the barriers must be treated with a certain sensitivity to sustainably benefit from the potentials of digitization. In summary, the use of digitization is essential in the context of the COVID-19 crisis. Digitization can profitably change and stabilize internal processes and business models to help small- and medium-sized companies to position themselves sustainably in the market regarding the ongoing COVID-19 crisis. Accordingly, the pandemic-crisis makes it obvious that on-site conversations or word-of-mouth communication (storytelling) is still indispensable and plays a major role in successful corporate communication. But online communication tools such as Zoom or Microsoft Teams will be an integral part of the communication philosophy in the future.

5   Limitations and Further Research

Even though the study shows many aspects of digitization in its context, several limitations need to be explained. The results of this study are explicitly aimed at small- and medium-sized enterprises with 10–250 employees. This study excluded corporations and sectors of the economy not surveyed in our investigations. Furthermore, an existing qualitative survey was used as a basis. However, the opportunities and barriers of digitization should be continuously followed up during the COVID-19 crisis. Especially in SMEs, digitization is a fast-moving process with constant innovations and changing application scenarios at short intervals, focusing the topic of restrictive and political measures. In addition, the study only refers to small- to medium-sized companies mostly located in Southern Germany. To gain an improved overall impression, it would be helpful to conduct the study in an international context, e.g., in other European countries or in countries outside Europe. Within the scope of our quantitative survey, 114 experts were interviewed. Therefore, the elaboration of this study is subject to a certain limitation with an empirical investigation in form of quantitative research due to the rather small sample size. The quality criteria used for the model represent certain limitations as well. The measurement of the constructs' reliability to the research question showed some inaccuracies since the reliability according to Cronbach's alpha and composite reliability defer significantly. Regarding the agreement between indicators and the constructs, the measurement model has got lower quality. Only one of the three measured constructs can properly be described by their indicators, which is the reason for using the general single item in the constructs Efficiency and IT Architectures. Only Changed Company Culture is described by multi-items. The results for hypothesis concerning the impaired communication between co-workers differ from the literature and the feedback received in course of the study.

The limitations of this study show the necessity for further studies on the topic "Opportunities and barriers of digitization during the COVID-19 crisis." Especially, the determinants Efficiency and IT Architectures can be extended and examined in more detail.

Acknowledgements This work was supported by Lisa Linnenfelser, Melissa Moldaschl, and Marcel Weber. We would like to take this opportunity to thank you all for your great support.

Appendix

Excerpt of Questionnaire:

- Efficiency:

  The potentials of digitalization are increased in our company...

  – due to faster work thanks to software solutions.
  – due to faster decision-making processes.
  – generally due to more efficiency.

- IT-architecture:

  The potentials of digitalization are increased in our company...

  – due to a simplified way of communication.
  – due to a higher level of data data security.
  – generally due to the IT architecture.

- Changed behavior:

  The potentials of digitalization are increased in our company...

  – due to a changed meeting culture.
  – due to changed customer needs.
  – due to a better communication between coworkers.
  – generally due to a changed company culture.

- New business models:

The potentials of digitalization are increased in our company…

– due to new business models.

References

1. Garzoni A, De Turi I, Secundo G, Del Vecchio P (2020) Fostering digital transformation of SMEs: a four levels approach. Manag Decis 58(8):1543–1562. https://doi.org/10.1108/MD-07-2019-0939
2. Häfner F, Härting R, Kaim R (2020) Potentials of digital approaches in a tourism industry with changing customer needs—a quantitative study. Proceedings of the FedCSIS/ACSIS 21:553–557
3. Härting R, Reichstein C, Lämmle P, Sprengel A (2019) Potentials of digital business models in the retail industry—empirical results from European experts. In: Proceedings of KES-2019, Elsevier B.V., vol 159, pp 1053–1062
4. Schmidt R, Möhring M, Härting R, Reichstein C, Neumaier P, Jozinović J (2015) Industry 4.0—potentials for creating smart products: empirical research results. In: Abramowicz W, Kokkinaki A (eds) Lecture notes in business information processing, vol 208. Springer, pp 16–27
5. Pedauga L, Sáez F, Delgado-Márquez BL (2021) Macroeconomic lockdown and SMEs: the impact of the COVID-19 pandemic in Spain. Small Bus Econ. https://doi.org/10.1007/s11187-021-00476-7
6. Priyono A, Moin A, Putri VNAO (2020) Identifying digital transformation paths in the business model of SMEs during the COVID-19 pandemic. J Open Innov Technol Mark Complex 6(4):104. https://doi.org/10.3390/joitmc6040104
7. Kuckertz A et al (2020) Startups in times of crisis–a rapid response to the COVID-19 pandemic. J Bus Ventur Insights 13:e00169 (Stuttgart, p 3). https://doi.org/10.1016/j.jbvi.2020.e00169
8. Härting R, Reichstein C, Jozinovic P (2017) The potential value of digitization for business—insights from German-speaking experts (p. 7.) in INFORMATIK 2017. Lecture notes in informatics (LNI), Gesellschaft für Informatik, Bonn
9. Mouzas S (2006) Efficiency versus effectiveness in business networks. J Bus Res 59(10–11):1124–1132. ISSN 0148–2963. https://doi.org/10.1016/j.jbusres.2006.09.018
10. Bührer C, Hagist C (2017) The effect of digitalization on the labor market. In: Ellermann H, Kreutterund P, Messner W (eds) The Palgrave handbook of managing continuous business transformation. Palgrave Macmillan UK, London, pp 115–137
11. Holmström J, Holweg M, Lawson B, Pil F, Wagner S (2019) The digitalization of operations and supply chain management: theoretical and methodological implications. J Oper Manag 65(8):728–734. https://doi.org/10.1002/joom.1073
12. Trum C (2020) Realizing the benefits of digitalization through standards. Nat Gas Electricity 36(7):9–17. https://doi.org/10.1002/gas.22157
13. Nissen V (2018) Digital transformation of the consulting industry. Extending the traditional delivery model. Springer, Cham, Switzerland
14. Wu H, Zuopeng Z, Wenzhuo L (2021) Information technology solutions, challenges, and suggestions for tackling the COVID-19 pandemic. Int J Inf Manag 57:102287. https://doi.org/10.1016/j.ijinfomgt.2020.102287
15. Lindeblad P, Voytenko Y, Mont O, Arnfalk P (2016) Organizational effects of virtual meetings. J Cleaner Prod
16. Martin K, Abhishek B, Palmatier R (2017) Data privacy: effects on customer and firm performance. J Mark 81(1):36–58. https://doi.org/10.1509/jm.15.0497

17. Study: impact of COVID-19 on consumer behavior (2020) Available: https://www2.deloitte. com/de/de/pages/consumer-business/articles/consumer-behavior-study-covid-19.html.   Last accessed 27 July 2020
18. Ubinger L, Gazendam A, Ekhtiari S, Nucci N, Payne N, Johal H, Khanduja H, Bhandari M (2020) Maximizing virtual meetings and conferences: a review of best practices. Int Orthop 44(8)
19. Celesti A, Leitner P (2016) Advances in service-oriented and cloud computing. Work-shops of ESOCC 2015, Taormina, Italy, September 15–17, 2015: revised selected papers. Communications in computer and information science. Springer, Switzerland, p 567
20. Härting R, Buck M, Geiger K, Brune M, Franzrahe C, Tesche N (2020) Digitization in the Covid-19 crisis—opportunities and barriers for SMEs. http://www.kmu-aalen.de/kmu-aalen/? page_id=169
21. Hennig-Thurau T, Dyckhoff H (2021) VHB-JOURQUAL 3 vhbonline.org/en/vhb4you/vhb-jourqual/vhb-jourqual-3. Last accessed 19 Sep 2021
22. Hair J, Hult G, Ringle C, Sarstedt M (2017) A primer on partial least squares structural equation modeling (PLS-SEM), 2nd edn. Sage Publications, Thousand Oaks, CA
23. Garson GD (2016) Partial least squares: regression and structural equation models. NC, Statistical Associates Publishers, Asheboro
24. Chin WW (1998) Issues and opinion on structural equation modeling. Manag Inf Syst Q 22:7–16
25. Chin W (1998) The partial least squares approach to structural equation modeling. In: Marcoulides GA (ed) Modern methods for business research, pp 295–336

Chapter 5
New Urban Mobility Strategies After the COVID-19 Pandemic

Domenico Suraci

Abstract The intent of this work is to investigate the measures taken in the various cities of the world due to COVID-19 crisis with regard to urban mobility and evaluate which have been the most effective, verifying their effects in both the short and long term. The cities considered are located in Europe and North and South America. The first part of the study tried to classify them by their similarity. Then, the data of modal split changement, mostly found on municipality official Web site, have been used to study the most effective measures taken. From this, generally, conclusion about the phenomenon and suggestions for future development are made.

1 Introduction and Background

The COVID-19 pandemic had, among the countless and tragic effects, that of putting in crisis the entire urban transport system of cities, showing how this is a strategic need, which must be satisfied for the very survival of cities. The new conditions forced a change in the strategy with which the problems had been solved up to now. Generally, the answer was the same, albeit with different tactical variations depending on the context: the increase in the offer of individual mobility, and in particular the sustainable one. However, the crisis did not represent a point of discontinuity as much as a point of ascending inflection: it did not generate any real news, and it was a catalyst for phenomena already underway [1]. In fact, the measures implemented almost entirely derive from proposals already presented or even approved for the next decade. The attempt, common all over the world, to use the response to the health crisis as a tool to change mobility within the city once it is over, is evident. The tragic evolution of the epidemic into a global pandemic has forced a large number of cities to take action, offering a vast case study.

D. Suraci (✉)
Graduated at Politecnico di Milano in Civil Engineering and Transport Engineering, Politecnico di Milano, Piazzale Dateo 5, 20129 Milan, Italy

© The Author(s), under exclusive license to Springer Nature Singapore Pte Ltd. 2022   61
R. J. Howlett et al. (eds.), *Smart and Sustainable Technology for Resilient Cities and Communities*, Advances in Sustainability Science and Technology,
https://doi.org/10.1007/978-981-16-9101-0_5

2 The Problem of Public Transport

The main effect of the epidemic on urban mobility was the collapse of public transport [2], both from the point of view of supply and from the point of view of demand. In fact, three phenomena went to compete.

The most obvious is the drastic decrease in the number of trips, due to the suspension of many work activities, smart working, the closure of schools, and many non-essential services. The decrease was around 90% compared to the average of the movements of previous years in the most stringent lockdown moments and then stood at around 40% less when the activities were able to reopen (Fig. 1 and Fig. 2).

Decline in mobility in European countries during the first phase of the pandemic [3].

Then, there was the need to reduce the maximum capacity of the means of transport—generally by 50% compared to the normal situation—to ensure distancing. Finally, there has been a general change in the choice of moving users toward individual mobility for fear of contagion on public transport, regardless of the safety levels declared by the health and transport authorities.

The problem on the part of the planner is therefore to be able to guarantee a safe service from a health point of view and as efficient as possible. The greatest criticality is found during peak hours, which are those of maximum overcrowding of vehicles. We have tried to solve the problem in two ways. On the one hand, the offer has increased through the enhancement of the service compared to normal during peak hours. However, this path has proved to be complex: the recruitment and training of new staff, the purchase of new vehicles, and the upgrading of lines often already at maximum capacity require costs and timing that cannot be reconciled with the emergency nature of an epidemic. In this way, private companies were used, which provided their own means to strengthen the network. However, the serious problem of lack of knowledge of this alternative on the part of users has been highlighted, which has largely continued to use the means they previously used [4].

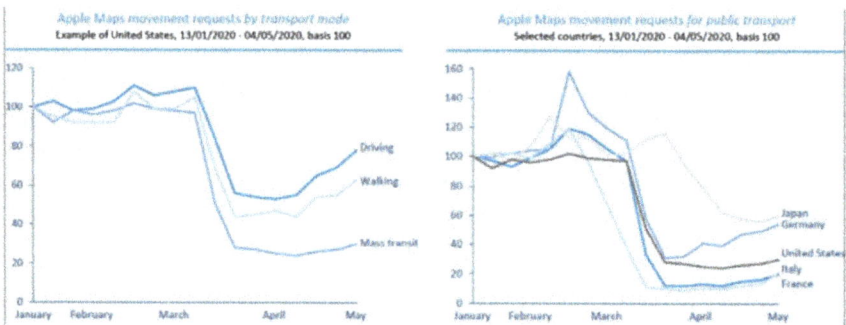

Fig. 1 Apple Mobility trends during the first lockdown in selected countries (February - March 2020)

On the other hand, an attempt was made to spread the demand in order to flatten the rush hour, especially the morning one which is the most loaded. The hourly flow was therefore completely rethought, staggering the openings and entrances of schools, services, and companies. In particular, the focus was on dividing the opening of schools into two time slots. However, it was not sufficient: the peak in demand was found outside the half hour before the two school entrance hours because the entrance to offices, shops, and companies was not staggered [4]. In fact, this operation turned out to be very complicated. Often it took months just to start negotiations for an agreement with the interested parties [5] and in the few places where it was reached it was often not respected.

From the experience gained, we can therefore see three ways to follow to be ready in the event of a new epidemic. First, the preparation of a pandemic plan that brings together companies, public bodies, social partners, and citizens on how to stagger the entrances, so that in the event of an emergency, immediate action can be taken with the participation of all. We can then think of staggering the entrances not in two but in three bands, anticipating the entry of half of the students in high schools and universities, which are the ones who use public transport the most. In this way, a band that is soft outside the emergency would be used. In this way, the opening hours of companies, shops, and services could be left almost unchanged with a consequent greater collaboration on the part of these. Finally, it is necessary to effectively communicate the enhancement of the service, so that users are aware of the options available to them. The dissemination of information must not only be carried out by the public service, but also by companies, schools, and trade associations before and especially during the emergency.

A final note should be made regarding the use of public transport itself during an epidemic. SARS-CoV-2 has among its greatest dangers a high contagiousness, which increases even more with the variants that have spread in 2021. However, in the future, we may have to deal with epidemics that have a high level of contagiousness at point of not being able to guarantee health safety on vehicles [6]. To cope with an emergency of this gravity, an adequate mobility plan must therefore be developed, which takes into account the inevitable and even more radical changes in people's lives and which is focused on individual travel, the only ones that can be safely practiced in this situation.

For these reasons, the work of the thesis focused on individual mobility, in particular the soft one, which turned out to be the most effective and the only possible response in the case of more serious epidemics than that caused by SARS-CoV-2.

3 Classification of Cities

Once the measures taken by each city were cataloged, criteria were sought to find points of contact between them, in order to subsequently verify which measures were effective in certain contexts and which were not. Population, infrastructure, previous modal split, and service levels were taken into consideration. The sources from which they were searched are the national statistical institutes of the respective countries and the mobility studies of the local and national transport departments. For the modal split, it was decided instead to use a database provided by the European Platform on Mobility Management [7], so that in this phase of comparison, data collected in the same way were used. The modal splits were all subsequently checked on the sites of the respective transport departments and were all consistent with the platform data. American cities, which were not present in the database and for which data from the transport departments and those from the Deloitte Insights Web site, were used as an exception [8].

However, no similarities were found when evaluating each of the parameters individually. Mobility is in fact such a complex and integrated system that it does not allow one to be decisive compared to the others outside the pandemic context. It was therefore necessary to make an overall assessment, which would go to evaluate the urban transport system and the cities.

This has brought to light how the most similar cities in terms of morphology, citizens' habits, infrastructures, and problems are found within the same state or a common cultural region. The same similarity was found, in the measures that were introduced, despite the absence of central national directives that would guide the choices.

Therefore, the "national" criterion is the one chosen to analyze the interventions before and the effects after.

4   Data Analysis

The data analysis was the longest and most complex part for many factors. The first difficulty was encountered in their own research. First, because they are scarce, almost none of the cities that have adopted the measures have been able to measure their effectiveness in a capillary manner, a fundamental condition for understanding what worked and what did not within the same context.

In addition, the work of aggregation was also complex: The data are calculated with different criteria, refer to different periods, and are represented differently. The help that can come from the giants of "big tech" must be used with great caution. In fact, the data made available show many shortcomings: They are almost never available for the years preceding the pandemic, they do not present all the most significant transport alternatives, and they are almost never data of actual movements but data on Internet searches on how to reach certain locations and surveys with a non-representative statistical sample. Given all these highly uncertain factors, these data are taken into consideration only to signal trends when there are variations with very significant orders of magnitude.

A happy exception is the Strava Metro app, which has made a dashboard available online [9] with data shared by users anonymously. They are trusted by the United States Centers for Disease Control and Prevention, which conducted a study [10] showing the strong correlation between the location of people using the app to track their movements and the location of commuters on bicycles and feet in the total population. The active commuting rate in each area of a city according to the census is therefore similar to the active commuting rate in the same area among Strava users. Their reliability is also evidenced by the consistency with the results of statistical studies and with the counting of steps, where these are available. It was therefore decided to refer to this database in the absence of "real" data.

Another problem that makes it difficult to make a comparison is the general decrease in mobility compared to the standard situation, oscillating on average between a minimum of 40% and a maximum of 90% in the most severe lockdown period. It is therefore not possible to reason in absolute terms but only in percentage terms. The evaluation of the "time" of the virus was also complex: The pandemic affected different states and geographic areas in different ways, with a different perception by the population of the seriousness of the situation even for comparable levels of contagion. The only period in which the pandemic situation would have allowed a confrontation would have been the summer, when in Europe it seemed that the virus had stopped. However, even in this circumstance, the sharp decline in mobility prevents us from making adequately precise comparisons.

Finally, the last and perhaps most significant difficulty was caused by the change in the process of choosing the mode of transport. The user is considered a rational decision maker who seeks to maximize his utility (or in other words to minimize the generalized cost) which he evaluates through a series of attributes [11]. In the presence of such a contagious epidemic, the attribute of health safety, previously so obvious that it was not considered, assumes a predominant importance over all the others. The demand for individual transport increases very strongly, and another element of uncertainty is generated when analyzing the data. The conditions of choice during the emergency are therefore radically different from the normal ones, which are those for which the measures introduced were designed: Particular attention must be paid to the difference between the response to the same interventions depending on whether one is inside or outside the pandemic, without making a single assessment.

For all these reasons, the exact effectiveness outside the pandemic context of the measures taken can only be verified when the epidemic ends. For now, trends can be underlined, taking advantage of those moments in which the contagion was low and those changes that were too sensitive to be considered momentary.

5 Study Cases

The contexts reported—France, Spain, United Kingdom, Latin America, USA, Northern Europe, and Italy—are those that were considered most interesting. Others have been omitted for various reasons: for China, insufficient elements have been found to make a description; Irish cities basically behaved like the English ones.

In reporting the cases studied, it was decided to follow the workflow described above. The context was described for each, briefly with regard to the spread of the epidemic and more in depth with regard to the characteristics and modal split of the cities. The measures taken and the reasons that led to their realization were then analyzed, focusing on the topography of the various interventions, considered a key element for a complete understanding. The construction techniques were among the most varied, and we focused on the most interesting ones. Finally, changes in mobility were analyzed, again in relation to the evolution of the epidemic and the consequent confinement measures.

6 Conclusions

The COVID-19 pandemic has hit the whole world hard, albeit at different times and in different ways. In the field of study of this thesis, it has been noted how many cities have reacted, but according to their respective possibilities and needs (Fig. 2).

Some of the cities that have most experienced an increase in cycling mobility [12–15].

The available data are not abundant, and largely focus on cycling, almost completely neglecting walking. However, it can be said that all measures have worked, often beyond the best expectations. They worked both in the most emergency period and when the lockdown was eased. The citizens' response was immediate and continued over time. It can therefore be said that the strategy implemented by all cities, of exploiting the response to the epidemic to solve the mobility problems prior to it, has been a success.

The full assessment of how mobility has changed will only be possible when the virus is eradicated and the health situation is back to normal, but the numbers available so far show that sustainable mobility has taken a leap forward decades.

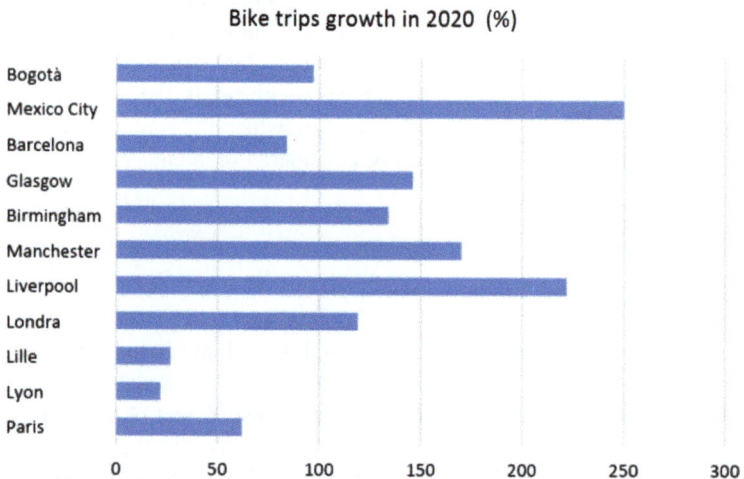

Fig. 2 Bike trips growth in 2020 compared to 2019 in some of the city observed during the study

6.1 *A Diversified Transport System*

An element of strong reflection is given by the importance of having a diversified mobility system, so that the emergency response is effective. In fact, the best response occurred in Spain and Central Europe, that is, in those contexts where transport choices were more evenly distributed between motorized transport, cycling, walking, and public transport. In fact, it was not necessary to build new infrastructures or upgrade existing ones, because in the absence of saturated alternatives, the system has an internal compensation capacity that is automatically activated with the different choices of citizens, allowing the emergency to be overcome without serious problems. Proof of this is the fact that in cities that did not present this balance, citizens began to change modes of transport before the measures discussed during the thesis were implemented, automatically recognizing in the movement a strategic need.

In planning urban mobility, public transport has probably been taken too much into consideration, which has always proved insufficient despite continuous improvements. Individual soft mobility is therefore inserted as an alternative to partially satisfy the demand for transport and as a need to avoid that the transport system is too fragile because focused solely on a solution. In the future, however, the same mistake must not be made, by neglecting collective transport in favor of individual transport: In order that urban transport to be efficient in periods of normality and ready to change in an emergency, it must be diversified and balanced.

6.2 *Example for the Future*

As the epidemic has affected the whole world, a huge number of cities have found themselves facing the same problems. The contexts dealt with here were the most disparate, from medium-sized European cities to megalopolises in South America. The construction methods were the most diverse and original, even when it was a question of similar measures. The requests for transport that were wanted to be satisfied were different, from that of the neighborhood to the commuters of medium-long distance.

We have seen that there are no universal solutions. Each city has an articulated, integrated, and detailed mobility system to such an extent that a schematic and functional answer cannot be found for each case. It is up to the planner's experience to be thoroughly familiar with the mobility of the city and to take appropriate measures. One of the reasons for the success of the measures was precisely to have implemented measures that had already been planned and therefore were already inserted in the city context.

Therefore, those who want to intervene in a similar way in the future will have to go and see what has been done in the same context in which they are operating. As shown throughout this work, the greatest similarities in context, morphology, mobility problems, and citizen habits are found in the national context. Looking at the nearby cities and copying what has been done and has proved successful (obviously adapting it to the case) is probably the best way to follow for all those cities that have not used the months of confinement to solve their previous mobility problems. The cases studied in this work, starting from the huge global mobility laboratory that was created with the epidemic, can precisely provide guidelines precisely because this is the approach we wanted to give.

6.3   The Trigger Problem

However, for the cities that want to take this path in the future, there is a big pitfall: the emergency context that triggered the explosion of individual sustainable mobility in 2020 and 2021 will be lacking (hopefully). The situation is very similar to that of the 1973 energy crisis in Amsterdam [16]. The city had congestion and security problems that it had not been able to solve for years despite being more than evident and much debated. Protected bicycle and cycle paths allowed mobility in emergencies and, when oil prices were lowered, they continued to be a very popular transport alternative.

Referring to the present, the case of Dunkerque is emblematic: having established the "30 zones" throughout the city has not increased cycling mobility for two years. During the epidemic, on the other hand, research on health safety increased bicycle trips, which, however, did not decrease when infections drastically decreased.

To change mobility with existences comparable to those measured during the epidemic, another push will therefore be needed, which represents a strong discontinuity. A solution could be represented by large economic incentives that make the use of bicycles, electric scooters, and mobility in general extremely advantageous for an initial period compared to other alternatives, going into direct competition with the mode that is most unbalanced in the respective context. For example, in a city where the car is the most used vehicle, a bike-sharing service could be provided at a very low cost.

The investment can be expensive, but still limited in time. As demonstrated by both crises, oil and health, the problem of users about soft mobility is mainly of a prejudicial nature. When it is possible to make this alternative perceive as real, users continue to benefit from it extensively even when the conditions that led them to change have ceased to exist.

6.4   Tactical Urbanism

A decisive factor for the success of all these experiences was the pop-up form with which they were implemented, which fall within the concept of tactical urbanism. This form of planning has in fact the best features to respond to the emergency according to the will of the administrations. Building the infrastructures in a flexible, rapid, and economical way was fundamental in the initial phase, when an immediate solution had to be found to the new mobility needs. Furthermore, in the long run, it is a low-cost form of experimentation that lends itself easily to improvements and can be made permanent later. Administrations can thus decide to invest the much larger sums necessary for permanent infrastructures when they are now sure of their necessity and effectiveness [17].

Further help to the planner is given by the spread of digitization: the immediacy with which each user, thanks to digital devices, can report defects on special portals allows to immediately highlight many critical issues, which precisely with pop-up infrastructures can be resolved quickly simply.

The opportunities provided by tactical urbanism are therefore very vast, and it is likely that this way of thinking and acting will be increasingly widespread in the future.

6.5 Accidental Safety

The analysis of the measures introduced allows to find a common denominator which, as evidenced in practically all the cases handled, was decisive: the accidental safety of the movement. In fact, if health safety was the decisive factor in the change of modality during the epidemic, the lack of intrinsic safety in the promiscuity of soft and motorized mobility was the basis of the lack of attention that users reserved for it. Accidental safety was sought through two different principles, depending on the types of movements that were wanted to be protected. For short trips, the solution identified was the integration of spaces. Since micro-mobility is not by its nature flexible, no dedicated infrastructures have been built, but the existing spaces have been redistributed and weak users have been protected with restrictions on circulation. It is in this direction that initiatives such as the widening of sidewalks, shared streets with speed limitations, and filtering of users who can travel the streets have gone. They are therefore measures that have been introduced in residential districts and historic centers.

However, to make long journeys safe, integration does not work. Segregation was required through the construction of infrastructures dedicated to soft mobility, practically everywhere in the pop-up form. In fact, the demand they must serve is very high and develops right on the road axes with the highest and busiest speeds, making the mixed location impracticable.

The construction of cycle paths in South America deserves a separate comment: having a transport infrastructure available even in the case of catastrophic earthquakes that, unfortunately, periodically hit that part of the world proves to be decisive in an urban environment to allow rescue in first few hours after the shock.

6.6   Future Developments

The work described here can be a starting point for deepening the topics covered. Three fields of interest are identified.

The first is the refining of the work itself when more statistical studies on soft mobility will be carried out during the epidemic and when, hopefully soon, the pandemic will be over and the long-term effects of the measures introduced can be assessed. In particular, the major limitation of the work done is the lack of data regarding the pedestrian mode. Helping the development of this modality was certainly not the first goal of the administrations, probably trusting that the change would happen automatically. This is probably why we have not even focused on monitoring the phenomenon.

The second concerns the safety of the measures introduced. We must go and study how mortality and accident rates changed during the epidemic to resolve any problems not identified by the designers as soon as possible. In general, it has been seen that in 2020, they have increased considerably compared to 2019. However, it is not yet possible to make an accurate study of the problem. First of all, because the data on the increase in soft mobility are still partial and therefore no comparison can be made between the increase in users and that of accidents: the mobility situation on the streets during the lockdown is absolutely unique. Secondly, because the measures introduced, even if they share the same aims, have such a constructive variety that they cannot be schematized in a single pattern. Therefore, a dedicated study is needed which analyzes what happened on a case-by-case basis, highlighting critical situations.

Finally, it is considered necessary to study a remodeling of the models that represent the transport supply and demand system of a city. The epidemic highlighted how the classic service-level attributes with which the generalized cost of the trip and the usefulness perceived by the user are defined (monetary cost, duration—and uncertainty of duration—of the trip, travel comfort, risk of accident), take on considerably less weight during an epidemic. The health risk, which until now had not been considered among the attributes, in this situation becomes a fundamental parameter that probably assumes the greatest weight of all. In particular, it would be interesting to highlight the difference between the level of health service actually offered by collective transport and the perceived level of users. For example, the authorities have established thresholds for the capacity of the vehicles below which it is believed there is safety from contagion. But public transport almost never reached the capacity limit that had been declared, because users still considered them dangerous.

References

1. Caracciolo L (2020) «L'ora più Chiara» Limes. Rivista Italiana di geopolitica, vol Il Mondo Virato, n 3(2020):7–32

2. Moovit (2020) «Coronavirus & your commute: how COVID-19 is affecting public transportation around the world», 19 Mar 2020 [Online]. Available: https://moovit.com/blog/coronavirus-effect-public-transportation-worldwide/
3. «Apple Maps Mobility Trends Reports» (2020) [Online]. Available: https://covid19.apple.com/mobility
4. Giambattista Attanasio SBNP «Milano zona gialla: il sistema tra sporti regge, sotto esame l'orario dei negozi» [Online]. Available: https://www.ilgiorno.it/milano/cronaca/trasporti-studenti-1.5973963
5. «Amplio acuerdo para aplanar la hora punta en Barcelona» (2020) lavanguardia, 6 Oct 2020 [Online]. Available: https://www.lavanguardia.com/local/barcelona/20201006/483882299578/amplio-acuerdo-aplanar-hora-punta-barcelona.html
6. Fondazione veronesi «Le pandemie nella storia: dal vaiolo del '500 al Covid-19» [Online]. Available: https://www.fondazioneveronesi.it/magazine/articoli/lesperto-risponde/le-pandemie-nella-storia-dal-vaiolo-del-500-al-covid-19
7. «European Platform on Mobility Management» (2020) [Online]. Available: http://epomm.eu/tems/index.phtml
8. «The 2020 Deloitte City Mobility Index» (2020) [Online]. Available: https://www2.deloitte.com/xe/en/insights/focus/future-of-mobility/deloitte-urban-mobility-index-for-cities.html
9. «Strava Metro» (2020) [Online]. https://metro.strava.com/
10. Whitfield GP, Ussery EN, Riordan B, Wendel AM «Centers for disease control and prevention». Available: https://www.cdc.gov/mmwr/volumes/65/wr/mm6536a4.htm#contribAff. Accessed 25 Feb 2021
11. Maja R (2011) Modellizzazione e simulazione dei sistemi di trasporto, Milano
12. Gobierno de la Ciudad de México (2020) «MOVILIDAD CICLISTA EN LA CIUDAD DE MÉXICO», 10 dicembre 2020 [Online]. Available: https://semovi.cdmx.gob.mx/storage/app/media/Feria%20Transporte%20de%20Pasajeros/MovilidadCiclista_101220.pdf
13. «Bogotà Governement» (2020) [Online]. Available: https://bogota.gov.co/en/international/seven-months-we-doubled-number-bicycle-trips-bogota
14. «Fréquentation vélo et déconfinement—Bulletin n°10», 10 Ottobre 2020 [Online]. Available: https://www.velo-territoires.org/wp-content/uploads/2020/10/2020-10-14-Bulletin_Velo_Deconfinement_10.pdf
15. «Strava Metro». https://metro.strava.com/
16. van der Zee R «How Amsterdam became the bicycle capital of the world», 5 Maggio 2015 [Online]. Available: https://www.theguardian.com/cities/2015/may/05/amsterdam-bicycle-capital-world-transport-cycling-kindermoord
17. Campo D (2015) Tactical urbanism: short-term action for long-term change. Island Press, Washington, D.C., United States

Chapter 6
Integration of Indoor Air Quality Concerns in Educational Community Through Collaborative Framework of Campus Bizia Laboratory of the University of the Basque Country

Iñigo Rodriguez-Vidal, **Xabat Oregi**, **Jorge Otaegi**, **Gaizka Vallespir-Etxebarria**, **José Antonio Millán-García**, and **Alexander Martín-Garín**

Abstract Nowadays, indoor air quality (IAQ) of occupied spaces has become a major concern for many stakeholders due to the current health situation caused by the COVID-19 pandemic. This problem is even more pressing in the educational context due to the high density of occupation, reduced distance between students, scarce awareness, coexistence of multiple mechanical and natural ventilation systems and the lack of training of users on the use of the building. These reasons, along with the difficulties in recommending a simple, clear criterion of action, generate the need to offer guidance on hands-on, effective improvement and prevention measures for lecturers, staff and students. In this sense, this chapter aims to analyse the evolution of indoor air quality (IAQ) of several educational facilities in the Gipuzkoa Campus of the University of the Basque Country (UPV/EHU). This study examines IAQ in several types of classrooms and offices with different characteristics in order to assess the performance of each one of them. To do this, based on a monitoring campaign, this research used CO_2 concentration as a key indicator of air quality that defines ventilation needs and displayed real-time information to occupants. Results have shown the clear effect that the visualisation of the data has on the users. It can be concluded that the information-based decision-making achieved through the employment of monitoring panels could provide an effective and easy-to-use solution to achieve IAQ goals. The research is also framed within the Campus Bizia Laboratory program where the collaborative process between academic staff, service and administrative

I. Rodriguez-Vidal (✉) · X. Oregi · J. Otaegi
CAVIAR Research Group, Department of Architecture, University of the Basque Country UPV/EHU, Plaza Oñati, 2, 20018 Donostia-San Sebastián, Spain
e-mail: inigo.rodriguez@ehu.eus

G. Vallespir-Etxebarria · J. A. Millán-García · A. Martín-Garín
ENEDI Research Group, Department of Thermal Engineering, Faculty of Engineering of Gipuzkoa, University of the Basque Country UPV/EHU, Plaza Europa 1, 20018 Donostia-San Sebastián, Spain

staff and students in order to respond to sustainability challenges within the university itself.

Keywords COVID-19 · Ventilation · Indoor air quality · Infectious risk assessment · Data-informed risk prevention · School · IoT (Internet of Things)

1 Objective of the Study

Education for Sustainable Development (ESD), Sustainable Education, and Education for Sustainability (EIS) are different names for a single objective: to rethink education, its purposes, methodologies and content. ESD is based on transformative learning and empowers people to acquire knowledge, skills, values and attitudes in order to enable them to contribute to a more sustainable future. In this way, the UPV/EHU encourages experiential learning and collaboration to learn about sustainability by practicing and sharing from different knowledge areas, in order for the teaching–learning environment of the UPV/EHU to foster crucial reflection and an integral vision of future scenarios, along with other skills needed to generate a change in thought and in real practice for sustainability. For that, the university created a program named Campus Bizia Laboratory (CBL) that serves as a tool to promote progress in sustainability in the university itself through the collaboration of teaching and research staff, administration and services staff and students who carry out end-of-degree projects and master's theses of different disciplines and knowledge in innovative projects that use the campus as a laboratory in order to respond to the sustainability challenges of the university itself.

Considering on the one hand this new and innovative educational approach and on the other hand the sanitary situation we are facing, through this research, the authors show the works that are being developed in the field of indoor air quality control in educational spaces of the University of the Basque Country.

2 State of the Art

In Europe, 17% of non-residential buildings are educational [1]. According to the census of the Cadastre in Spain (excluding the Basque Country and Navarre), there are 51,349 cultural buildings [2], among which the majority are educational buildings. Many educational buildings are older than the entry into force of the current regulations (In Spain, between 90 and 95% of educational buildings were built before the Technical Building Code of 2006, which promoted effective requirements on ventilation, air quality and indoor comfort. Some 60% of these centres were built before any related regulations existed, before 1980). From the recently issued Strategy for Energy Rehabilitation in the Spanish Building Sector [3], the lack of controlled ventilation in these buildings is evident. In the Basque Country, there are 1209 primary and secondary education buildings (600 in Bizkaia, 426 in Gipuzkoa and 183 in Alava) [4]. As for the university sphere, the University of the Basque Country alone has 32 schools and faculties distributed in different parts of the territory. In short, this means that these buildings are generally not equipped with mechanical ventilation systems, i.e., they manage IAQ by means of traditional natural ventilation, windows to the outside or courtyards. This has a few important effects:

- Indoor air quality is not always adequate. Current national regulations require, for school buildings, a fresh air flow rate of 12.5 litres/second per occupant, high values that cannot always be guaranteed with natural ventilation with a more irregular operation of windows and dependent on the user's behaviour [5].
- Natural ventilation often implies considerable energy losses.
- Insufficient air renewal has proven implications for student learning and work performance in general.
- Poor IAQ encourages the development of airborne diseases.

The standard solution to achieve the current legislative performance standards is the installation of mechanical ventilation systems in old buildings. This measure should logically follow the improvement of both building envelope and airtightness. However, the spatial requirements and the cost of implementation mean that such interventions are not always feasible.

As a result, good training in the use of natural ventilation (opening and closing windows and doors) is essential to achieve adequate air renewal rates and therefore proper IAQ. The current situation caused by the COVID-19 pandemic has accentuated this problem, as it is considered that adequate ventilation of teaching spaces may be the best way to prevent infection of students and teachers. The goals of this project are:

- To assess the capacity of natural ventilation systems in university spaces;
- To evaluate the impact of the implementation of real-time monitoring data with the use of screens and calibrate user response;
- To generate a culture of good practices among students, teachers and staff regarding ventilation of shared spaces;
- To determine whether it is necessary or not to make future investments in mechanical ventilation equipment under a scientific logic;
- To generate a methodology that allows for the replicability of the decision-making in all university spaces with natural ventilation or of insufficient air quality;
- To promote, in short, an improvement in the health conditions of the workspaces of all the groups present in the university and build healthy habits regarding IAQ in the UPV/EHU community.

Our initiative seeks to position all users of university spaces as active players in improving the health of their workplace and to provoke their learning supported by the information generated in the implemented IAQ monitoring system.

3 IAQ in Schools and College Campuses

In Europe, numerous studies assess indoor air quality in naturally ventilated buildings [6–8]. Generally, they report CO_2 concentration values, and the levels of renovation (ACH) achieved by acting on ventilation are evaluated. In Spain, a relevant monitoring study of 36 schools throughout the country during one school year [9] concluded that if hygrothermal comfort alone was considered (ambient temperature and relative humidity), only 68.06% of the time the classrooms were in comfortable conditions. The results for CO_2 levels (ppm) showed that only 32.40% of the time acceptable levels of CO_2 are available. Overlaying both data, it concludes that only 16.16% of the time the schools are in use, there are adequate comfort conditions. In the Atlantic coastal climate, where the campus of UPV/EHU is located, this study concludes from the monitored data that the hygrothermal comfort in the classrooms is satisfactory 74.93% of the time, but CO_2 levels (ppm) are only acceptable 27.27% of the occupied time. In the Basque Country, a study carried out in six schools in Bajo Nervión with different outdoor air qualities showed a poor indoor air quality overall, with all the studied classrooms exceeding the recommended values of 800 ppm CO_2 [10].

3.1 Relationship Between CO_2 Concentration and COVID-19. Prevention Protocols

CO_2 is co-exhaled with aerosols containing SARS-CoV-2 particles by COVID-19 infected people and can be used as a proxy of SARS-CoV-2 concentrations indoors. Scientific studies confirm that aerosols are one of the routes of transmission of SARS-CoV-2, so the possibility of airborne transmission increases in high-occupancy indoor spaces such as classrooms. Therefore, international protocols and guidelines have established the requirement for educational buildings to be overventilated with an outdoor fresh air supply [11–18] using natural or mechanical ventilation as a means of maintaining low risk rates for airborne virus transmission. In the case of natural ventilation, different ventilation procedures have been studied. Depending on the CO_2 values reached in ppm, the efficacy of the different systems can be measured.

In a first stage (September 2020), the UPV/EHU established a provisional ventilation protocol based on intermittent ventilation. Thus, a 10-min period of open windows was mandatory every 20 min during lectures. It was later found that for intermittent ventilation schedules, CO_2 concentration increased rapidly after the windows were closed again, so the Spanish Ministry of Health recommended in February 2021 to keep doors and windows constantly open as the most effective measure to prevent airborne COVID-19 transmission [19].

This recommendation conflicts with hygrothermal comfort in classrooms, especially during cold days. Other factors that have hindered the implementation of

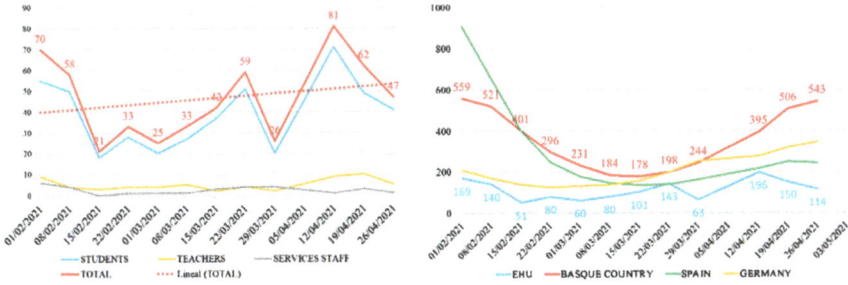

Fig. 1 Left, number of COVID-19 cases detected in the UPV/EHU on a weekly basis since February 2021. Right, weekly cumulative incidence of COVID-19 per 100,000 inhabitants detected in the UPV/EHU on a weekly basis since February 2021 compared to the Basque Country, Spain and Germany. *Source* UPV/EHU Monitoring Committee and the Spanish Ministry of Health

continuous ventilation protocols have been outside noise, the need to darken classrooms for teaching and in some cases the absence of good natural ventilation potential in some teaching spaces (lack of operable windows, suboptimal layouts for cross-ventilation, etc.).

However, it is not easy to establish a direct relationship of contagion within the classroom and IAQ. In the case of the UPV/EHU, in May 2021, teacher and student unions reported 1750 COVID-19 cases out of a total of 41,389 users (34,000 students, 5496 teachers and 1893 workers). A surveillance committee reports every 15 days on the cases detected [20]. The week of 12–18 April, for example, 81 new positive cases were reported (71 students, 9 teachers and 1 administrative staff), but no further information is given on the sites of infection. The evolution of the cases detected at the university shows that the different groups are affected in a similar way (Fig. 1, left).

It can be seen that the cumulative incidence per 100,000 inhabitants detected each week in the UPV/EHU is below the average for Spain and the Basque Country and even below that reported by Germany (Fig. 1, right).

3.2   Characterisation of UPV/EHU Buildings

The University of the Basque Country is made up of different faculties and schools mainly concentrated in three campuses: Leioa in Bizkaia, Donostia-San Sebastian in Gipuzkoa and Vitoria-Gasteiz in Araba. These buildings correspond to very different building periods, from the beginning of the twentieth century to the most modern ones (Fig. 2).

Natural ventilation is predominant in all campuses except for the most modern buildings, which are equipped with mechanical ventilation systems. In general, this type of equipment was only retrofitted in older buildings in auditoriums and libraries,

Fig. 2 Campus of the UPV/EHU in Gipuzkoa with the School of Architecture and a sample of different buildings of the UPV/EHU campuses. *Source* UPV/EHU

which due to their characteristics did not have natural ventilation. In these spaces, ventilation is associated with the HVAC system.

3.3 Campus Bizia Laboratory (CBL) Research Initiative

This research project is framed within the 2020–2021 call of the vice-rectorate for Innovation, Social Commitment and Cultural Action under the coordination of the Directorate of Sustainability under the title of Campus Bizia Laboratory. CBL consists of a research and action process that aims to develop a high-impact practice among students (transdisciplinary learning based on challenges related to sustainability) in which teaching staff acts as researchers of their own practice. This program has a curricular nature and is being implemented through the students' final degree and final master's theses. The objectives are:

- To create a transdisciplinary community that works cooperatively in the resolution of challenges and problems of unsustainability that are detected on the UPV/EHU campuses themselves;
- To design, develop and evaluate a working device that allows to successfully carry out high-impact learning processes linked to sustainability within the UPV/EHU campuses;
- To articulate and make visible an institutional CBL project that, in stages, extends its scope of action to all UPV/EHU degrees and generates multilevel sustainable practices.

4 Methodology

The methodology carried out during the project consisted of the development of certain actions and organisation of work, which are mainly reflected in the following items:

4.1 Management and Teacher-Student Coordination for Autonomous Work

The first step consists on orienting the student on the subject to be addressed. For this purpose, it was proposed to establish a series of driving questions that would allow the student to progress in a structured and methodical way during the learning process. In addition, this allows the student to achieve a significant understanding of the concepts, principles and practices of the subject to be addressed. To this end, the following questions were proposed: What is the purpose of the project and why are we analysing it? What is the framework of the project? How can I achieve the objectives of the project? What are the methods or tools that will enable the work to be carried out? Is there any previous experience on the subject or similar work? This allows the student's interest to be maintained throughout the learning process; thanks to the fact that they are working on the search for a solution to a real and current problem. For the development of the driving questions, we have taken into consideration attitudes such as: being proactive to motivate the student, inviting to analyse an attractive topic for the student, presenting a challenge and promoting discussion and debate on the same topic.

4.2 Involvement of Other Staff (General Student Body, Teaching Staff and Administrative and Service Staff)

This involves the necessary actions to involve the university community in the project. The idea of the project and of CBL is to take the campus as a laboratory in which to carry out the proposed research. For this reason, on the one hand, two centres of the Gipuzkoa Campus of the UPV/EHU were chosen, the Gipuzkoa School of Engineering and the School of Architecture. The choice of both case studies is due to the fact that the study will make it possible to analyse the effect and the differences obtained in the IAQ of two buildings with different ventilation systems. The first one having mechanical ventilation systems and the second relying on natural ventilation only.

Next, in order to involve the staff, a series of direct contacts were made with the personnel involved. In the classrooms where the lectures are taught, the students were informed by the teachers involved in the project, guiding them on the measures to be adopted for the correct ventilation of the spaces. Based on the information collected by the monitoring systems, students can make decisions based on the collected data and analyse the effect of their interaction with respect to the indoor conditions of the spaces.

In relation to the administrative and service staff, meetings were organised with the staff responsible for the operation and maintenance of the building in order to present the project and analyse their involvement. In this sense, the first group, along with the students, is participating in the monitoring of the spaces where they

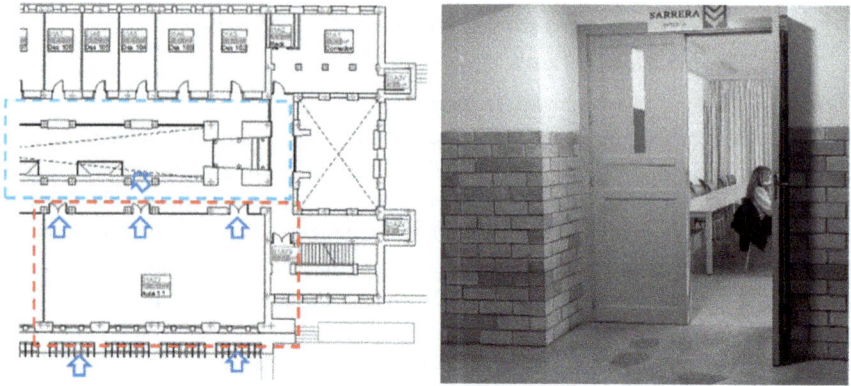

Fig. 3 Example of a classroom used in the experiment. Classroom 1.1, School of Architecture. The blue arrows indicate the natural ventilation flows. On the right, access doors open during a lecture

carry out their activities in order to follow up the IAQ conditions. The second group participated in the installation of screens at the classrooms in order to inform about the IAQ conditions of the different parameters that are being analysed and thus be able to offer a global vision of the project as a whole.

4.3   Case Study Selection

The cases analysed cover several classrooms and administrative spaces of the School of Architecture of the UPV/EHU, inaugurated in 1992 and built with brick facades with an intermediate air chamber. The windows were improved in a recent refurbishment and it only has mechanical ventilation in the basement, conference room and drawing room on the top floor (former library). As a sample, the floor plan of one of the classrooms selected for the experiment is shown. It is a 160 m² classroom with a capacity reduced to 50% of 65 students. It has natural ventilation through windows and doors that open onto large common atriums of the school that facilitate proper cross-ventilation (Fig. 3). The windows are large and allow for proper ventilation when fully opened. The classroom is darkened by interior curtains. The rest of the spaces analysed have similar characteristics.

4.4   Methods and Tools

Based on previous experiences [21–23], two monitoring campaigns were deployed at the same time. The first of them has made it possible to use the campus itself as a

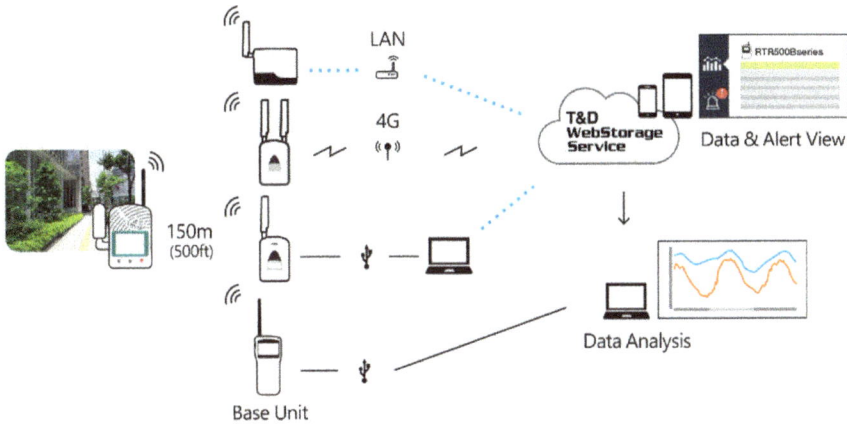

Fig. 4 Commercial monitoring system based on RTR-576. Connection diagram to the network and sending data to the server for control and management. *Source* T&D Corporation

laboratory where the proposed analysis can be implemented, specifically in interior spaces of the School of Architecture and the Faculty of Engineering of Gipuzkoa.

Regarding the monitoring systems, the T&D RTR-576 device has been used for the measurement of CO_2 concentration, temperature and relative humidity. The system consists of a CO_2, temperature and relative humidity sensor (RTR-576) and a receiver connected via the University's LAN network (RTR-500). T&D's FTP service allows remote data visualisation, as well as data downloading and alert programming (Fig. 4).

In parallel, also a novel approach was implemented based on open-source platforms (OSP) and the Internet of Things (IoT) aiming to offer alternative and low-cost monitoring systems. The aim is to compare both commercial and OSP systems and verify the suitability of the latter for future deployments in other buildings of the campus. Both systems allow not only the data recording and storing, but also remote reading of the monitored values in order to be shown to users being this last characteristic the most important for the object of this study. In addition, magnetic switch sensors have been implemented for detecting the opening status of windows. Greater use of windows has a direct effect on ventilation, so these sensors will allow us to analyse and verify the effect of users on the IAQ of the spaces and therefore the reduction of the risk of COVID-19 infection.

For the visualisation of the data, a screen connected to the Internet has been set up using either the commercial platform of T&D Corporation or the company's own platform (see Fig. 5).

The displays have been arranged in a way that is visible to the greatest number of users in order to trigger effective action on natural ventilation based on the data displayed and associated with an informative legend on how to act on the ventilation based on the CO_2 concentration (Fig. 6).

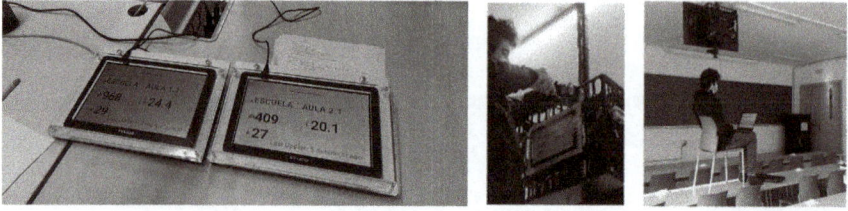

Fig. 5 Calibration of display screens, installation and programming of the monitoring equipment

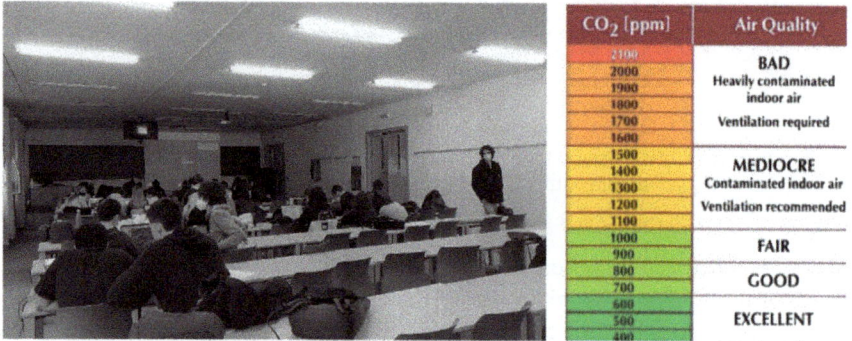

Fig. 6 Display of data on the screen and legend and instructions poster of the actions to be taken with the ventilation according to the measured CO_2 concentration

Regarding the alternative open-source IoT monitoring system, its development is based on previous work carried out in the field of building monitoring [24–26]. This methodology is based on the implementation of open software and hardware such as Arduino and Raspberry Pi that allows for the acquisition, storage, management and visualisation of the environmental variables of interest. In this project, one of the cornerstones was the deployment of a LoRaWAN network on the university campus that offers a long-range and low-power IoT network infrastructure for sending the data collected by the measurement nodes (Fig. 7).

Subsequently, the data is collected on the Raspberry server, being incorporated into a TSDB-type database, specifically InfluxDB, to finally be represented on a dashboard through Grafana. Access to said panel is done through a Web browser by entering the IP address of the server and offering direct visualisation of the generated panel in order to be able to track the measurements made on indoor air quality (Fig. 8).

The main highlights of the open-source monitoring system are:

– Cost-effective and open-source system to allow mass deployment solution;
– Based on an IoT technology to allow RTD;
– Wireless communication technology to avoid disturbance to the property;
– Use of an independent network so as not to depend on the existence or not of a connection on the site;
– Employ a separate power supply to avoid relying on site electrical supply;

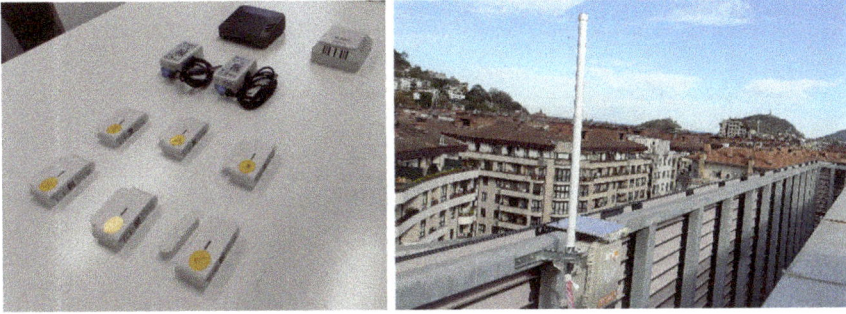

Fig. 7 Deployed open- source IoT monitoring system in the Campus of Gipuzkoa of the University of the Basque Country (UPV/EHU). Left, monitoring nodes and server. Right, deployed IoT LoRaWAN gateway

Fig. 8 Designed dashboard for RTD visualisation

- Data ownership and independence;
- Lower cost.

5   Results

The effect of the availability of information in classrooms is assessed by studying air quality during two school weeks from Monday to Friday with samples every 5 min. The IAQ is controlled using CO_2 concentration as a proxy. For schools, the Spanish RITE standard sets an IDA 2 requirement, allowing a maximum CO_2 concentration of 900 ppm. Lectures run from 8:00 am to 8:00 pm with breaks from approximately 2:30 pm to 3:00 pm. The analysis of a classroom with a somewhat lower natural ventilation capacity than others monitored in which a generally better IAQ has been recorded is shown in Fig. 9.

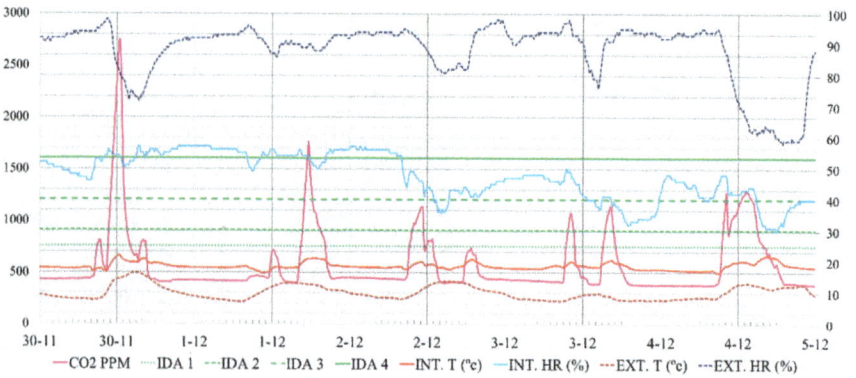

Fig. 9 Monitoring results for classroom 1.1 during the week 30/11/2020 to 06/12/2020

In the first week, from 30 November to 4 December 2020, no real-time data is available for classroom users. It was a cold week outside (8–13 °C), and the heating was partially out of order and working at 30% of its full power. The evolution of indoor and outdoor temperatures and relative humidity is also shown.

The maximum values occur at the peak times of the morning or evening classes. On Monday 30 December, for example, the peak CO_2 concentration occurs at 12:30 am (2738 ppm), on Tuesday at 5:40 pm (1759 ppm), on Wednesday at 11:10 am (1132 ppm), on Thursday at 4:20 pm (1135 ppm) and on Friday at 1:00 pm (1282 ppm). During the week, air quality remains within the standard's limits (<900 ppm) 75% of the occupied hours.

The second week analysed runs from 26 April to 2 January 2021. In this case, the classroom users have live data available on the screens installed. It is a warm week outside (12–17 °C). In this case, the heating is running at full power. It can be seen that the maximum values of CO_2 concentration have been considerably reduced (see Fig. 10).

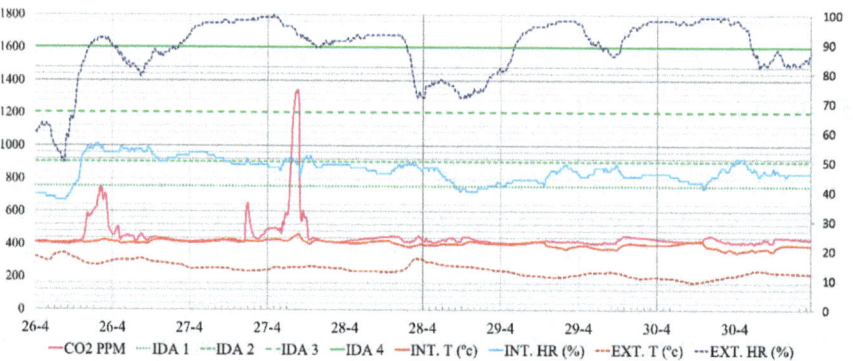

Fig. 10 Monitoring results for classroom 1.1 during 26/04/2021 to 02/05/2021

On Monday 26 April, there is a peak CO_2 concentration at 15:40 h (742 ppm), on Tuesday at 16:40 h (1339 ppm), on Wednesday at 13:500 h (448 ppm), on Thursday at 13:50 h (450 ppm) and on Friday at 13:50 h (449 ppm). During the week, air quality remains within the standard (<900 ppm) 98% of the time, marked with the IDA 2 limits.

The peak observed on Tuesday at 16:40 h, which occurs on Tuesday afternoon during a class with high occupancy and group work, is noteworthy. The remote visualisation of the data by the school staff enabled the teacher and students to be alerted. CO_2 values could be quickly lowered in less than 10 min by fully opening doors and windows. Similar effects are produced in the other spaces used as case studies over the 30 weeks of the 2020/2021 academic year.

In terms of thermal comfort, the more precise management of natural ventilation logically allows for greater energy savings by being able to carry out more precise ventilation only at times when the regulatory CO_2 concentration values are exceeded. However, as strict permanent ventilation protocols are in place, it is difficult to assess the overall ventilation system in terms of its energy effects. In the comparison between the two mentioned weeks, thermal comfort is assessed by plotting temperature and relative humidity pairs in the week of 30th of November. It can be seen that during school hours from Monday to Friday (in colour) it is difficult to reach the comfort ranges of the EN 7730 standard [27] and national regulations (RITE [5] and INSHT [28]). The same thing occurs during unoccupied hours and at the weekend (in black in the graph). The effect is amplified by the poor functioning of the heating system. The maximum temperature reached during school hours is 22.0 °C with a minimum of 16.2 °C. The RH ranges between 30 and 57% (see Fig. 11).

In the week of 26 April, we see the effect of warmer outdoor temperatures and the full operation of the heating system. In this case, we can see that almost all school hours are within a reasonable comfort range in all the models. The maximum temperature reached during school hours is 25.5 °C with a minimum of 19.3 °C. The RH ranges between 40 and 56% (see Fig. 12).

It is important to note that the actual perceived comfort must be lower as the windows are open and cold draughts from outside increase local discomfort.

6 Discussion and Conclusions

The results obtained show a logical evolution of CO_2 concentration throughout the duration of the classes. CO_2 concentration peaks are produced in the start of the morning lectures (9:00 am) and evening classes (4:00 pm). The target set is that established by the national regulation RITE IDA 2 (CO_2 level < 900 ppm). Permanent ventilation can lower these values in most of the cases or keep them stable. The large volume of the studied spaces allows for a better dissolution of CO_2. The display of real-time monitoring data to users allows for a better, more rational use of natural ventilation, allowing a better compromise between IAQ and thermal comfort and reducing unnecessary energy losses. This is a first step towards the implementation

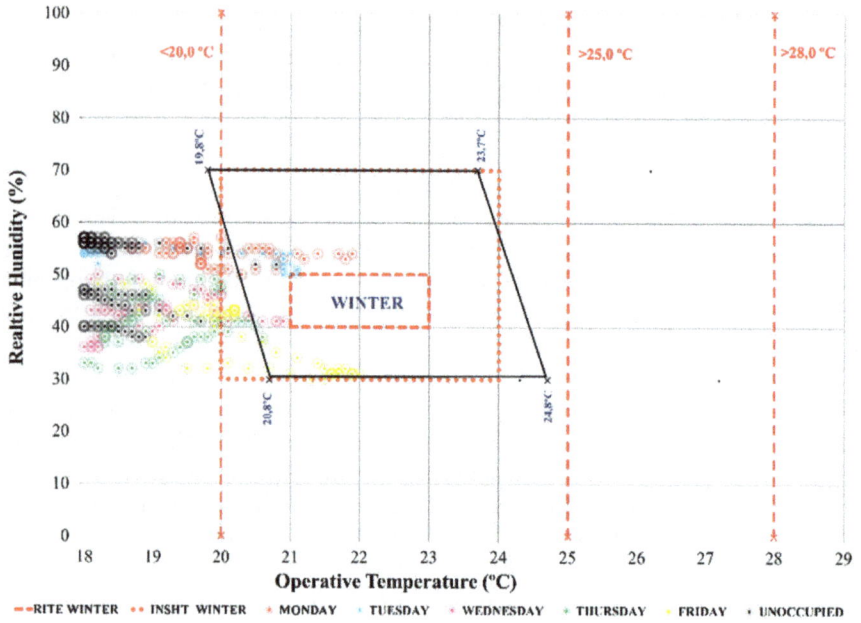

Fig. 11 Thermal comfort monitoring results in classroom 1.1, 30/11/2020 to 06/12/2020 according to EN 7730, RITE y INSHT standards

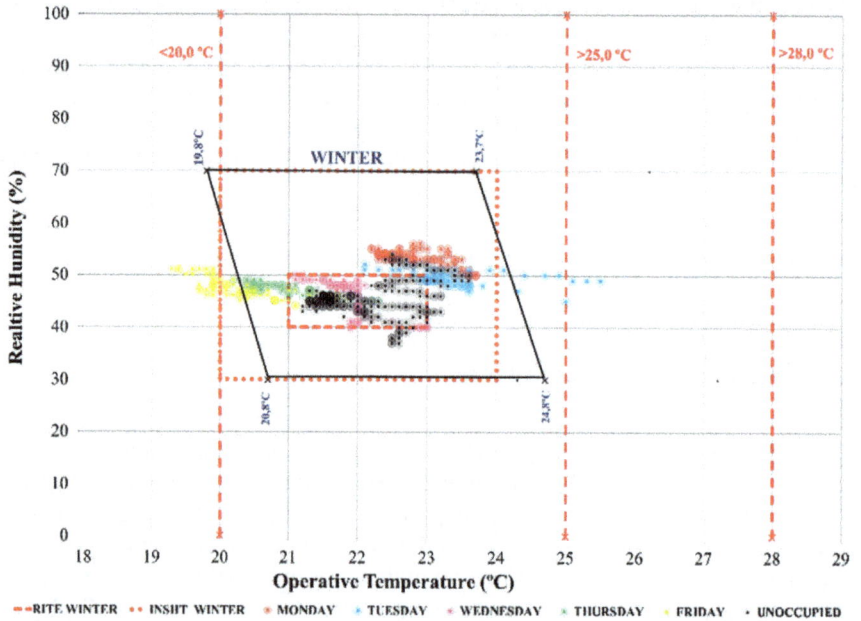

Fig. 12 Thermal comfort monitoring results for classroom 1.1, 26/04/2021 to 02/05/2021 according to EN 7730, RITE y INSHT standards

of hybrid ventilation systems with a greater or lesser degree of mechanisation and automation in an IoT environment.

Acknowledgements This study was financed by the Territorial Planning and Housing Department of the Basque Government and the Department of Architecture of the University of The Basque Country (UPV/EHU). It was also financed under the title "*Proyecto Piloto Sobre Calidad Del Aire En Espacios Interiores Universitarios*" in the CBL program 2020–2021, promoted by the Directorate of Sustainability, Vice-Rectorate for Innovation and Social Commitment of the University of the Basque Country.

References

1. BPIE (2011) Europe's buildings under the microscope. A country-by-country review of the energy performance of buildings. Last accessed 10 May 2021
2. Cuerdo-Vilches T COVID-19: La ventilación en centros educativos, una asignatura pendiente. http://hdl.handle.net/10261/227119. Last accessed 10 May 2021
3. ERESEE 2020 (2020) Actualización 2020 de la estrategia a largo plazo para la rehabilitación energética en el sector de la edificación en españa. Ministerio de Transportes, Movilidad y Agenda Urbana. Last accessed 10 May 2021
4. Departamento de Educación del Gobierno Vasco. Directorio de Centros educativos del País Vasco 2020/2021. https://www2.hezkuntza.net. Last accessed 10 May 2021
5. RITE (Regulation of Thermal Installations) (2013) Ministerio de Industria, Energía y Turismo: Reglamento de Instalaciones Térmicas En Los Edificios. Boletin Oficial Del Estado
6. Dorizas PV, Assimakopoulos MN, Helmis C, Santamouris M (2015) An integrated evaluation study of the ventilation rate, the exposure and the indoor air quality in naturally ventilated classrooms in the Mediterranean region during spring. Sci Total Environ 502:557–570
7. Stabile L, Dell'Isola M, Russi A, Massimo A, Buonanno G (2017) The effect of natural ventilation strategy on indoor air quality in schools. Sci Total Environ 595:894–902
8. Schibuola L, Scarpa M, Tambani C (2016) Natural ventilation level assesment in a school building by CO_2 concentration measures. Energy Procedia 101:257–264
9. Gutiérrez Cuevas B, Manso Villalaín JM (2018) Proyecto de monitorización de colegios en el curso 2017–2018. Plataforma de Edificación Passivhaus/Universidad de Burgos, Burgos
10. Campo Díaz VJ, Mendivil Martínez A (2006) Calidad del aire interior en los Centros de Educación Infantil del País Vasco. El Instalador. ISSN 0210–4091(427):34–42
11. REHVA (Federation of European Heating, Ventilation and Air Conditioning Associations) REHVA COVID-19 guidance document, 3 April 2020
12. Education & skills funding agency: guidelines on ventilation, thermal comfort and indoor air quality in schools. Building Bulletin, p 101 (2018)
13. Gobierno de España: Medidas de prevención, higiene y promoción de la salud frente a Covid-19 para centros educativos en el curso 2020–2021 (2021)
14. CSIC (Superior Council of Scientific Investigations) Guide for classroom ventilation. https://www.csic.es/sites/default/files/guia_para_ventilacion_en_aulas_csic-mesura.pdf
15. Gobierno del Principado de Asturias: Recomendaciones preventivas de ventilación en centros educativos para reducir las probabilidades de contagio de Covid-19 (2020)
16. Generalitat Valenciana: Guía para la ventilación en los centros educativos de la Comunitat Valenciana (2020)
17. Gobierno Vasco: Guía para reducir el riesgo de transmisión del SARS-CoV-2 por aerosoles en centros educativos (2020)
18. Asociación Madrileña de Empresas Privadas de Enseñanza (CECE-Madrid): Guía para la ventilación en centros escolares (2020)

19. Muelas A, Remacha P, Tizné E, Ballester J (2021) Ventilación natural en las aulas. Universidad de Zaragoza, CSIC, Ventilación Continua vs. Intermitente
20. Comité de Vigilancia de la COVID-19 de la Universidad del País Vasco. https://www.ehu.eus/es/informacion-del-comite-de-vigilancia
21. Rodríguez Vidal I, Otaegi J, Oregi X (2020) Thermal comfort in NZEB collective housing in Northern Spain. Sustainability 12(22):9630. https://doi.org/10.3390/su12229630
22. Rodriguez Vidal I, Oregi X, Otaegi J (2020) Thermal comfort evaluation of offices integrated into an industrial building. Case study of the Basque Country. Environ Clim Technol 24(2):20–31. https://doi.org/10.2478/rtuect-2020-0051
23. Rodriguez I, Oregi X, Otaegi J (2021) Thermal comfort assessment in an administrative area of an industrial building in Spain. In: Littlewood J, Howlett RJ, Jain LC (eds) Sustainability in energy and buildings 2020. Smart innovation, systems and technologies, vol 203. Springer, Singapore. https://doi.org/10.1007/978-981-15-8783-2_2
24. Martín-Garín A, Millán-García JA, Hernández-Minguillón RJ, Prieto MM, Alilat N, Baïri A (2021) Open-Source framework based on LoRaWAN IoT technology for building monitoring and its integration into BIM models. In: Hussain CM, Di Sia P (eds) Handbook of smart materials, technologies, and devices. Springer Nature. https://doi.org/10.1007/978-3-030-58675-1_9-1
25. Martín-Garín A, Millán-García JA, Baïri A, Gabilondo M, Rodríguez A (2020) IoT and cloud computing for building energy efficiency (Ch. 10), Start-Up creation 2nd edition: the smart eco-efficient built environment. Woodhead Publishing Ser Civ Struct Eng 235–265. https://doi.org/10.1016/B978-0-12-819946-6.00010-2
26. Martín-Garín A, Millán-García JA, Baïri A, Millán-Medel J, Sala-Lizarraga JM (2018) Environmental monitoring system based on an open-source platform and the Internet of Things for a building energy retrofit. Autom Constr 87:201–214. https://doi.org/10.1016/j.autcon.2017.12.017
27. ISO (2005) ISO 7730: ergonomics of the thermal environment analytical determination and interpretation of thermal comfort using calculation of the PMV and PPD indices and local thermal comfort criteria. Management
28. INSHT (National Institute for Safety and Health at Work), website of the INSHT (2020): https://www.insst.es/. Last accessed 10 May 2021

Part II
Smart Techniques for Monitoring a Pandemic and Forecasting its Course

Chapter 7
An Overview of Methods for Control and Estimation of Capacity in COVID-19 Pandemic from Point Cloud and Imagery Data

Jesús Balado⊙, Lucía Díaz-Vilariño⊙, Elena González⊙, and Antonio Fernández⊙

Abstract The main actions to control the COVID-19 pandemic and prevent the spread of the virus have focused on population control and social distancing. Over the years, many applications of sensing technologies have shown their effectiveness in solving problems related to the acquisition, identification and modelling of the environment, although not always from a human-centred approach. This chapter compiles sensing techniques from point cloud and imagery data related to population control and estimation of the capacity: people counting, biometric identification, monitoring of activities, distance measurement and 3D modelling. The current state-of-the-art techniques and the most common algorithms are summarized. Finally, the advantages and disadvantages of point cloud data and imagery are discussed, as well as the current trends of the predominant technology in each field.

Keywords LiDAR · Human detection · 3D modelling · Photogrammetry · Pose estimation · Biometric identification

1 Introduction

The world is currently facing an unprecedented situation owing to the coronavirus pandemic (COVID-19). The pandemic has impacted all aspects of life, and this impact has been more pronounced in cities, where most people live. Indeed, it is estimated that 55% of the world's population currently lives in cities, and cities just represent 3% of Earth's area [1].

Many cities are in the process of transition to become 'smart cities' mainly through smart solutions in waste management, traffic and public transport. But even these cities have been unable to implement similar solutions to effectively deal with the pandemic. Social or physical distance is one of the aspects that have been proved to reduce the spread of coronavirus disease. This requirement has been imposed

J. Balado (✉) · L. Díaz-Vilariño · E. González · A. Fernández
Universidade de Vigo, CINTECX, GeoTech Group, 36310 Vigo, Spain
e-mail: jbalado@uvigo.es

© The Author(s), under exclusive license to Springer Nature Singapore Pte Ltd. 2022
R. J. Howlett et al. (eds.), *Smart and Sustainable Technology for Resilient Cities and Communities*, Advances in Sustainability Science and Technology,
https://doi.org/10.1007/978-981-16-9101-0_7

by governments through rules, plans and protocols, adopting different measures according to the cultural criteria of each country. However, controlling social distance is being difficult and mostly depends on the surveillance of people responsible for public spaces.

Measured data from remote sensing techniques is becoming increasingly popular for realistic planning by optimizing variables such as the number of people, time and space, and it offers an insight into the automatic control of social distance, capacity estimation and crowd simulation. The use of cameras and laser scanners for mapping the real environment is consolidated, but the effectiveness of those systems depends on the success of processing the captured data to extract useful information for the applications they intend. Over the last decade, much effort has been devoted to applications such as infrastructure monitoring [2], indoor modelling [3] and face recognition [4].

This work offers a general perspective of the potential use of remote sensing techniques in the context of the COVID-19 pandemic, both in indoor and outdoor environments. This study discusses recent methods dealing with automatic point cloud and imagery data processing with a special emphasis on their use in applications aimed at tackling the pandemic, such us social distance control and capacity control/estimation. The strengths and weaknesses of each method and data source, as well as their principles of operation, are briefly explained.

2 Imagery Data

Due to the COVID-19 pandemic, governments try to implement tools to accurately monitor the occupancy of public spaces as well as compliance with social distance. Computer vision is one of the most promising technologies to achieve that endeavour. The widespread availability of imagery data together with recent advances in machine learning makes computer vision an ideal candidate to implement those tasks and are likely to be the underlying reason why research on those topics has experienced a new impetus in the last few months.

Image processing methods do not make a clear distinction between indoor and outdoor application. Although the spread of the virus may be significantly different in these two environments [5], regulations in certain countries do not always make such a distinction and often extend indoor amendments to outdoor applications when crowds of people are present, so image processing methods must be applicable to both environments.

2.1 People Counting

Understanding and monitoring crowd behaviour have been a challenge for security agencies across the world. Thus, a huge research effort has gone into finding a proper solution to such a problem. People counting/density estimation is one of the major subfields of crowd analysis and monitoring. This task can be accomplished on still images or video sequences in both outdoor and indoor environments. People counting can be approached through three different strategies, namely detection, clustering and regression, each of them having its pros and cons [6]. The current trend is to replace handcrafted image features such as Histograms of Oriented Gradients (HOG), Grey-Level Co-occurrence Matrices (GLCM), Scale-Invariant Feature Transform (SIFT) and so on by Deep Convolutional Neural Networks (DCNN) [7], which has become the dominant paradigm in most computer vision applications. Besides, it is worth pointing out that several commercial systems for occupancy monitoring have come onto the market [8–10].

2.2   Biometric Identification

Identifying people from imagery data is one of the most successful application fields of computer vision. There are different biometric modalities, which can be classified into physiological and behavioural. When it comes to occupancy monitoring, the most important physiological modality is face recognition. In the past, Local Binary Patterns (LBPs) were the dominant approach to facial recognition [11]. As in many other computer vision areas, the features on which face recognition systems are based have evolved over the years from handmade visual features to deep learning [12]. This also applies to the most important behavioural modality: gait recognition. The main advantage of gait recognition over face recognition is that it does not require observed subject attention and cooperation. Besides, gait recognition offers a great potential for recognition of low-resolution videos [13].

2.3   Monitoring of Activities

Regarding health protocols related to COVID-19, a set of activities such as disinfection of hands or attitude of approach to other people are treatable from images.

Activities are related to the position of the body's limbs, trunk and head, from which the understanding of human behaviour and the analysis of the activity are predicted [14, 15]. Monitoring of activities is a classification and recognition task based on a set of features extracted from image. Shape, silhouette or pose is adopted as features, the extraction of which is difficult because the human body can adopt multiple geometries [16].

In literature, feature extraction is usually enfaced as Human Pose Estimation (HPE). A group of techniques estimates key points of the body for the skeleton approach. Their invariance to clothing, lighting and background makes these techniques useful and popular [17–21]. Other techniques estimate the pose by decomposing the human body into connected rectangular parts at different orientations [22,

23]. Not only based on pose approaches have been proposed, but also others such as Unstructured Feature Point [24], where glimpses, RGB data, play the role of key points.

Human Activity Recognition (HAR), classification and recognition step, has been reviewed by [25, 26], outlined in [27], as well as evaluated the different techniques in [22]. In this area, there is a great variety of proposals both in classic methods, Deep Neural Networks (DNN) and Convolutional Neural Networks (CNN).

2.4   Distance Measurement Using Images

The problem of measuring distances using images is solved from photogrammetry [28]. Several basics must be considered. By itself, the image does not give a real measurement; it is necessary to fix an equivalence between the number of pixels per unit of reality, mm, cm and m. This ratio is established by associating a reference entered in the image, which can be a rule, even an average value obtained from a set of images [29]. Likewise, this ratio can be derived by knowing the focal length of the camera and the distance at which it is a reference object. Images taken using low-cost cameras and webcams may appear radial distortion. This phenomenon can make peripheral measurements erroneous. Therefore, the most reliable measurements are those in the centre of the image.

2.5   3D Modelling from Imagery

Capacity estimation leads to the need for 3D models. In addition, these 3D models can also be used for crowd simulation [30]. The generation of 3D models from images is currently carried out through a combination of photogrammetry and computer vision [31]. While photogrammetry provides position and shape, computer vision, as defined by Szeliski [32], 'is a mathematical technique for recovering the three-dimensional shape and appearance of objects in imagery'. To carry out this modelling, the first phase of imaging is carried out following a careful orientation and camera calibration protocol from different angles. Subsequently, the set of images is processed using mathematical algorithms based on epipolar geometry [33] to generate the 3D model. As a result, capacity can be estimated. In addition, this model can be viewed at any angle ranging from 0° to 360°. Correct visualization is guaranteed by matching the pixel and measurement images, as well as the visual aspects of colour and shadow.

3 LiDAR Data

The light detection and ranging (LiDAR) technology is a method for determining ranges by targeting an object with a laser and measuring the time for the reflected light to return to the receiver. Through LiDAR scanning, a 3D environment can be acquired quickly and accurately in a point cloud. Point clouds are unobstructed 3D vector data whose processing requires a neighbour search or distance search to establish relations between points. These search processes are time-consuming and computationally expensive.

3.1 People Detection

The literature on people detection and tracking from LiDAR data is scarce. The main use of Terrestrial Laser Scanning (TLS) and Mobile Laser Scanning (MLS) data in urban areas, both indoors and outdoors, is the mapping of objects and modelling of the built environment. People are not considered as an object to be modelled, since their existence in the scene is temporary. The few papers that detect people in mapping-modelling consider people as a noise class rather than a target class [34]. With Airborne Laser Scanning (ALS) data, the number of points per square metre acquired with ALS is insufficient to detect a person. Human detection with LiDAR data focuses on the fields of autonomous driving and robotics. Low-cost LiDAR is mounted in robots and vehicles to keep the cost of these devices as affordable as possible. Consequently, the generated point clouds are less dense, and they have a limited range to recognize objects and can be processed in real time.

In autonomous driving, pedestrian detection is performed in the immediate vicinity of the vehicle to avoid collisions. In [35], different classifiers are evaluated in the detection of pedestrians from LiDAR data, showing a difference of 30% in the precision detection of people located up to/more than 20 m away from the vehicle. Detection of distant pedestrians by LiDAR is hindered by the low point density of pedestrians, which interferes with feature extraction. Pedestrian tracking is also implemented in autonomous driving as the different pedestrian positions between consecutive scans. The tracking is also analysed in order to ensure that the pedestrian's path does not intersect with the vehicle's path [36]. By combining pedestrian detection, location and tracking with the detection of other built elements, such as curbs, it is possible to know whether the pedestrian is on the road or the sidewalk and whether the pedestrian intends to cross the street.

In robotics, there is no need to process such a large environment because the movement speed of most support robots is much lower than in autonomous driving, and therefore, they do not need to analyse distant data for safe interaction with the surrounding environment. Most LiDAR applications are based on human tracking [37, 38]. The robot is located closer to the person, so the point clouds generated are denser, even with low-cost LiDAR, and the point clouds allow to distinguish people according to objects they are carrying (walkers, baggage, etc.) [39] or specific situations, such as one person accompanying another [40].

The use of static LiDAR for human monitoring has been explored in [41], where a fixed LiDAR is installed on a traffic light pole. The authors claim that this device obtains accurate and real-time information on the presence, position, velocity, and direction of pedestrians and vehicles. In [42], a platform that integrates LiDAR for crowd control and detection in smart cities is presented. This early works on the use of LiDAR to replace or complement cameras are attracting more and more interest, and new datasets have been recently developed for the monitoring of people in point clouds [43].

The process of human detection in all previous applications has common steps. In most of them, detection is done in two steps. First, the objects are segmented and clustered by removing the ground from the point cloud [44], and second, persons are recognized from other objects, either through feature extraction [45] or comparison with models [46]. Recently, with the incursion of artificial intelligence in point cloud processing, detection is performed through 3D detectors directly on point clouds [47], or through cloud-to-image transformation to apply CNNs [48]. Other more specific techniques focus on leg detection, based on movement and geometry, although this is mostly employed in tracking robots located near the person [49, 50].

Therefore, human detection is concentrated on a very specific low-cost LiDAR in real-time, allowing both monitoring and tracking at distances close to the positive.

3.2   3D Modelling from LiDAR Data

One of the main uses of point clouds is for the modelling of buildings, both indoors and outdoors, since the geometry provided by the point cloud facilitates the recognition of building elements based on the point distribution in space. The generated models are used to analyse the available space or to obtain measurements of the environment [51], so that the capacity and the space available to accommodate people can be calculated. Since the geometrical conditions of indoor and outdoor are different, different LiDAR and techniques are used for acquisition and modelling.

Indoor environment has traditionally been provided with TLS devices, as TLS allows for greater mobility and is adapted for indoor use. Recently, it is also possible to purchase handled devices or backpacks, which allow faster acquisition at the cost of some loss of data accuracy [52]. Indoor modelling focuses on structural elements such as floors, ceilings, walls, and columns. These elements are easily recognizable by their typically vertical or horizontal geometry [53, 54]. Other relevant elements are transit or connecting elements or spaces, such as doors or windows, which allow movement or illumination [55]. Finally, furniture is also relevant indoors, because it occupies and defines the use of each space [56, 57]. The detection and modelling of these elements is more complex, due to their great variety of shapes and uses. It usually relies on neural networks for detection and classification. For a correct model generation, topology (relationships between elements) needs to be considered, so that each element fits coherently with its neighbours [58].

Outdoor can be understood from an aerial or street perspective. Point clouds with aerial perspective are obtained by ALS acquisitions and the methods used focus on roof modelling, because there is an unequivocal relationship between the shape of the roof and the shape of the building [59]. The most common method of building modelling with ALS data is the recognition of the building's watershed, or roof shape, and projection, resulting in smooth façades [60, 61]. MLS is used for the modelling of façades over large streets, although partially occluded facades can be supplemented with TLS or UAV data if necessary [62]. The detection and segmentation of façade elements are conducted based on the geometric features or by applying AI techniques, similar to that mentioned indoors [63].

Since the movement of people occurs on the ground, one option to estimate the capacity is through the measurement of the navigable ground. Both indoors and outdoors, the navigable ground (floor) is the main horizontal element of the environment with little or no inclination [64]. On the ground, there are static objects that occupy a fixed space [65]. In addition, dynamic objects acquired do not require space on the navigable ground. The navigable ground is the one that allows the mobility of people and is not occupied by static objects [66]. Since ground and objects are identifiable in point clouds, people can be positioned according to the ground geometry while maintaining the social distance. This alternative to the capacity calculation based on the space enclosed by walls or buildings is more accurate, as it considers the geometry of the surroundings and the ground occupancy.

From the different LiDAR data sources, it is possible to measure the free space indoors or outdoors, both in empty and furnished areas and the occupied space.

4 Discussion

4.1 Comparative Imagery and LiDAR Data for Human Monitoring

It clearly emerges from the literature review that image-based detection and tracking of people are much more developed than point cloud-based counterparts. This is because imagery data for scientific use were available long before point clouds. Moreover, computer vision has been one of the principal driving forces for the extraordinary development of artificial intelligence. An example can be found in the support robot shown in [67], where LiDAR is used for person tracking, but when the interaction is needed, human poses are obtained from imagery data and robots can follow instructions. Similarly, in autonomous driving, pedestrians can be detected with LiDAR, but the estimation of poses and knowing pedestrian intentions is accomplished through image processing [68]. Another example of the superiority of imagery over LiDAR is shown in [41], where a LiDAR is installed in a traffic light for detection and tracking of vehicles and pedestrians. The accuracy of the equipment is measured by taking images as ground truth. The success rate with LiDAR is 96% for pedestrian detection and 95% for tracking. Therefore, pedestrian detection and tracking in point clouds can be catching up with imagery data.

An advantage of LiDAR data over images is the measure of distances. If the positioning of the person in each consecutive LiDAR acquisition and the time between acquisitions are known, the direction vectors and velocity of the person can be estimated accurately and directly, without resorting to transformations and distances extraction from the images.

Another relevant factor in the usefulness of LiDAR versus camera is the dependence of atmospheric conditions and illumination. Although most tests are conducted under ideal conditions, the devices should work in any environment or situation. The operation of LiDAR and cameras is completely different in data acquisition. As LiDAR is an active sensor, the illumination of the scene does not influence the point cloud acquisition, and therefore, there is no difference between night and daytime acquisition procedures [69]. In contrast, although daytime images are of higher quality than LiDAR data for pedestrian detection, it is not clear that detection is better than with LiDAR in night-time images. LiDAR has limitations of operation in environments with rain, fog or airborne dust, which produces a lot of ambient noise and false returns [70]. Although the same problems affect images, they are not so relevant and can be corrected [71, 72].

The question of whether LiDAR data is more suitable for detecting people than images or vice versa remains controversial, but there is a broad consensus among different authors that the fusion of both types of data improves the detection accuracy. In autonomous driving, numerous studies are showing that point cloud and image-based detection yield better hit rates when combined. LiDAR helps camera to detect occluded objects, and camera helps LiDAR to detect wrong segmented objects [73]. There are different ways to embroider the joint use of both data: some authors choose to paint the point cloud according to images, and others to use first one data type to over detect people and the other data type to eliminate false positives [74–76].

Undoubtedly, the most important reason for choosing between the installation of LiDAR or cameras is price, as long as both technologies provide good results. Cameras are considerably cheaper than LiDAR devices, although the price varies between models, in general, a camera can be considered two orders of magnitude cheaper than a LiDAR. For human monitoring, where wide area coverage has to be provided and many devices need to be installed, cameras outperform LiDAR, as the price per monitored area is considerably lower.

The LiDAR and camera comparison for human detection is summarized in Table 1.

Table 1 Comparison between imagery and LiDAR data for human detection and tracking

Data type	Imagery	LiDAR
Main field	Surveillance	Autonomous driving and robotics
Algorithms	Image processing and CNN	Geometric processing
Quality results	Very high, it can be taken as a reference	Not as high as in images
Measurement of distances	Photogrammetry and metric units in pixels	Direct
Atmospheric conditions	Limited visibility in fog and rain	Noise introduced due to rain, fog and dust
Illumination	Passive sensor, needed of external illumination	Active sensor, not affected by changes in illumination
Price/coverage	Low	High

4.2 Comparative Imagery and LiDAR Data for 3D Modelling and Capacity Estimation

Unlike human detection, point clouds and images are used in the same field when the acquisition of the built environment is required. Numerous commercial software packages are available for the generation of 3D point clouds from photogrammetric images. LiDAR point clouds are intrinsically 3D. However, if several acquisitions are made, each one must be registered to complete the model. The generation of the 3D mesh from images as well as registration leads to added errors in the final model, which are more relevant in the imagery case [77, 78]. Therefore, models generated from images have lower precision than those generated from LiDAR data.

In addition to the 3D geometry, each data type provides different attributes. Images contain colour and texture information, while LiDAR provides reflectivity. This information is useful to identify objects or to obtain a more realistic final model. Many LiDAR devices integrate cameras to perform colour acquisitions and generate coloured point clouds [79]. However, for gauge calculations, the measurements are obtained from the 3D geometry, where colour and intensity are not relevant.

The lighting of the scene strongly influences the generation of a 3D model from the images. While the LiDAR can be acquired day and night, cameras for 3D modelling require the best possible lighting conditions. However, data acquisition for modelling is schedulable, and the best conditions can be selected for both LiDAR and camera use [80]. The same consideration applies to weather conditions.

Finally, as mentioned in human detection, the price of the LiDAR is higher than cameras. But in modelling, there is no need for constant fixed monitoring of the environment, and therefore, no fixed devices need to be installed and the purchase of a LiDAR can be amortized over the acquisition of several environments, so the price per modelled environment decreases. Considering the above advantages of LiDAR in precision and processing speed over imagery and the price to coverage ratio, LiDAR seems to be a better choice than imagery when it comes to modelling.

Table 2 Comparison between imagery and LiDAR data for 3D modelling

Data type	Imagery	Lidar
Attributes	Colour and textures	Geometry and reflectivity
Algorithms	Image processing and CNN	Geometric processing
Measurement of distances	Photogrammetry	Direct
Precision	High, errors caused by photogrammetry	Very high, errors caused by registration
Illumination	Passive sensor, needed of external illumination	Active sensor, not affected by changes in illumination
Price/coverage	Low	High

The LiDAR and camera comparison for 3D modelling has been summarized in Table 2.

4.3 Privacy

Apart from the technological aspect, privacy is a concern of society as massive data is acquired. LiDAR and imagery data have completely opposite privacy considerations. From a point cloud, it is not possible to recognize an individual, so the privacy of the citizen is not violated. Point clouds have no colour, and moving objects are deformed. The opposite is the case with images where they are often used to identify people. However, privacy is an aspect closely related to the regulations of each country, and it is the authorities who are responsible for ensuring that the technology is used in accordance with the regulations in force. Nowadays, it is common to see cameras in areas of special security and crowds of people: airports, train stations, shopping centres and main city streets.

There is a current trend to try to respect users' privacy as far as possible. In certain cases, such as biometric identification, this is not possible, since its objective is to identify the person directly, either with or without their consent. In activity monitoring and in people counting, there is significant published research on alternative less intrusive techniques, such as the analysis of infrared imagery to avoid facial or personal recognition in the imagery [81] and thus potentially enhance privacy. However, thermographic sensors tend to be more expensive and of lower resolution than conventional cameras. Another alternative is the use of Bluetooth or Wi-Fi sensors [82] to estimate the number of people in a place or the distance between them; however, the accuracy of these two technologies is low, and the consent of the person to be detected is required (through the activation of sensors on their smartphone) and also involves the detection of the person not through their face, which also constitute sensible data from a privacy point of view.

5 Conclusion

This paper gathers and comments on some articles that show how geospatial data can help in the control of pandemics by monitoring people and calculating capacity. The main conclusion that can be drawn from this work is that LiDAR and imagery data have their pros and cons. So far, the dominant paradigm has been to use a single data type for each particular application. Concretely, in human detection and tracking, images are commonly preferred over LiDAR data, although it should be pointed out that low-cost LiDAR devices with similar functionality are slowly beginning to appear. It is also worth mentioning that—to the best of our knowledge—LiDAR-based systems for pose detection or biometric identification have not yet been developed. Although more expensive than imaging hardware, LiDAR devices are widely used in 3D modelling because they provide superior precision, while the imagery is only used when LiDAR equipment is not available or colour information is required.

Acknowledgements This work has received funding from Xunta de Galicia through grant ED481B-2019-061 and ED431C 2020/01, and from the Government of Spain through project PID2019-105221RB-C43 funded by MCIN/AEI/10.13039/501100011033 and through human resources grant RYC2020-029193-I funded by MCIN/AEI/10,13039/501100011033. The statements made herein are solely the responsibility of the authors.

References

1. The World Bank (2020) Urban and disaster risk management responses to COVID-19
2. Soilán M, Sánchez-Rodríguez A, del Río-Barral P et al (2019) Review of laser scanning technologies and their applications for road and railway infrastructure monitoring. Infrastructures 4
3. Khoshelham K, Díaz Vilariño L, Peter M et al (2017) The ISPRS benchmark on indoor modelling. ISPRS Int Arch Photogramm Remote Sens Spat Inf Sci XLII-2/W7:367–372. https://doi.org/10.5194/isprs-archives-XLII-2-W7-367-2017
4. Masi I, Wu Y, Hassner T, Natarajan P (2018) Deep face recognition: a survey. In: 2018 31st SIBGRAPI conference on graphics, patterns and images (SIBGRAPI), pp 471–478
5. Greenhalgh T, Jimenez JL, Prather KA et al (2021) Ten scientific reasons in support of airborne transmission of SARS-CoV-2. Lancet 397:1603–1605. https://doi.org/10.1016/S0140-6736(21)00869-2
6. Lamba S, Nain N (2017) Crowd monitoring and classification: a survey BT—advances in computer and computational sciences. In: Bhatia SK, Mishra KK, Tiwari S, Singh VK (eds) Springer Singapore, Singapore, pp 21–31
7. Khan A, Ali Shah J, Kadir K et al (2020) Crowd monitoring and localization using deep convolutional neural network: a review. Appl Sci 10
8. Fluke SafeCount by FLUKE (2021). https://www.fluke.com/en-us/products/building-infrastructure/occupancy-monitoring
9. Senstar (2021) Crow detection by SENSTAR. https://senstar.com/products/video-analytics/crowd-detection
10. Gradiant (2021) Occupancy monitoring by gradiant. https://www.gradiant.org/wp-content/uploads/2020/08/2020.08.07_BeachOccupancy-TS_EN.pdf

11. Ahonen T, Hadid A, Pietikainen M (2006) Face description with local binary patterns: application to face recognition. IEEE Trans Pattern Anal Mach Intell 28:2037–2041. https://doi.org/10.1109/TPAMI.2006.244

12. Wang J, She M, Nahavandi S, Kouzani A (2010) A review of vision-based gait recognition methods for human identification. In: 2010 International conference on digital image computing: techniques and applications, pp 320–327

13. Sundararajan K, Woodard DL (2018) Deep learning for biometrics: a survey. ACM Comput Surv 51. https://doi.org/10.1145/3190618

14. Gupta A, Gupta K, Gupta K, Gupta K (2020) A survey on human activity recognition and classification. In: 2020 international conference on communication and signal processing (ICCSP), pp 915–919

15. Gawande U, Hajari K, Golhar Y (2020) Pedestrian Detection and tracking in video surveillance system: issues, comprehensive review, and challenges. Recent Trends Comput Intell

16. Nguyen DT, Li W, Ogunbona PO (2016) Human detection from images and videos: a survey. Pattern Recognit 51:148–175. https://doi.org/10.1016/j.patcog.2015.08.027

17. Dhiman C, Saxena M, Vishwakarma DK (2019) Skeleton-based view invariant deep features for human activity recognition. In: 2019 IEEE fifth international conference on multimedia big data (BigMM), pp 225–230

18. Raj NB, Subramanian A, Ravichandran K, Venkateswaran N (2020) Exploring techniques to improve activity recognition using human pose skeletons. In: 2020 IEEE winter applications of computer vision workshops (WACVW), pp 165–172

19. Ghazal S, Khan US (2018) Human posture classification using skeleton information. In: 2018 international conference on computing, mathematics and engineering technologies (iCoMET), pp 1–4

20. Ke L, Qi H, Chang M, Lyu S (2018) Multi-scale supervised network for human pose estimation. In: 2018 25th IEEE international conference on image processing (ICIP), pp 564–568

21. Halim AA, Dartigues-Pallez C, Precioso F et al (2016) Human action recognition based on 3D skeleton part-based pose estimation and temporal multi-resolution analysis. In: 2016 IEEE international conference on image processing (ICIP), pp 3041–3045

22. Serpush F, Rezaei M (2020) Complex human action recognition in live videos using hybrid FR-DL method

23. Andriluka M, Roth S, Schiele B (2009) Pictorial structures revisited: people detection and articulated pose estimation. In: 2009 IEEE conference on computer vision and pattern recognition, pp 1014–1021

24. Baradel F, Wolf C, Mille J, Taylor GW (2018) Glimpse clouds: human activity recognition from unstructured feature points. In: Proceedings of the IEEE conference on computer vision and pattern recognition, pp 469–478

25. Aggarwal JK, Ryoo MS (2011) Human activity analysis: a review. ACM Comput Surv 43. https://doi.org/10.1145/1922649.1922653

26. Angeleas A, Bourbakis N, Tsihrintzis G (2016) Categorization of research surveys and reviews on human activities. In: 2016 7th international conference on information, intelligence, systems & applications (IISA), pp 1–6

27. Boualia SN, Amara NEB (2019) Pose-based human activity recognition: a review. In: 2019 15th international wireless communications & mobile computing conference (IWCMC), pp 1468–1475

28. Kraus K (2007) Photogrammetry: geometry from images and laser scans/Karl Kraus; translated by Ian Harley, Stephen Kyle. Photogramm. Geom. from images laser scans

29. Chaudhary P, D'Aronco S, de Vitry M et al (2019) Flood-water level estimation from social media images. ISPRS Ann Photogramm Remote Sens Spat Inf Sci 4:5–12

30. Musse SR, Ulicny B, Aubel A, Thalmann D (2005) Groups and crowd simulation. In: ACM SIGGRAPH 2005 courses. Association for Computing Machinery, New York, NY, USA, pp 2–es

31. Aicardi I, Chiabrando F, Lingua AM, Noardo F (2018) Recent trends in cultural heritage 3D survey: the photogrammetric computer vision approach. J Cult Herit 32:257–266

32. Szeliski R (2010) Computer vision: algorithms and applications. Springer Science & Business Media
33. Zhang Z (1998) Determining the epipolar geometry and its uncertainty: a review. Int J Comput Vis 27:161–195
34. Balado J, Sousa R, Díaz-Vilariño L, Arias P (2020) Transfer learning in urban object classification: online images to recognize point clouds. Autom Constr 111:103058. https://doi.org/10.1016/j.autcon.2019.103058
35. Fürst M, Wasenmüller O, Stricker D (2020) LRPD: long range 3D pedestrian detection leveraging specific strengths of LiDAR and RGB. In: 2020 IEEE 23rd international conference on intelligent transportation systems (ITSC), pp 1–7
36. Wang H, Wang B, Liu B et al (2017) Pedestrian recognition and tracking using 3D LiDAR for autonomous vehicle. Rob Auton Syst 88:71–78. https://doi.org/10.1016/j.robot.2016.11.014
37. Álvarez-Aparicio C, Guerrero-Higueras ÁM, Rodríguez-Lera FJ et al (2019) People detection and tracking using LIDAR sensors. Robot 8
38. Yan Z, Sun L, Duckctr T, Bellotto N (2018) Multisensor online transfer learning for 3D LiDAR-based human detection with a mobile robot. In: 2018 IEEE/RSJ international conference on intelligent robots and systems (IROS), pp 7635–7640
39. Yan Z, Duckett T, Bellotto N (2020) Online learning for 3D LiDAR-based human detection: experimental analysis of point cloud clustering and classification methods. Auton Robots 44:147–164. https://doi.org/10.1007/s10514-019-09883-y
40. Koide K, Miura J, Menegatti E (2019) A portable three-dimensional LIDAR-based system for long-term and wide-area people behavior measurement. Int J Adv Robot Syst 16:1729881419841532. https://doi.org/10.1177/1729881419841532
41. Zhao J, Xu H, Liu H et al (2019) Detection and tracking of pedestrians and vehicles using roadside LiDAR sensors. Transp Res Part C Emerg Technol 100:68–87. https://doi.org/10.1016/j.trc.2019.01.007
42. Yamaguchi H, Hiromori A, Higashino T (2018) A human tracking and sensing platform for enabling smart city applications. In: Proceedings of the workshop program of the 19th international conference on distributed computing and networking. Association for Computing Machinery, New York, NY, USA
43. Romero-González C, Villena Á, González-Medina D et al (2017) InLiDa: a 3D Lidar dataset for people detection and tracking in indoor environments. In: VISIGRAPP
44. Liu K, Wang W, Wang J (2019) Pedestrian detection with Lidar point clouds based on single template matching. Electron 8
45. Lin T, Tan DS, Tang H et al (2018) Pedestrian detection from Lidar data via cooperative deep and hand-crafted features. In: 2018 25th IEEE international conference on image processing (ICIP), pp 1922–1926
46. Yamamoto T, Kawanishi Y, Ide I et al (2018) Efficient pedestrian scanning by active scan LIDAR. In: 2018 international workshop on advanced image technology (IWAIT), pp 1–4
47. Ali W, Abdelkarim S, Zidan M et al (2018) Yolo3d: end-to-end real-time 3d oriented object bounding box detection from lidar point cloud. In: Proceedings of the European conference on computer vision (ECCV) workshops
48. Jansen L, Liebrecht N, Soltaninejad S, Basu A (2020) 3D object classification using 2D perspectives of point clouds BT—smart multimedia. In: McDaniel T, Berretti S, Curcio IDD, Basu A (eds) Springer International Publishing, Cham, pp 453–462
49. Guerrero-Higueras ÁM, Álvarez-Aparicio C, Calvo Olivera MC et al (2019) Tracking people in a mobile robot From 2D LIDAR Scans using full convolutional neural networks for security in cluttered environments. Front Neurorobot 12:85. https://doi.org/10.3389/fnbot.2018.00085
50. Duong HT, Suh YS (2020) Human gait tracking for normal people and walker users using a 2D LiDAR. IEEE Sens J 20:6191–6199. https://doi.org/10.1109/JSEN.2020.2975129
51. Xie L, Zhu Q, Hu H et al (2018) Hierarchical regularization of building boundaries in noisy aerial laser scanning and photogrammetric point clouds. Remote Sens 10
52. Otero R, Lagüela S, Garrido I, Arias P (2020) Mobile indoor mapping technologies: a review. Autom Constr 120:103399. https://doi.org/10.1016/j.autcon.2020.103399

53. Oh S, Lee D, Kim M et al (2021) Building component detection on unstructured 3D indoor point clouds using RANSAC-based region growing. Remote Sens 13
54. Yang F, Li L, Su F et al (2019) Semantic decomposition and recognition of indoor spaces with structural constraints for 3D indoor modelling. Autom Constr 106:102913. https://doi.org/10.1016/j.autcon.2019.102913
55. Previtali M, Diaz-Vilariño L, Scaioni M (2018) Towards automatic reconstruction of indoor scenes from incomplete point clouds: door and window detection and regularization. ISPRS Int Arch Photogramm Remote Sens Spat Inf Sci 624:507–514. https://doi.org/10.5194/isprs-archives-XLII-4-507-2018
56. Charles RQ, Su H, Kaichun M, Guibas LJ (2017) PointNet: deep learning on point sets for 3D classification and segmentation. In: 2017 IEEE conference on computer vision and pattern recognition (CVPR), pp 77–85
57. Poux F, Neuville R, Nys G-A, Billen R (2018) 3D point cloud semantic modelling: integrated framework for indoor spaces and furniture. Remote Sens 10
58. Tran H, Khoshelham K, Kealy A, Díaz-Vilariño L (2017) Extracting topological relations between indoor spaces from point clouds. ISPRS Ann Photogramm Remote Sens Spat Inf Sci 42W4:401–406. https://doi.org/10.5194/isprs-annals-IV-2-W4-401-2017
59. Xiong B, Oude Elberink S, Vosselman G (2014) A graph edit dictionary for correcting errors in roof topology graphs reconstructed from point clouds. ISPRS J Photogramm Remote Sens 93:227–242. https://doi.org/10.1016/j.isprsjprs.2014.01.007
60. Li M (2019) A voxel graph-based resampling approach for the aerial laser scanning of urban buildings. IEEE Geosci Remote Sens Lett 16:1899–1903. https://doi.org/10.1109/LGRS.2019.2910575
61. Li Y, Hu Q, Wu M et al (2016) Extraction and simplification of building façade pieces from mobile laser scanner point clouds for 3D Street view services. ISPRS Int J Geo-Information 5
62. Andriasyan M, Moyano J, Nieto-Julián JE, Antón D (2020) From point cloud data to building information modelling: an automatic parametric workflow for heritage. Remote Sens 12
63. Baik A (2019) From point cloud to existing bim for modelling and simulation purposes. ISPRS Int Arch Photogramm Remote Sens Spat Inf Sci 42W2:15–19. https://doi.org/10.5194/isprs-archives-XLII-5-W2-15-2019
64. Poux F, Mattes C, Kobbelt L (2020) Unsupervised Segmentation of indoor 3d point cloud: application to object-based classification. ISPRS Int Arch Photogramm Remote Sens Spat Inf Sci 44W1:111–118. https://doi.org/10.5194/isprs-archives-XLIV-4-W1-2020-111-2020
65. Balado J, Díaz-Vilariño L, Arias P, Lorenzo H (2019) Point clouds for direct pedestrian pathfinding in urban environments. ISPRS J Photogramm Remote Sens 148:184–196. https://doi.org/10.1016/j.isprsjprs.2019.01.004
66. Balado J, Díaz-Vilariño L, Arias P, Frías E (2019) Point clouds to direct indoor pedestrian pathfinding. In: International archives of the photogrammetry, remote sensing and spatial information sciences—ISPRS archives
67. Islam MM, Lam A, Fukuda H et al (2019) A Person-following shopping support robot based on human pose skeleton data and LiDAR sensor BT—intelligent computing methodologies. In: Huang D-S, Huang Z-K, Hussain A (eds) Springer International Publishing, Cham, pp 9–19
68. Fang Z, López AM (2018) Is the pedestrian going to cross? Answering by 2d pose estimation. In: 2018 IEEE intelligent vehicles symposium (IV), pp 1271–1276
69. Lindner P, Richter E, Wanielik G et al (2009) Multi-channel lidar processing for lane detection and estimation. In: 2009 12th international IEEE conference on intelligent transportation systems, pp 1–6
70. Tang L, Shi Y, He Q et al (2020) Performance test of autonomous vehicle Lidar Sensors under different weather conditions. Transp Res Rec 2674:319–329. https://doi.org/10.1177/0361198120901681
71. Tan RT (2008) Visibility in bad weather from a single image. In: 2008 IEEE conference on computer vision and pattern recognition, pp 1–8
72. Tarel J, Hautiere N, Caraffa L et al (2012) Vision enhancement in homogeneous and heterogeneous fog. IEEE Intell Transp Syst Mag 4:6–20. https://doi.org/10.1109/MITS.2012.2189969

73. Wu T, Tsai C, Guo J (2017) LiDAR/camera sensor fusion technology for pedestrian detection. In: 2017 Asia-Pacific signal and information processing association annual summit and conference (APSIPA ASC), pp 1675–1678

74. Fei J, Chen W, Heidenreich P et al (2020) SemanticVoxels: sequential fusion for 3D pedestrian detection using LiDAR point cloud and semantic segmentation. arXiv Prepr arXiv200912276

75. Treméau A, El Ansari M, Lahmyed R (2018) A hybrid pedestrian detection system based on visible images and LIDAR data

76. Matti D, Ekenel HK, Thiran J-P (2017) Combining LiDAR space clustering and convolutional neural networks for pedestrian detection. In: 2017 14th IEEE international conference on advanced video and signal based surveillance (AVSS), pp 1–6

77. He Y, Liang B, Yang J et al (2017) An iterative closest points algorithm for registration of 3D laser scanner point clouds with geometric features. Sensors 17

78. Tushev S, Sukhovilov B (2017) Photogrammetric system accuracy estimation by simulation modelling. In: 2017 international conference on industrial engineering, applications and manufacturing (ICIEAM), pp 1–6

79. Remondino F (2011) Heritage recording and 3D modeling with photogrammetry and 3D scanning. Remote Sens 3:1104–1138. https://doi.org/10.3390/rs3061104

80. Chiabrando F, Spanò A, Sammartano G, Teppati Losè L (2017) UAV oblique photogrammetry and lidar data acquisition for 3D documentation of the Hercules Fountain. Virtual Archaeol Rev 8. https://doi.org/10.4995/var.2017.5961

81. Tateno S, Meng F, Qian R, Hachiya Y (2020) Privacy-preserved fall detection method with three-dimensional convolutional neural network using low-resolution infrared array sensor. Sensors 20

82. Longo E, Redondi AEC, Cesana M (2019) Accurate occupancy estimation with WiFi and bluetooth/BLE packet capture. Comput Networks 163:106876. https://doi.org/10.1016/j.comnet.2019.106876

Chapter 8
Modeling and Evaluating the Impact of Social Restrictions on the Spread of COVID-19 Using Machine Learning

Mostafa Naemi⬤, **Amin Naemi**⬤, **Romina Zarrabi Ekbatani**⬤, **Ali Ebrahimi**⬤, **Thomas Schmidt**⬤, and **Uffe Kock Wiil**⬤

Abstract COVID-19 has influenced different aspects of human life, such as working and socializing, over the past year. Authorities in different countries have imposed various levels and forms of social restrictions to control the outbreak of this disease. This chapter investigates the effect of social restrictions, including restrictions on schools, workplaces, public events, gatherings, and internal and international flights, on the control of virus spread. For this aim, three machine learning models, including random forest (RF), extreme gradient boosting (XGB), and long short-term memory neural network (LSTM), are applied to simulate the number of infected cases per day in Denmark under different levels of restrictions. Different scenarios of social restrictions are simulated to study the impact of decisions on social restrictions and imposing more strict ones. The results show that LSTM has superior performance in detecting temporal and long dependencies. Also, it is shown that school and workplace closures are the two most decisive restrictions, and therefore, considering more strict restrictions on them can lead to a more significant decline in the number of infected cases. The results of this study also provide an insight on the importance of timely decisions by government and health authorities and quantify their effect on virus spread, which can be very useful for managing future pandemics.

Keywords COVID-19 · Machine learning · Deep learning · LSTM · Social restrictions

M. Naemi
The University of Melbourne, Melbourne, Australia

A. Naemi (✉) · A. Ebrahimi · T. Schmidt · U. K. Wiil
Center for Health Informatics and Technology, The Maersk Mc-Kinney Moeller Institute,
University of Southern Denmark, Odense, Denmark
e-mail: amin@mmmi.sdu.dk

R. Z. Ekbatani
Swinburne University of Technology, Melbourne, Australia

1 Introduction

Coronavirus Disease 2019 (COVID-19) is a contagious disease induced by severe acute respiratory syndrome coronavirus 2 (SARS-CoV-2) [1] discovered in Wuhan, China, in December 2019. The World Health Organization (WHO) announced COVID-19 as a global pandemic on March 11, 2020 [2]. During this period, the rapid spread of COVID-19 has influenced various aspects of human life significantly. For decades, quarantine has been considered as one of the oldest and most successful techniques for controlling communicable disease spread. However, from the citizens' point of view, quarantine is one of the most obscure and somewhat frightening issues in controlling the spread of a disease, which could have serious psychological, emotional, and financial consequences for quarantined patients and families [3, 4].

During the COVID-19 pandemic, authorities of different countries employed various quarantine strategies at different levels to manage the spread of the virus and minimize its impact on the economy, healthcare systems, human well-being, etc. Several researchers have investigated the efficacy of quarantine strategies on the containment of outbreaks and come up with different conclusions about the impact of restrictions on the spread of COVID-19. Jia et al. stated that strict restrictions in China such as home quarantine, travel bans, and delay in returning to work decreased the spread of infection in the community [5]. Li et al. showed that a one-week delay in applying restrictions would increase the number of infected people by approximately 10%. On the other hand, they estimated that with the early establishment of restrictions by one and two weeks the number of infected people would decrease by about 25% and 57%, respectively [6]. Moreover, Qiu and Xiao stated that by applying lockdown in Wuhan only seven days earlier, the total number of patients would decline by 72% [7]. Zhang et al. also concluded that social distancing and school and festivals closings could significantly mitigate the spread of infection [8]. Even so, some studies have argued that lockdowns or some restrictions such as travel restrictions are not highly effective in such situation [9].

Recent studies show that machine learning algorithms have performed well in various applications, leading to higher accuracy, reliability, and scale up ability [10, 11]. As data on applied restrictions and the number of infected and dead people along with other information about the spread of COVID-19 for a year is available from approximately 180 countries, by using this historical data and machine learning techniques, we can simulate and evaluate the impact of different strategies employed to control the outbreak. Moreover, we can find the correlation and relationship between different strategies as well as complicated patterns and effects of those variables together.

This paper therefore analyzes and models the impact of employed restrictions on the infection spread in Denmark using machine learning techniques. The data of eight social restrictions, including school closing, workplace closing, limitations on public events, gathering restrictions, public transport closing, stay-at-home requirement, and restrictions on internal and international travels for Denmark, is used.

Three machine learning algorithms, including random forest (RF), extreme gradient boosting (XGB), and long short-term memory neural networks (LSTM), are trained on historical data. Denmark experienced an outbreak around end of December 2020, so to evaluate the impact of these restrictions, different scenarios of more strict restrictions than actual ones are also simulated to investigate the importance of decisions on the level and forms of restrictions and its timing. Comparison of the results of simulated scenarios with historical data provides an insight about the most effective forms of restrictions and the required time for managing and control of future pandemics.

2 Data

Data used in this study was obtained from the Oxford COVID-19 Government Response Tracker (OxCGRT) [12]. This dataset contains the number of confirmed cases and deaths per day for about 180 countries, along with information about government policies related to closure and containment, health, and economic policies from January 1, 2020. It includes eight policy indicators, such as school closing (C1), workplace closing (C2), cancelation of public events (C3), restrictions on gathering size (C4), public transport closing (C5), stay-at-home requirements (C6), and restrictions on internal (C7) and international travels (C8) (Table 1). Moreover, this dataset contains four and seven indicators for economic and health policies, respectively.

In this study, Denmark was chosen because of the credibility of the data and also the government's success in controlling the outbreak by introducing different forms of social restrictions [13]. Seven social restrictions, including school closing, workplace closing, public events cancelation, gathering size restrictions, restrictions on public transport, and restrictions on internal and international travels, were enforced in Denmark (C6 was not used). Data from February 1, 2020, to January 1, 2021, was used in this study.

3 Model Development

In this study, three machine learning algorithms, including RF, XGB, and LSTM, were implemented. RF and XGB belong to ensemble learning algorithms, which are among the state-of-the-art machine learning algorithms. Ensemble learning refers to the process of constructing a model by merging several individual models to build a model that outperforms the primary models. RF is a bagging-based algorithm that considers homogenous weak learners such as decision trees that are trained independently on specific subsets of data, while XGB is a boosting-based algorithm that considers homogenous weak learners sequentially in an adaptative way. Ensemble

Table 1 Description of data variables

ID	Indicator name	Type	Possible values
C1	School closing	Ordinal	0—no measures 1—recommend closing 2—require closing (only some levels or categories, e.g., just high school) 3—require closing all levels
C2	Workplace closing	Ordinal	0—no measures 1—recommend closing (or work from home) 2—require closing (some sectors or categories of workers) 3—require closing (or work from home) for all-but-essential workplaces (e.g., grocery stores, doctors)
C3	Public events cancelation	Ordinal	0—no measures 1—recommend cancelling 2—require cancelling
C4	Restrictions on gathering	Ordinal	0—no restrictions 1—restrictions on large gatherings (above 1000 people) 2—restrictions on gatherings between 101 and 1000 people 3—restrictions on gatherings between 11 and 100 people 4—restrictions on gatherings of 10 people or less
C5	Public transports closing	Ordinal	0—no measures 1—recommend closing (or significantly reduce volume, route, and means of transport available) 2—require closing
C6	Stay-at-home requirements	Ordinal	0—no measures 1—recommend not leaving house 2—require not leaving house with exceptions, e.g., daily grocery shopping, and essential trips 3—require not leaving house with minimal exceptions
C7	Restrictions on internal travels	Ordinal	0—no measures 1—recommend not to travel between regions/cities 2—internal movement restrictions in place

(continued)

Table 1 (continued)

ID	Indicator name	Type	Possible values
C8	Restrictions on international travels	Ordinal	0—no restrictions 1—screening arrivals 2—quarantine arrivals from some or all regions 3—ban arrivals from some regions 4—ban on all regions or total border closure

learning has many benefits, including local optimal prevention, avoiding overfitting, and eliminating the curse of dimensionality [14].

LSTM is a recurrent deep neural network architecture which has connections for feedbacking data. Hidden layers are replaced by LSTM cells, which enable the network to manage input flow. LSTM cells are made up of three gates, including input gate, forget gate, and output gate [15]. Figure 1 shows a LSTM cells.

Equation 1 shows the equations of the forward pass of a LSTM unit, where w represents the recurrent relationship between the previous and current hidden layers. Also, U and $\tilde{C}$ are the weight matrix that connects the input layer to the hidden layer and a candidate hidden state calculated from the present input and the previous hidden state, respectively. Additionally, C denotes the unit's internal memory that is made up of the previous memory multiplied by the forget gate and the newly calculated hidden state multiplied by the input gate [16].

$$i_t = \sigma\left(x_t U^i + h_{t-1} W^i\right)$$
$$f_t = \sigma\left(x_t U^f + h_{t-1} W^f\right)$$
$$o_t = \sigma\left(x_t U^o + h_{t-1} W^o\right)$$

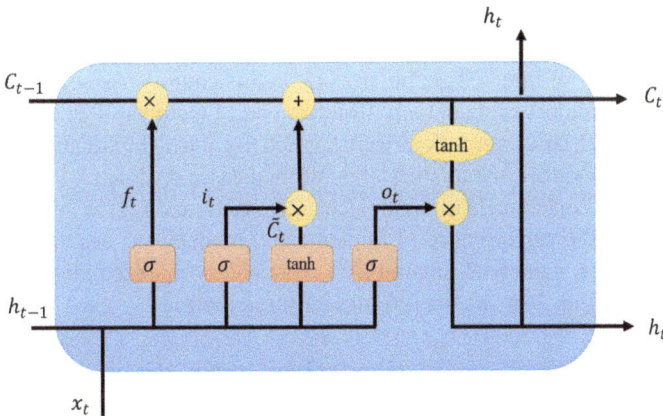

Fig. 1 LSTM cell structure

Table 2 Predictive models hyperparameters

Models	Hyperparameters
RF	Number of decision trees = 50, max depth = 30
XGB	Learning rate = 0.1, number of decision trees = 50
LSTM	LSTM cells = 64, batch size = 14, epochs = 1000, optimizer = Adam

$$\tilde{C}_t = \tanh\left(x_t U^g + h_{t-1} W^g\right)$$
$$C_t = \sigma\left(f_t * C_{t-1} + i_t * \tilde{C}_t\right)$$
$$h_t = \tanh(C_t) * o_t \tag{1}$$

To investigate the effect of different social restrictions on the control of virus spread, the whole data samples were used for training and validation of models. To do this, cross-validation on a rolling basis which is suitable for time series was applied.

To model the selected restrictions, dummy variables were considered for each level of every restriction individually, i.e., as we selected 24 different forms and levels of restrictions, therefore, there were 24 dummies in our models. In addition, as other studies and medical findings showed, the incubation period of COVID-19 is 14 days [17], and the impact of these restrictions on the number of cases appear in the data with 14 days delay was investigated. This delay was considered in assigning values to these dummy variables. For instance, if a restriction was applied on January 1, its corresponding dummy was set to one on January 15. The hyperparameters of machine learning models are shown in Table 2.

4 Results and Discussion

As mentioned in Sect. 2, COVID-19 data for Denmark was used in this study. Figure 2 shows the number of infected people per day in this country. As seen, the numbers were increasing almost exponentially from the end of October 2020, and finally, on December 9, 2020, strict restrictions were applied to control the situation. As a result, by the end of December, the number of confirmed cases declined.

Three selected machine learning models (RF, XGB, and LSTM) were trained on the dataset, and the performance of implemented models was evaluated using two regression metrics, namely R-squared (R^2) score and normalized mean square error (NRMSE). Equation 2 shows the formula for these metrics.

$$R^2 = 1 - \frac{\sum_i \left(CC_i - \widehat{CC_i}\right)^2}{\sum_i \left(CC_i - \overline{CC}\right)^2}$$

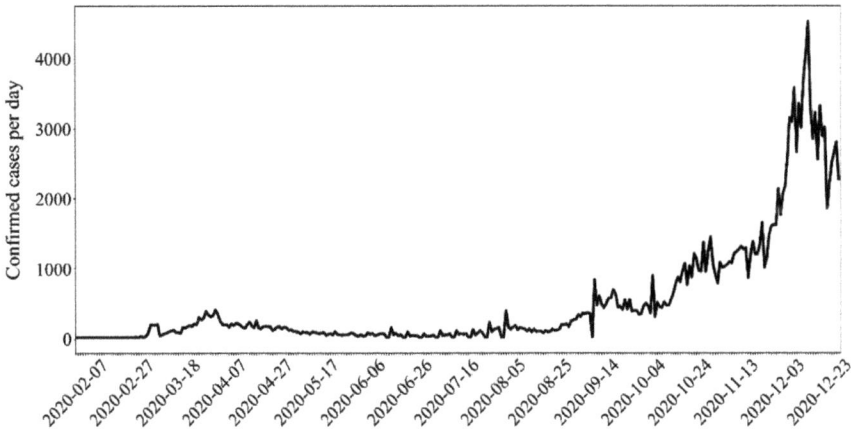

Fig. 2 Daily confirmed cases for Denmark

$$\text{NRMSE} = \frac{\sqrt{\frac{\Sigma_i \left(cc_i - \widehat{CC}_i\right)^2}{T}}}{CC_{\max}} \tag{2}$$

where CC_i and $\widehat{CC}_i$ are the actual confirmed cases and the predicted confirmed cases at day i, respectively. Also, $\overline{CC}$ and $CC_{\max}$ are the actual average and maximum number of confirmed cases in validation period; T denotes the validation period. It is worth mentioning that R^2 score is a number between zero and one, which indicates how much of the variation of a time series can be modeled by the model. A model is better at predicting a time series if its R^2 score is closer to one. Models' performance is shown in Table 3. According to this table, the LTSM has superior performance compared with the other two models. This is because of the better capability of LSTM in detecting temporal and long-time dependencies.

Table 4 shows the fortnightly transition of applied social restrictions in Denmark from the first of October to the end of December. According to this table, the strictness levels of C1, C2, C3, C5, and C7 are lower than other restrictions. Therefore, four scenarios which consider more strict levels are considered as follows:

- *Scenario 1*: More severe restrictions on C1 and C2, i.e., C1 = 2, and C2 = 3. In this scenario, schools are required to close, except at some levels, such as high

Table 3 Performance of models

Models	R^2 score	NRMSE
RF	0.48	0.27
XGB	0.51	0.26
LSTM	0.66	0.13

Table 4 Imposed restriction from October 1 to December 31, 2020, for Denmark

	1 Oct–15 Oct	15 Oct–31 Oct	1 Nov–15 Nov	15 Nov–30 Nov	1 Dec–15 Dec	15 Dec–31 Dec
C1	1	$1 \to 0$	$0 \to 2$	$2 \to 1$	$1 \to 2$	2
C2	$2 \to 1$	1	1	1	$1 \to 2$	2
C3	1	1	1	1	$1 \to 2$	2
C4	$3 \to 2$	$2 \to 4$	4	4	4	4
C5	$1 \to 0$	0	0	0	0	0
C7	0	0	$0 \to 1$	$1 \to 0$	0	0
C8	3	3	3	3	3	3

schools. Workplaces are also required to close, or people are asked to work from home, except for essential occupations, e.g., grocery stores and doctors.

- *Scenario 2*: More severe limitations on C5, meaning C5 = 1. In this case, it is recommended to significantly reduce volumes or means of available transport.
- *Scenario 3*: More serious restrictions on internal movement, meaning C7 = 1. In this scenario, it is recommended not to travel between cities and regions.
- *Scenario 4*: Considering all the above-mentioned restrictions simultaneously, i.e., C1 = 2, C2 = 3, C3 = 2, C5 = 1, C7 = 1.

Considering the growth in the number of confirmed cases from October to December 2020 and its peak on December 18, November 15 was therefore selected as the start date for simulating the aforementioned scenarios. This enables us to investigate the possible outcome of imposing more strict restrictions during this peak period and illustrates the importance of timely and crucial decisions by authorities. It is worth mentioning that C4 for the test period has the strictest value, so it was not considered in designing scenarios. Similarly, the C8 value has not been changed for the whole test period, and it was set to 3, which is a strict restriction; a stricter value, which is border closure, was not considered for this variable.

The parameters of these scenarios were fed into the three mentioned machine learning models, and the results are shown in Fig. 3. This figure shows that restrictions in Scenario 1 (C1 and C2) can reduce the number of infected cases significantly. In scenarios 2 and 3, although the number of confirmed cases is around the actual, the overall trend of predictions does not decrease, which implies that the restrictions on internal travels and transport do not impact the number of confirmed cases significantly. Among all the simulated scenarios, scenario 4 (combination of all other scenarios) for all three models shows the lowest peak, after which the trend is decreasing. Statistics for different scenarios using the three machine learning models are presented in Table 5. It should be noted that for real data, the maximum number of infected cases occurred on December 18, 2020, with 4527 cases, and for the simulation period, mean and standard deviation of confirmed cases were 2179 and 946, respectively. It should be noted that the peak values are calculated based on the results of the simulation period (November 15, 2020, to January 1, 2021).

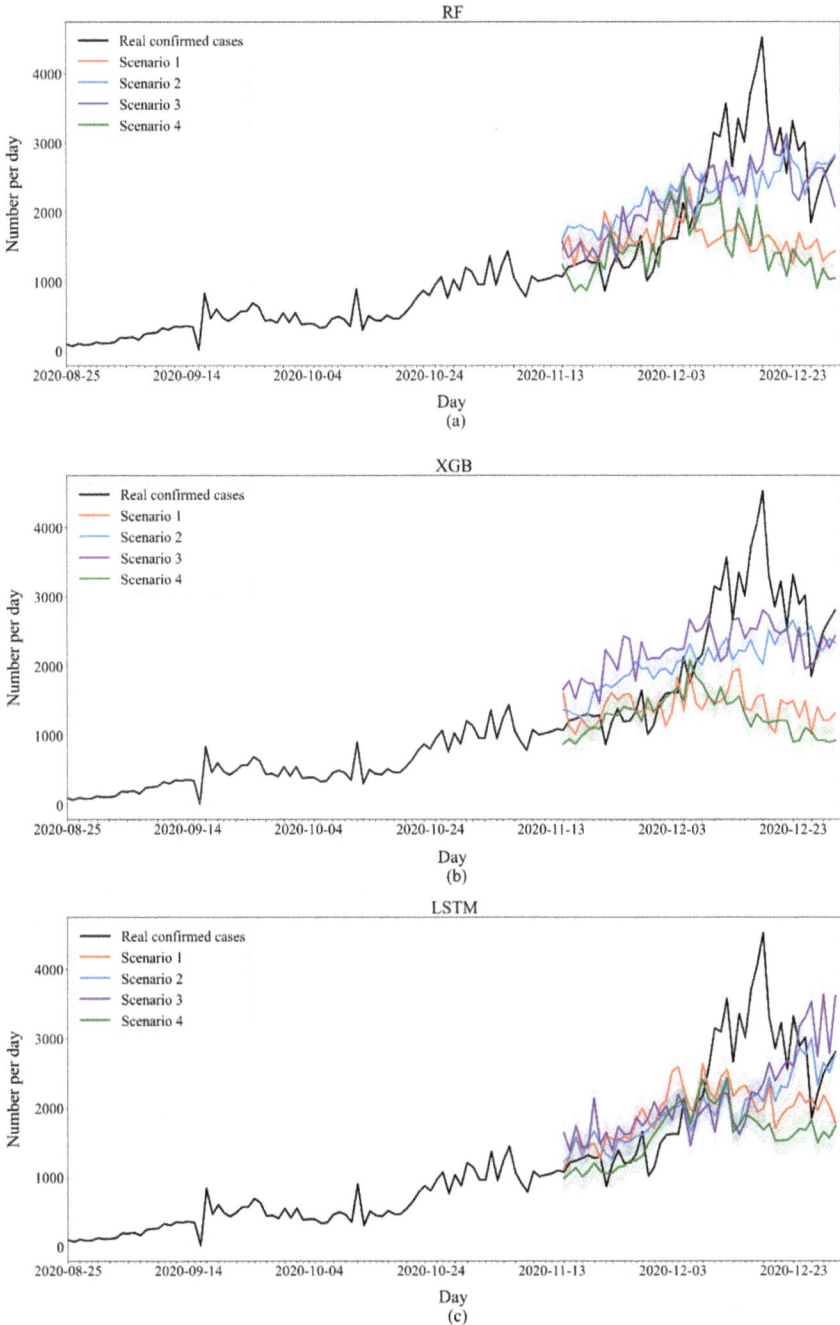

Fig. 3 Predicted confirmed cases per day under different scenarios **a** RF, **b** XGB, and **c** LSTM

Table 5 Statistics of predicted confirmed cases for different scenarios for three machine learning models

	Statistics	Peak date	Peak value	Mean (SD)
Scenario 1	RF	8/12/2020	2129	1606 (271)
	XGB	14/12/2020	1991	1555 (238)
	LSTM	8/12/2020	2628	1838 (347)
Scenario 2	RF	24/12/2020	2971	2431 (234)
	XGB	23/12/2020	2743	1965 (209)
	LSTM	26/12/2020	2994	2110 (379)
Scenario 3	RF	20/12/2020	3243	2575 (281)
	XGB	18/12/2020	2882	2184 (255)
	LSTM	28/12/2020	3628	2278 (436)
Scenario 4	RF	6/12/2020	2526	1533 (264)
	XGB	6/12/2020	2063	1411 (249)
	LSTM	12/12/2020	2430	1506 (344)
Base case	Actual	18/12/2020	4527	2179 (946)

[*] SD: standard deviation

According to Table 5, in Scenario 1 of the LTSM model, the average number of confirmed cases was reduced by 16% as compared to historical value. This reaches 30% in Scenario 4, which implies that in these scenarios, the load on the healthcare system will be reduced considerably. Based on the LTSM model in scenarios 2 and 3, the total number of cases was almost equal to the number of infected cases in historical data, which shows a minimal impact of the corresponding restrictions individually on healthcare system load. Moreover, as can be seen in Fig. 3 and Table 5, the peak time shifted to earlier dates for the two scenarios with the most effective restrictions, scenarios 1 and 4. This important finding indicates that by imposing the right restrictions on time, the peak will be lower and occur sooner, which will help society to return to normal conditions sooner. Furthermore, the comparison between scenarios in Fig. 3 shows that the number of cases in scenarios 1 and 4 is very similar, which suggests that C1 and C2 are the most crucial restrictions to enforce as they have the highest impact on the number of confirmed cases. Therefore, it is recommended to take action regarding these restrictions in similar future pandemics. In addition, our results showed that some restrictions, such as internal and international travel restrictions, cannot be effective individually, but considering social restrictions together can lead to better control of spread. The findings of this study can be very useful for managing and controlling future pandemics.

5 Limitations

This study has limitations. First, we considered seven variables related to social restrictions to investigate the impact of social restrictions on the control of outbreaks. As we know, there are more variables in health, economy, etc., policies that can affect the spread of the virus. Second, it is also possible that authorities did not apply stricter restrictions because of some other considerations, such as society mental health and economic issues. Third, C6, which indicates a restriction on staying at home requirement, seems to be one of the strictest restrictions, but it has not been applied in Denmark to consider it in this study.

6 Conclusion

In this study, we explored the impact of social restrictions on controlling the spread of COVID-19 in Denmark. To do this, the OxCGRT dataset was used, and seven social restriction variables, including school closing, workplace closing, public event cancelation, restrictions on gathering, public transport closing, restrictions on internal travels, and restrictions on international travels, were considered.

To consider the effect of restrictions on the number of infected cases per day, three machine learning algorithms including RF, XGB, and LSTM were considered. Moreover, we simulated four scenarios stricter than the real one applied by Danish authorities. Data from November 15 to January 1, 2021, was used as a simulation period to evaluate the effects of proposed scenarios. The results showed that LSTM had a better performance in prediction of confirmed cases. It was also shown that restrictions on schools (C1) and working places (C2) had a high impact on the control of the pandemic. Also, by applying scenario 4, which considers more strict restrictions on the aforementioned categories simultaneously, the maximum number of confirmed cases was reduced for all models significantly. For instance, in the LSTM model, the maximum number and mean number of infected cases for the simulation interval decreased to 2430 and 1506, respectively, compared to the historical values of 4527 and 2179, respectively. Furthermore, the results showed that by imposing stricter restrictions, the peak time shifted to earlier dates, meaning that society could potentially return to its normal conditions sooner. Such a system that can conclude based on historical data and show the effect of different authorities' decisions can be very beneficial. By using such a system, authorities can get an insight about the situation and detect the most effective restrictions variables.

In our future work, more variables in different domains, such as economic and health indicators, will be taken into account. Also, data from several countries with different strategies for controlling virus spread will be used to conduct a comparative study.

References

1. Gorbalenya AE, Baker SC, Baric R, de Groot RJ, Drosten C, Gulyaeva AA et al (2020) Severe acute respiratory syndrome-related coronavirus: the species and its viruses—a statement of the Coronavirus study group. https://doi.org/10.1038/s41564-020-0695-z

2. Mækelæ MJ, Reggev N, Dutra N, Tamayo RM, Silva-Sobrinho RA, Klevjer K et al (2020) Perceived efficacy of COVID-19 restrictions, reactions and their impact on mental health during the early phase of the outbreak in six countries. Roy Soc Open Sci 7(8). https://doi.org/10.1098/rsos.200644

3. Taghrir MH, Akbarialiabad H, Marzaleh MA (2020) Efficacy of mass quarantine as leverage of health system governance during COVID-19 outbreak: a mini policy review. Arch Iran Med 23(4):265–7. https://doi.org/10.34172/aim.2020.08

4. Brooks SK, Webster RK, Smith LE, Woodland L, Wessely S, Greenberg N et al (2020) The psychological impact of quarantine and how to reduce it: rapid review of the evidence. Lancet 395(10227):912–920. https://doi.org/10.1016/S0140-6736(20)30460-8

5. Jia J, Ding J, Liu S, Liao G, Li J, Duan B et al (2020) Modeling the control of COVID-19: impact of policy interventions and meteorological factors. arXiv preprint arXiv: 200302985. Preprint

6. Li R, Lu W, Yang X, Feng P, Muqimova O, Chen X et al (2020) Prediction of the epidemic of COVID-19 based on quarantined surveillance in China. medRxiv. https://doi.org/10.1101/2020.02.27.20027169

7. Qiu T, Xiao H (2020) Revealing the Influence of national public health response for the outbreak of the SARS-CoV-2 epidemic in Wuhan, China through status dynamic modeling. China Through Status Dynamic Modeling (Preprint)

8. Zhang Y, Jiang B, Yuan J, Tao Y (2020) The impact of social distancing and epicenter lockdown on the COVID-19 epidemic in mainland China: a data-driven SEIQR model study. MedRxiv. https://doi.org/10.1101/2020.03.04.20031187

9. Read JM, Bridgen JRE, Cummings DAT, Ho A, Jewell CP (2020) Novel coronavirus 2019-nCoV: early estimation of epidemiological parameters and epidemic predictions. 101101/202001

10. Zhang X, Zhang T, Young AA, Li X (2014) Applications and comparisons of four time series models in epidemiological surveillance data. PLoS One 9(2). https://doi.org/10.1371/journal.pone.0088075

11. Lalmuanawma S, Hussain J, Chhakchhuak L (2020) Applications of machine learning and artificial intelligence for Covid-19 (SARS-CoV-2) pandemic: a review. Chaos Solitons Fractals 110059. https://doi.org/10.1016/j.chaos.2020.110059

12. Hale T, Angrist N, Goldszmidt R, Kira B, Petherick A, Phillips T et al (2021) A global panel database of pandemic policies (oxford covid-19 government response tracker). Nat Hum Behav 1–10. https://doi.org/10.1038/s41562-021-01079-8

13. da Silva JG (2020) A healthy, innovative, sustainable, transparent, and competitive methodology to identify twenty benchmark countries that saved people lives against Covid-19 during 180 days. Int J Innov Educ Res 8(10). https://doi.org/10.31686/ijier.vol8.iss10.2710

14. Naemi A, Mansourvar M, Schmidt T, Wiil UK (2020) Prediction of Patients severity at emergency department using NARX and Ensemble learning. In: 2020 IEEE international conference on bioinformatics and biomedicine (BIBM). IEEE, pp 2793–9. https://doi.org/10.1109/BIBM49941.2020.9313462

15. Sagi O, Rokach L (2018) Ensemble learning: a survey. Wiley Interdisc Rev Data Min Knowl Discovery 8(4):e1249. https://doi.org/10.1002/widm.1249

16. Naemi A, Schmidt T, Mansourvar M, Wiil UK (2020) Personalized predictive models for identifying clinical deterioration using LSTM in Emergency departments. Stud Health Technol Inf 275:152–156. https://doi.org/10.3233/SHTI200713

17. Chinazzi M, Davis JT, Ajelli M, Gioannini C, Litvinova M, Merler S et al (2020) The effect of travel restrictions on the spread of the 2019 novel coronavirus (COVID-19) outbreak. Science 368(6489):395–400. https://doi.org/10.1126/science.aba9757

Chapter 9
Forecasting the COVID-19 Spread in Iran, Italy, and Mexico Using Novel Nonlinear Autoregressive Neural Network and ARIMA-Based Hybrid Models

Amin Naemi⬤, Mostafa Naemi⬤, Romina Zarrabi Ekbatani⬤,
Thomas Schmidt⬤, Ali Ebrahimi⬤, Marjan Mansourvar⬤,
and Uffe Kock Wiil⬤

Abstract This paper analyzes single and two-wave COVID-19 outbreaks using two novel hybrid models, which combine machine learning and statistical methods with Richards growth models, to simulate and forecast the spread of the infection. For this purpose, historical cumulative numbers of confirmed cases for three countries, including Iran, Italy, and Mexico, are used. The analysis of the Richards models shows that its single-stage form can model the cumulative number of infections in countries with a single wave of outbreak (Italy and Mexico) accurately while its performance deteriorates for countries with two-wave outbreaks (Iran), which clarifies the requirement of multi-stage Richards models. The results of multi-stage Richards models reveal that the prevention of the second wave could reduce the outbreak size in Iran by approximately 400,000 cases, and the pandemic could be controlled almost 7 months earlier. Although the cumulative size of outbreak is estimated accurately using multi-stage Richards models, the results show that these models cannot forecast the daily number of cases, which are important for health systems' planning. Therefore, two novel hybrid models, including autoregressive integrated moving average (ARIMA)-Richards and nonlinear autoregressive neural network (NAR)-Richards, are proposed. The accuracy of these models in forecasting the number of daily cases for 14 days ahead is calculated using the test data set shows that forecast error ranges from 8 to 25%. A comparison between these hybrid models also shows

A. Naemi (✉) · T. Schmidt · A. Ebrahimi · M. Mansourvar
Center for Health Informatics and Technology, The Maersk Mc-Kinney Moeller Institute,
University of Southern Denmark, Odense, Denmark
e-mail: amin@mmmi.sdu.dk

M. Naemi · U. K. Wiil
The University of Melbourne, Melbourne, Australia

R. Z. Ekbatani
Swinburne University of Technology, Melbourne, Australia

© The Author(s), under exclusive license to Springer Nature Singapore Pte Ltd. 2022   119
R. J. Howlett et al. (eds.), *Smart and Sustainable Technology for Resilient Cities and Communities*, Advances in Sustainability Science and Technology,
https://doi.org/10.1007/978-981-16-9101-0_9

that the machine learning-based models have superior performance compared with statistical-based ones and on average are 20% more accurate.

Keywords COVID-19 · Hybrid modeling · Machine learning · ARIMA · NAR neural network · Richards growth models

1 Introduction

The fast spread of the novel coronavirus infection, known as COVID-19, has risen worldwide concerns regarding the healthcare systems planning and human well-being [1]. Modeling the spread of this pandemic can play a significant role in the strategy development to assessment of the pathways to securely remove restrictions [2]. These modelings, which require the time-series data of the infection, can be utilized to quantify medicinal services required for COVID-19 patients and make decisions considering public well-being measures [3, 4]. We can also use these models to forecast the growth of COVID-19 outbreaks, which enables us to investigate future situations and the effect of various immunization techniques. These forecasts also provide some insights, which can potentially help the authorities and decision-makers setting up the medical consideration and different assets needed to remove restrictions [5–7].

Several methods have been used to model the epidemiological time evolution, among which there are two major groups: collective models and networked models. Collective models, such as Richards models and susceptible-infected-recovered (SIR), describe the spread using a limited number of collective variables and are defined by their small number of factors [8, 9]. On the other hand, networked models describe the spreading process at individual level considering the population as a network of individuals that can interact. This type of model provides a more detailed explanation of the epidemic spread compared to the collective model; however, it needs a substantial amount of data, and its parameter identification is more difficult than collective models. Hence, collective models are easier to implement and can be used by the public health authorities even under data-scarce conditions [9].

The Richards model has also been used in epidemiology for outbreak prediction for illnesses like SARS [10] and swine flu pandemic [11]. The multi-stage Richards models have also been proposed to simulate the multi-wave infection outbreak [12]. Typically, this method produces reliable estimations of the outbreak size and ending time; however, it cannot simulate the fluctuations in the number of infections precisely [13].

Statistical models have also been used to forecast the epidemiological trend of the virus infections. Autoregressive integrated moving average (ARIMA) models have been used extensively in various studies to forecast a virus spread [14]. For example, Benvenuto et al. [15] used ARIMA time-series models to forecast COVID-19 infection spread and showed that their model can accurately forecast the total number of confirmed cases for 2 and 9 days ahead, respectively.

Recent studies show that machine learning techniques can be used by various programs due to their higher forecast accuracy, reliability, and ability to scale up [16]. Chimmula et al. [17] modeled COVID-19 spread to forecast its peak time in Canada using deep learning methods. Ribeiro et al. [18] developed a new model to forecast the number of patients for 1–6 days ahead in different states of Brazil using stacking ensemble and support vector regression algorithm. These studies showed that the machine learning-based models can potentially enhance the short-term forecasting of COVID-19 spread; however, similar to ARIMA models, their performance deteriorates significantly for long-term forecast.

Hybrid modeling, which combines two or more different methods, has also been used to model complicated infection spread behavior [19]. Hasan [20] proposed a hybrid model, which combined neural network (NN) and ensemble empirical mode decomposition (EEMD) to forecast the daily pattern of COVID-19 using the world-wide statistics. These studies revealed the potential use of machine learning-based hybrid models, but the combination of these models with epidemic growth models for long-term forecast of COVID-19 has not been addressed previously.

This paper therefore first presents the use of single and multi-stage Richards growth models for forecasting the cumulative size of COVID-19 outbreak and esti-mated end time of the pandemic in three different countries, including Iran, Italy, and Mexico. Then, we investigate the use of non-hybrid ARIMA and nonlinear autore-gressive neural network (NAR) to forecast the number of daily cases for at least one month ahead. Finally, two novel hybrid models combining ARIMA and NAR models with Richards growth models are proposed to forecast short- and long-term number of daily cases, and their accuracy is evaluated using available data.

2 Data

The data required for this study was obtained from World Health Organization (WHO) database [21]. Our data set includes the cumulative and daily number of confirmed cases from January 22 to August 1, 2020 (this is referred to as the data period) for Iran, Italy, and Mexico. These three countries are selected as they repre-sent three different categories of countries responding to the COVID-19 outbreak. According to Fig. 1, Iran successfully controlled the first wave of the COVID-19 outbreak by imposing restrictions; however, they experienced a second wave as restrictions were removed. Italy represents countries who controlled the fast-growing virus outbreak by introducing strict rules and at the end of the data period did not experience a second wave. Finally, Mexico is an example of countries who could not control the extreme growth of the infections by the end of the data period.

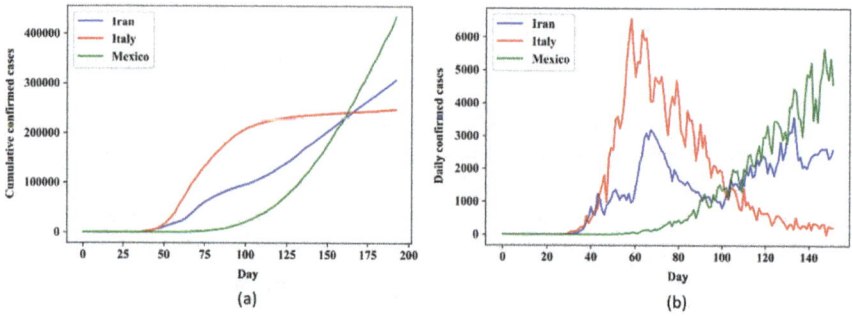

Fig. 1 **a** Cumulative and **b** daily number of confirmed cases for Iran, Italy, and Mexico from January 22 to August 1, 2020

3 Method

Our methodology is incremental which means, first we apply Richards growth models, then we evaluate machine learning models, and finally, we propose novel models combining both models to overcome the limitations of each type to have robust models.

3.1 Richards Models

Various methods can be used to model and forecast the spread of COVID-19, among which generalized logistic growth models, known as Richards models, are used in this study. Equation 1 shows the general formulation of this model,

$$I(t) = \frac{K}{(1 + e^{-rt})^{\frac{1}{v}}} \tag{1}$$

where I, K, and r denote the cumulative number of cases, final outbreak size, and growth rate, respectively, and v is a parameter which determines the point of inflection on the vertical axis. These parameters are determined using the least-square method to fit a curve to the data of selected countries. Since this paper aims to also model the second wave of COVID-19 outbreak and its impact on the control of the virus spread, the multi-stage Richards models are used. Stages are determined using the historical data to find the local minimum in cumulative size of the outbreak, which indicates a new wave of spread. For each stage, we calculate the Richards models parameters separately, and by combining all stages, the final outbreak model is achieved.

3.2 ARIMA Model

ARIMA and its simpler forms such as ARMA have been used to forecast a time series using its previous values. The general formulation of this method is,

$$\nabla^d Y(t) = \sum_{i=1}^{q} \beta_i \epsilon(t-i) + \sum_{i=1}^{p} \alpha_i \nabla^d Y(t-i) + \epsilon(t) \qquad (2)$$

where Y, ϵ are the modeled variable and the error, and p; q; d denote the autoregressive (AR), moving average (MA), and differencing orders of ARIMA(p, d, q). These orders are usually determined by analyzing the autocorrelation and partial autocorrelation function of the time series. Also, α; β are the coefficients of the ARIMA model, which are determined by fitting the model to the data. In Eq. 2, the ∇ is the differencing operator. In fact, the ARIMA model assumes that the value of a time series in the next interval is correlated to its value in previous intervals.

3.3 Nonlinear Autoregressive Neural Network

Nonlinear autoregressive neural network (NAR) can also forecast a time series based on its previous values $\{y(t-1), y(t-2), …, y(t-p)\}$ where p denotes the AR order. The architecture of a NAR can be generally described using the following equation,

$$Y(t) = F(Y(t-1), Y(t-2), …, Y(t-p)) + \epsilon(t) \qquad (3)$$

where Y is the desired parameter and F is the function that NAR aims to approximate during the training phase by optimization of weights and neuron bias of networks. Equation 3 can be written as:

$$Y(t) = \alpha_0 + \sum_{i=1}^{M} \alpha_i \emptyset \left(\sum_{j=1}^{N} \beta_{ij} Y(t-j) + \beta_{0i} \right) + \epsilon(t) \qquad (4)$$

where M is the number of hidden layers, N is the number of nodes per hidden layer, $\emptyset$ denotes the activation function, and α_i, β_{ij} are the weight between hidden unit j and the output and the connection weight between input j and hidden unit i, respectively. The constant values related to the connection between hidden unit i and the output unit are β_{0i} and α_0 [22].

3.4 Hybrid Models

In this study, hybrid models benefit from the combination of Richards growth models and statistical and machine learning techniques, discussed in the previous section, to improve the performance of the model in terms of the long-term forecasting of the number of cases. Figure 2 provides the flowchart of our novel hybrid models for simulation and forecasting of COVID-19 spread.

4 Results and Discussion

In this section, we first examine the performance of Richards models for the simulation of the COVID-19 outbreak using its single-stage and multi-stage forms. Then, we use the developed models to analyze the impact of second-wave prevention on the disease control in Iran and estimate the end time of the COVID-19 outbreaks in Mexico and Italy as well. In Sect. 4.2, we study non-hybrid models to forecast the number of daily cases. The performance of the novel hybrid models, which combine machine learning and statistical models with Richards models, in forecasting the number of the daily confirmed cases, is presented in Sect. 4.3 and their accuracy is compared.

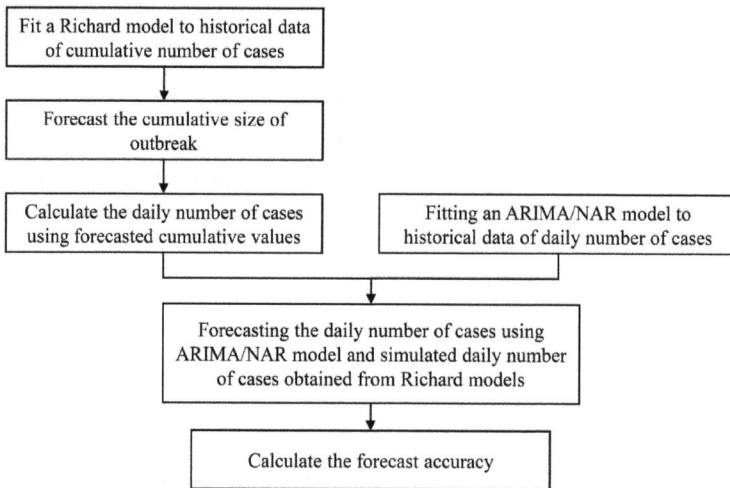

Fig. 2 Block diagram of novel hybrid models

4.1   Performance of Richards Growth Model

As explained in Sect. 3.1, Richards models are used to model the cumulative size of the outbreak. The single-stage model is first used to model and forecast the level of virus spread in our selected countries. Then, the two-stage model is used for Iran, experiencing the second wave of the COVID-19 outbreak, to examine the performance of this model and compare it to the single-stage one.

Single Wave Model

The parameters of Richards models have been calculated using the least-square method, which minimizes the distance between the fitted values and actual ones. Table 1 shows these parameters along with their 95% confidence interval (CI).

Figure 3 shows the historical and modeled daily number of confirmed cases in our studied countries using the Richards models. It is obvious that for Iran, which

Table 1 Richards growth model parameters along with their 95% confidence interval (CI) determined using the least-square method

Country	K	r	v	K-95% CI	r-95% CI	v-95% CI
Iran	473,833	0.0151	0.1703	±17,491	±0.0004	±0.0034
Italy	241,262	0.0586	0.0890	±365	±0.0004	±0.0017
Mexico	1,034,794	0.0175	0.0869	±10,892	±0.0001	±0.0005

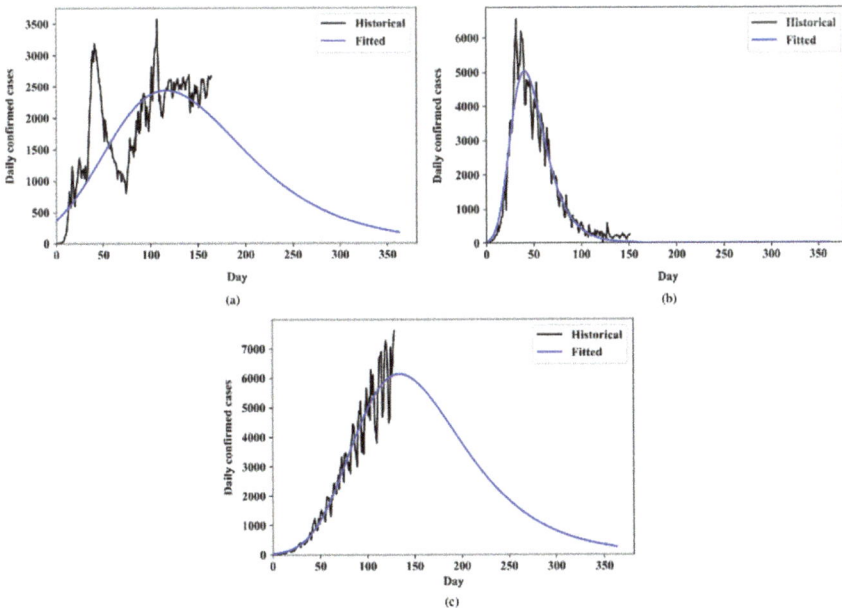

Fig. 3 Historical and modeled daily number of confirmed cases in **a** Iran, **b** Italy and **c** Mexico using the single-stage Richards model

has two waves of the virus outbreak, the single-stage Richards models cannot model the trend of daily cases. Figure 3b, c demonstrates that despite the fitted models are unable to capture fluctuation in the number of daily cases, they can simulate the trend very well. The end time for the virus spread can be seen in Fig. 3, where the number of cases reaches zero. For example, Italy could achieve almost zero number of cases after 4 months while this is approximately 12 months for Mexico.

Two Waves Model

In this section, we use multi-stage Richards models to simulate each wave of the infection spread and forecast the final size of the outbreak and the ending time. The focus of this section is therefore on Iran as it is the only country among our studied cases that experiences the second wave of COVID-19.

Table 2 shows the calculated parameters of the two-stage Richards model for Iran, along with their 95% confidence interval. It should be noted that the final outbreak size in this country is the summation of K values, which equals to approximately 514,000 cases. Comparing this number to what was calculated using single-stage model reveals that the final outbreak size increases by almost 8.5% using a more accurate modeling approach. Also, the ending time (zero daily cases) reduces significantly.

Figure 4 shows the cumulative and the daily number of confirmed cases in Iran using multi-stage Richards models. Comparing the result in Figs. 3 and 4 clarifies that the multi-stage model has a superior performance compared with the single-stage one. Figure 4a clearly exhibits the difference in the final outbreak size if restrictions were in place, and the second wave was prevented in Iran. For example, the outbreak

Table 2 Calculated parameters of two-stage Richards model for Iran along with their 95% confidence interval

	K	r	v	K-95% CI	r-95% CI	v-95% CI
First wave	100,132	0.0891	0.6374	±1411	±0.005	± 0.0941
Second wave	414,574	0.0198	0.3697	±15,842	±0.0016	±0.0981

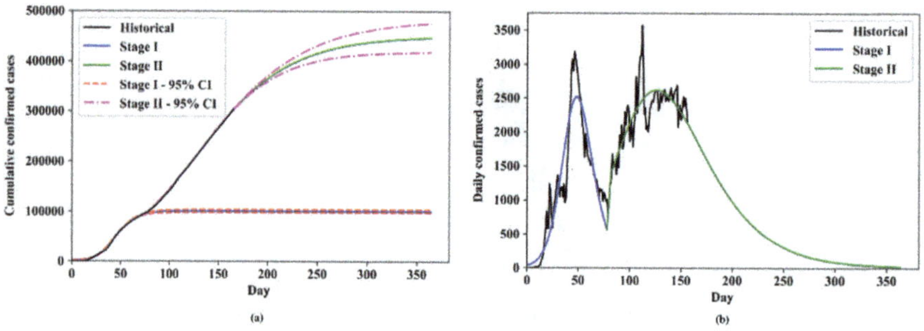

Fig. 4 Historical and modeled number of confirmed cases for Iran using two-stage Richards model: **a** cumulative number of cases **b** daily number of cases

size of the first wave is approximately 100,000 while this is approximately 514,000 for the second outbreak, i.e., 400% increase in the number of cases. Also, the ending time of the first wave was approximately 3 months; however, the second wave delayed this for almost 7 months and the number of daily cases achieves zero after more than 10 months since the start of the outbreak in Iran.

Despite two-stage Richards model has a superior performance compared with single stage, it cannot simulate the dynamics of the daily number of cases accurately. Figure 4b reveals this drawback as the fluctuations of the daily cases cannot be captured using Richards models. Therefore, we use time-series forecasting methods to simulate these dynamics of the virus spread as they are crucial for health system planning and decision making. Since it was demonstrated that multi-stage Richards model has a better accuracy, it is used for the rest of analysis conducted on Iran in this paper.

4.2 Non-hybrid Models Performance

This section examines the performance of non-hybrid ARIMA and NAR models for forecasting the number of daily cases. First, the historical data of each country is divided into two sets, known as training and testing sets. The training sets are used to build our models, and the testing data set enables us to calculate the accuracy of our model in forecasting the number of daily cases for next days. In this study, the training data set includes the data from the January 22, 2020, to the July 16, 2020, while the last 14 days are selected as the test data set.

ARIMA Models

As explained in Sect. 3.2, ARIMA models have three main components that need to be optimized to improve their accuracy. The optimal value of AR (p), MA (q), and differencing order for each country were calculated separately and are presented in Table 3.

Figure 5 shows the forecast of the daily number of cases using the ARIMA-only model for our three studied countries for one month after the end of the historical data. It can be seen that the ARIMA model has a higher accuracy for short-term (a few days) forecasting of the daily number of cases, but for long-term forecasting (a few months), its accuracy deteriorates rapidly. It should be noted that the ARIMA model considers the general trend of time series for forecasting the number of daily cases for periods after the test data; therefore, the numbers of daily cases for Iran

Table 3 Optimal order values of ARIMA models for selected countries

Country	p	q	d
Iran	5	0	2
Italy	5	0	1
Mexico	7	0	1

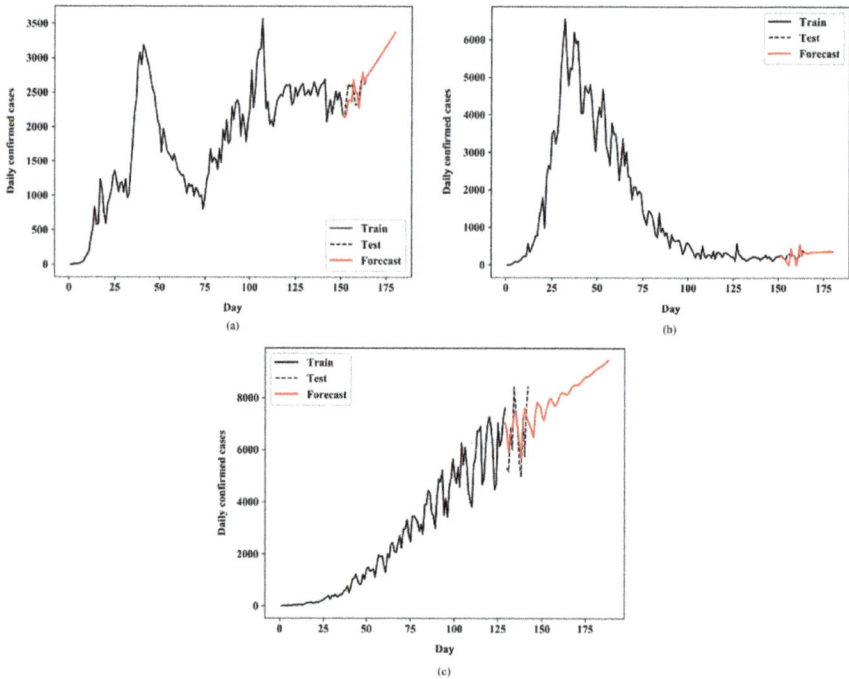

Fig. 5 Forecasting of the daily number of confirmed cases for one month ahead using ARIMA-only models for **a** Iran, **b** Italy, **c** Mexico

and Mexico increase almost monotonically, while for Italy it remains approximately constant. This highlights the necessity for other models or combination of different models, to forecast the number of daily cases more accurately.

NAR Models

To determine the optimal hyperparameters for the NAR model, grid search method is used. This method is an exhaustive search that is applied on a specified subset of hyperparameters of a learning algorithm to find their best values. Learning rate is a positive number between zero and one, which indicates the amounts of weights (αi, βij in Eq. 4) that are updated during the training phase. Moreover, epochs refer to the number of cycles for training a neural network on the entire train data set. Table 4 shows the optimal values of our NAR model hyperparameters.

Figure 6 shows the NAR model for forecasting the daily number of cases for our studied countries. It can be seen that the number of daily cases in Italy starts to rise

Table 4 Optimal hyperparameters of the NAR-only model, determined by grid search

Hidden layer (p)	Neurons	AR order	Learning rate	Epochs	Activation function
2	10, 10	5	0.01	500	Sigmoid

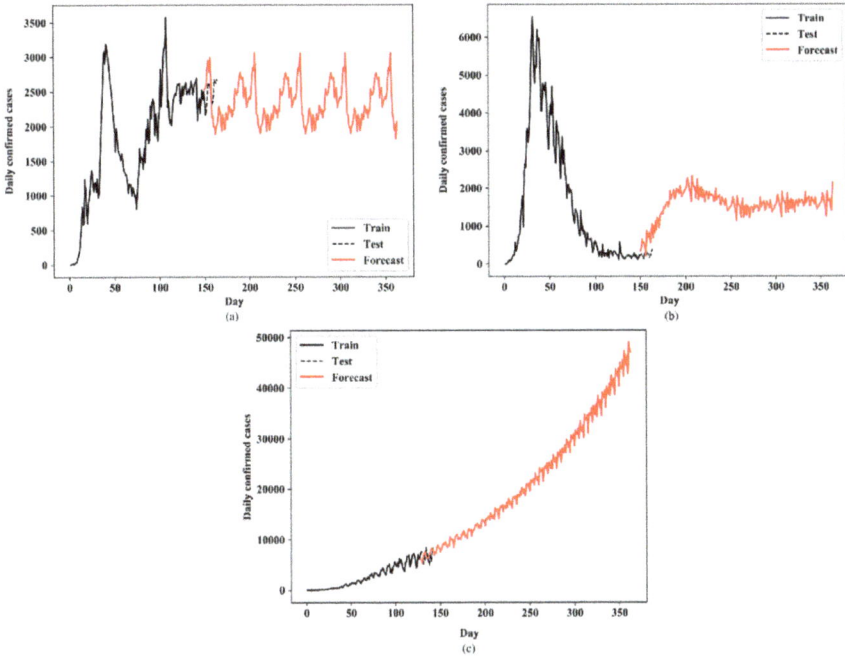

Fig. 6 Forecasting of the daily number of confirmed cases using NAR-only models for **a** Iran, **b** Italy, and **c** Mexico

after day 150, which is due to the historical trend and peak observed at approximately day 50. Also, for Iran, periodic behavior can be seen, which has a similar profile to the historical data. In fact, the NAR model forecasts the future number of infected people based on the overall trend existing in the historical time series. Table 5 shows the forecast error of the ARIMA and NAR models in terms of the normalized root mean square error (NRMSE) using the test data set and the following equation:

$$\text{NRMSE} = \frac{\sqrt{\frac{\sum_{t=1}^{T}\left(\text{DC}_{\text{test}}(t) - \text{DC}_{\text{pred}}(t)\right)^2}{T}}}{\text{DC}_{\text{test}}^{\text{max}}} \tag{5}$$

Table 5 Accuracy of ARIMA and NAR non-hybrid models for forecasting 14 days ahead		NRMSE (%) ARIMA	NRMSE (%) NAR
	Iran	16.71	12.33
	Italy	38.13	29.41
	Mexico	25.29	18.26

where DC denotes the number of daily cases, t is the day index, T is the test period (14 days here), and DC_{test}^{max} represents the maximum value in the test data set. Forecasting the number of daily cases for 14-days ahead.

Although the NAR model is not accurate for long-term forecasting of COVID-19 outbreak, it provides useful insights into the potential future of the outbreak if no restriction is applied and the historical trend continues. For example, Fig. 6a shows that Iran experiences a new peak with around 3000 every 60 days if dynamics of the virus spread remain the same. In addition, we can see that the spreading of virus in Mexico will grow exponentially, if no decision or action is undertaken by authorities in this country. Since the non-hybrid NAR models are not able to forecast the long-term profile of the number of daily cases, the requirement of hybrid models which combine the time-series forecasting methods with growth population models is inevitable.

4.3 Hybrid Models Performance

As demonstrated, non-hybrid models are not able to simulate the long-term behavior of an infection spread, which illuminates the necessity of another model, such as Richards model, to improve the performance of our forecasting system.

Figure 7 shows the block diagram of the novel, adaptive hybrid models used in this study. According to this figure, historical data of the number of daily cases is first used to train the ARIMA or NAR model. The trained models then forecast the number of daily cases for the next day $(t + 1)$. Next, the estimated number of cases obtained from Richards models for that interval $(t + 1)$ is used to tune the ARIMA or NAR model for forecasting the next interval $(t + 2)$. This prevents the behavior of the non-hybrid models observed in the previous section for forecasting the number of cases for a few months ahead. It should be noted that the ARIMA or NAR model as well as the Richards model can be tuned quickly once the historical data set is

Fig. 7 Block diagram of the proposed novel, adaptive hybrid models, which are trained with historical data and the estimations obtained from Richards model to forecast the daily number of cases for future intervals. In this figure, t denotes the time

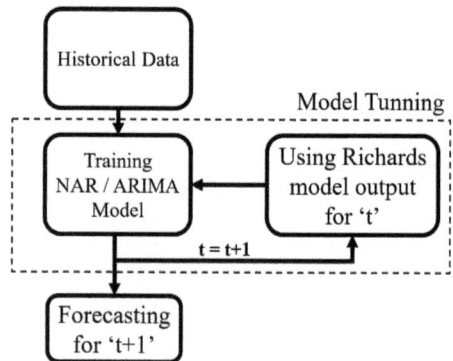

updated using newly measured data. This tuning allows us to have a better forecast of the possible future of the COVID-19 outbreak.

ARIMA-Richards Models

In this section, the performance of a hybrid model, combining Richards model and ARIMA method, is investigated. It should be noted that ARIMA models used in this section have the same order as those in Sect. 4.2. The use of Richards model informs the ARIMA model about the general long-term trend in the number of daily cases, which could potentially increase the accuracy of its forecast (Fig. 7).

Figure 8 presents forecasting the number of cases using the ARIMA-Richards hybrid model for the rest of the year since the start of an outbreak in Iran and Mexico and one month after the end of historical data for Italy. As expected, results indicate that the combination of the Richards and ARIMA model enables the forecasts to follow the general trend estimated by the growth models analysis. It should be noted that modeling of the historical fluctuations along with the white noise term in Eq. 2 causes the predicted number of the daily cases for Italy not to reach zero. Also, it is clear that in Fig. 8, fluctuations in the forecast of the number of cases are much smaller as compared with the historical values, which leads to its lower accuracy. Table 7 shows the accuracy of our proposed hybrid models. Results reveal that the ARIMA-based model has the best accuracy for forecasting the number of daily cases

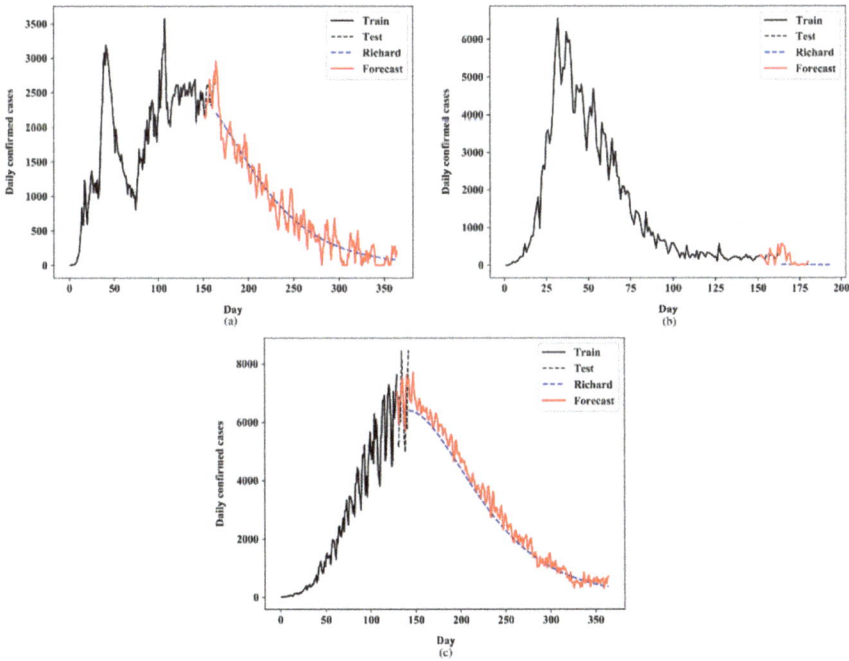

Fig. 8 Forecasting of the daily number of confirmed cases using ARIMA-Richards models for **a** Iran, **b** Italy, and **c** Mexico

Table 6 Optimal hyperparameters of the NAR-Richards model, determined by grid search

Hidden layer (p)	Nodes	AR order	Learning rate	Epochs
2	16, 8	7	0.01	500

in Iran, while its accuracy is the lowest for Italy. The reason is that for Italy the number of daily cases is close to zero and even small errors lead to high loss of accuracy (see Eq. 5). Also, for Iran, the amplitude of fluctuations in the number of daily cases is much lower than those of Mexico, which ultimately results in better accuracy for this country as forecasting of a noisy time series is more difficult.

NAR-Richards Models

Similar to Sect. 4.2, the optimal values for hyperparameters, shown in Table 6, are determined using the grid search method to build the best neural network for forecasting the daily number of cases. Figure 9 shows the forecasting of the number of daily cases using NAR-Richards hybrid model. Again, the utilization of the Richards model forces the machine learning model to follow the general trend of the number of daily cases approximated by fitted Richards models.

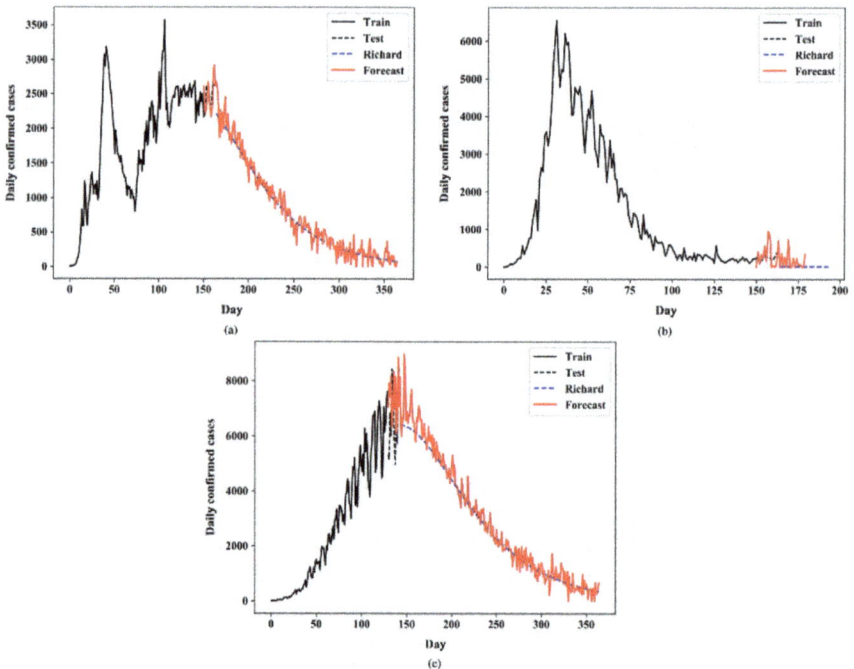

Fig. 9 Forecasting of the daily number of confirmed cases using NAR-Richards models for **a** Iran, **b** Italy, and **c** Mexico

Table 7 Accuracy of the ARIMA-Richards and NAR-Richards hybrid models for forecasting 14 days ahead in terms of NRMSE for selected countries

	NRMSE (%) ARIMA-Richards	NRMSE (%) NAR-Richards	Improvement (%)
Iran	10.91	8.19	24.93
Italy	25.61	21.21	17.18
Mexico	15.78	10.02	36.50

Comparison between the performance of our hybrid models in test period shown Figs. 8 and 9 indicates the NAR-Richards model can forecast the fluctuations more accurately. This can be clearly seen in Figs. 8c and 9c, which shows our forecasts for Mexico. Also, for Iran, the amplitude of fluctuations in the forecast of the number of cases is more consistent with the historically observed fluctuations in Fig. 9a while that is relatively higher than the historical values in Fig. 8a, especially between days 150 to 200.

The NRMSE of the NAR-Richards model is also presented in Table 7. The accuracy of the model is the highest for Iran while it has the worst accuracy for Italy. The reason for this is similar to what was explained in the previous section for ARIMA-Richards model. Furthermore, the comparison between the ARIMA-Richards and NAR-Richards model shows that our machine learning-based hybrid model outperforms the statistical-based hybrid model. On average, the accuracy of our forecasts has been improved by approximately 20% (Eq. 6) using the NAR-Richards model as compared with the ARIMA-Richards, while the computation time remained comparable. This reveals the potential of using machine learning methods in the field of epidemiology for modeling and forecasting the infections spread to assist the authorities to plan and control the outbreaks.

$$\text{Improvement} = \frac{\text{NRMSE}_{\text{ARIMA-Richards}} - \text{NRMSE}_{\text{NAR-Richards}}}{\text{NRMSE}_{\text{ARIMA-Richards}}} \times 100 \quad (6)$$

As shown in Fig. 7, these hybrid models are highly adaptable, meaning that they can be trained rapidly using more updated data in order to provide a more accurate forecast of the possible future of the COVID-19 pandemic.

5 Conclusion

This paper proposed novel experimentally determined hybrid models integrating logistic growth models into statistical and machine learning models to simulate and forecast COVID-19 outbreaks in three countries, Iran, Italy, and Mexico. Among these countries, Iran experienced a second wave of the outbreak due to lifting restrictions, while Italy successfully controlled the first wave and prevented the second wave by the end of the data period (August 1, 2020). Mexico also faces a rapid increase in the size of the first wave of outbreak.

The results showed that the conventional Richards models could simulate the infection growth in Italy and Mexico, which only have a single wave of virus outbreak while its accuracy reduced for Iran, having two waves of infection spread. Therefore, multi-stage Richards models, where each stage was associated with one outbreak wave, were used to model Iran infection growth that resulted in a higher model accuracy. The comparison between two stages of Richards models for Iran indicated that the prevention of the second-wave outbreak could reduce the number of incidences and the required time for controlling the virus spread by approximately 87% and 75%, respectively. This highlighted the importance of the prevention of the second-wave outbreak in countries.

In addition, the results indicated that Richards models are not able to capture and forecast the fluctuations in the daily number of cases, which is an important factor for health system planning. To address this issue, non-hybrid ARIMA and NAR model were initially studied. The results showed that these models can forecast the short-term profile of the daily number of cases accurately, but they were not suitable for long-term forecasting (a few months ahead). Therefore, two novel hybrid models, ARIMA-Richards and NAR-Richards, were proposed. The results revealed that these hybrid models can forecast the daily number of cases more accurately and reasonably for short and long forecast horizons. The comparison between the results of these hybrid models showed that machine learning-based hybrid models have a better performance than the statistical-based models and on average are 20% more accurate. This along with the high adaptability of the proposed hybrid models make them suitable for providing a better insight into the possible future of the COVID-19 pandemic and illuminates the potentials of using machine learning algorithms to investigate the short- and long-term impacts of pandemics on the healthcare systems, human health, and environment.

References

1. Xie M, Chen Q (2020) Insight into 2019 novel coronavirus—an updated interim review and lessons from SARS-CoV and MERS-CoV. Int J Infect Dis 94:119–124
2. McBryde ES, Meehan MT, Adegboye OA, Adekunle AI, Caldwell JM, Pak A et al (2020) Role of modelling in COVID-19 policy development. Paediatr Respir Rev
3. George DB, Taylor W, Shaman J, Rivers C, Paul B, O'Toole T et al (2019) Technology to advance infectious disease forecasting for outbreak management. Nat Commun 10(1):1–4
4. Chretien J-P, Riley S, George DB (2015) Mathematical modeling of the West Africa Ebola epidemic. Elife 4:e09186
5. Ren H, Zhao L, Zhang A, Song L, Liao Y, Lu W et al (2020) Early forecasting of the potential risk zones of COVID-19 in China's megacities. Sci Total Environ 729:138995
6. Cássaro FAM, Pires LF (2020) Can we predict the occurrence of COVID-19 cases? Considerations using a simple model of growth. Sci Total Environ 728:138834
7. Roosa K, Lee Y, Luo R, Kirpich A, Rothenberg R, Hyman JM et al (2020) Real-time forecasts of the COVID-19 epidemic in China from February 5th to February 24th, 2020. Infect Dis Model 5:256–263

8. Chowell G (2017) Fitting dynamic models to epidemic outbreaks with quantified uncertainty: a primer for parameter uncertainty, identifiability, and forecasts. Infect Dis Model 2(3):379–398
9. Calafiore GC, Novara C, Possieri C (2020) A time-varying SIRD model for the COVID-19 contagion in Italy. Annu Rev Control
10. Hsieh Y-H (2009) Richards model: a simple procedure for real-time prediction of outbreak severity. In: Modeling and dynamics of infectious diseases. World Scientific, pp 216–36
11. Hsieh Y (2010) Pandemic influenza A (H1N1) during winter influenza season in the southern hemisphere. Influenza Other Respir Viruses 4(4):187–197
12. Hsieh Y, Chen CWS (2009) Turning points, reproduction number, and impact of climatological events for multi-wave dengue outbreaks. Trop Med Int Health 14(6):628–638
13. Wang X-S, Wu J, Yang Y (2012) Richards model revisited: validation by and application to infection dynamics. J Theor Biol 313:12–19
14. Li Q, Guo N-N, Han Z-Y, Zhang Y-B, Qi S-X, Xu Y-G et al (2012) Application of an autoregressive integrated moving average model for predicting the incidence of hemorrhagic fever with renal syndrome. Am J Trop Med Hyg 87(2):364–370
15. Benvenuto D, Giovanetti M, Vassallo L, Angeletti S, Ciccozzi M (2020) Application of the ARIMA model on the COVID-2019 epidemic dataset. Data Brief 29:105340
16. Davenport T, Kalakota R (2019) The potential for artificial intelligence in healthcare. Future Healthc J 6(2):94
17. Chimmula VKR, Zhang L (2020) Time series forecasting of COVID-19 transmission in Canada using LSTM networks. Chaos Solitons Fractals 135:109864
18. Ribeiro MHDM, da Silva RG, Mariani VC, dos Santos Coelho L (2020) Short-term forecasting COVID-19 cumulative confirmed cases: perspectives for Brazil. Chaos Solitons Fractals 135:109853
19. Currie CSM, Fowler JW, Kotiadis K, Monks T, Onggo BS, Robertson DA et al (2020) How simulation modelling can help reduce the impact of COVID-19. J Simul 14(2):83–97
20. Hasan N (2020) A methodological approach for predicting COVID-19 epidemic using EEMD-ANN hybrid model. Internet Things 11:100228
21. WHO coronavirus Diesease (COVID-19) Dashboard [Internet] (2020). Available from: https://covid19.who.int/table
22. Naemi A, Mansourvar M, Schmidt T, Wiil UK (2020) Prediction of patients severity at emergency department using NARX and Ensemble learning. In: 2020 IEEE international conference on bioinformatics and biomedicine (BIBM). IEEE, pp 2793–9

Chapter 10
Spatial Statistics Models for COVID-19 Data Under Geostatistical Methods

S. Zimeras

Abstract Geostatistics provides the practitioner with a methodology to quantify spatial uncertainty. Statistics come into play because probability distributions are the meaningful way to represent the range of possible values of a parameter of interest. In addition, a statistical model is well suited to the apparent randomness of spatial variations. It must be noted that there is considerable variety of statistical methods that have been applied in the analysis of spatial variation in data, summarized by Dale (Spatial pattern analysis in plant ecology. Cambridge University Press, 1999) [1]. These include dispersal analysis, spectral analysis, wavelet analysis, kriging, and spatial Monte Carlo simulations, and many geostatistics methods. Kriging was developed for estimating thresholds of continuous variables. It has been used for interpolation and simulation of categorical variables and for spatial uncertainty analysis.

Keywords Spatial statistics · Geostatistical analysis · COVID-19 · Kriging · Kernel estimation

1 Introduction

Model evaluation assumes a certain general structure (e.g., multiple linear), and the model is built through adding terms (variables) which are significant or which aid in prediction (hierarchical modeling). Parameter uncertainty is defined as a problem of estimation. Stochasticity is often introduced through stochastic functions (e.g., weather) or random effects in parameter values.

Interactions between regions at different scales are characterized by their local dynamics, and the emergent spatial patterns are the outcome of different processes. The development of specific new, applied statistical techniques can be explained by the emerging field of specific regions of human body, which focuses on spatial processes operating over various spatial extents epidemiologistics are trying to collect

S. Zimeras (✉)
Department of Statistics and Actuarial-Financial Mathematics, University of the Aegean, Samos, Greece
e-mail: zimste@aegean.gr

© The Author(s), under exclusive license to Springer Nature Singapore Pte Ltd. 2022 137
R. J. Howlett et al. (eds.), *Smart and Sustainable Technology for Resilient Cities and Communities*, Advances in Sustainability Science and Technology,
https://doi.org/10.1007/978-981-16-9101-0_10

quantified information about geostatistic spatial pattern in order to answer questions regarding the underlying processes (e.g., competition).

Geostatistics is based upon the recognition that in the Earth sciences there is usually a lack of sufficient knowledge concerning how properties vary in space. Therefore, a deterministic model may not be appropriate. If we wish to make predictions at locations for which we have no observations, we must allow for uncertainty in our description as a result of our lack in knowledge. So, the uncertainty inherent in predictions of any property we cannot describe deterministically is accounted for through the use of probabilistic models.

Markov chain geostatistics (MCG) is a new non-kriging geostatistics [2, 3]. The basic idea of this geostatistics is to use Markov chains to perform multidimensional interpolation and simulation. Compared with the covariance-based (or variogram-based) geostatistics, MCG is transition probability-based. Compared with the kriging-based geostatistics, MCG is Markov chain-based. MCG directly uses Markov chains to accomplish conditional simulation. The basic idea of MCG is that an unknown location is related on its nearest known neighbors in different directions. With a Markov chain moving around in a space, its conditional probability distribution at any unknown point is entirely dependent on its nearest known neighbors in different directions [4–7]. The interaction between each nearest known neighbor and the unknown location is expressed by a transition probability at the corresponding distance. Therefore, transiograms are the explicit components of the conditional probability function.

Referring to conditional probabilities, due to the largeness of the configuration space it is impractical to sample from it by direct computation of the probabilities. Markov chains Monte Carlo (MCMC) methods have been investigated by various researchers as an alternative to exact probability computation [8, 9]. The general method is to simulate a Markov chain with the required probability distribution as its equilibrium distribution. If the chain is aperiodic and irreducible, the convergence is guaranteed.

Model selection techniques for spatial models need to include the correlation structure in determining the best set of predictors. By computing the AIC statistic for all possible sets of explanatory variables and autocorrelation functions, one can find a single "best" model or a set of models which fit the data well [6, 10]. This method attempts to strike a balance between the competing forces of large-scale variability, as modeled via the explanatory variables, and small-scale variability, as modeled through the correlation in the residuals.

Hierarchical models, whereby a problem is decomposed into a series of levels linked by simple rules of probability, assume a very flexible framework capable of accommodating uncertainty and potential a priori scientific knowledge while retaining many advantages of a strict likelihood approach. Estimation of the parameters of the models could be achieved by applying MCMC techniques [6–9].

In this work, illustration of the spatial modeling based on coordinates considering the epicenters of COVID-19 virus in a small region could be described. Statistical analysis under spatial point process methodology has been applied and estimation techniques have been proposed using kernel methods (Gaussian kernel estimation).

Finally application of the K-function to spatial background rate estimates for point spatial pattern using simulated and real data based on the position of COVID-19 virus.

2   Point Process Analysis

The region of interest is defined as a d-dimensional space D with $D \subset \mathfrak{R}^d$, where d is the applied dimension. The data location is defined as $\mathbf{s}$ where in 2-D space, two coordinates could be illustrated as (x, y). Based on the data location $\mathbf{s}$, the notation $z(s)$ is the value for each z in location s. The $Z(s)$ is defined as a random variable at each location, and the spatial model corresponding to the above random variable is denoted as $\{\mathbf{Z}(\mathbf{s}) : \mathbf{s} \in D\}$ [11–14].

In point process analysis, the investigation of a given pattern based on the data is analyzed The data set consists of n locations $s_1 \ldots s_n$, each based on an (x, y) coordinate in two-dimensional space. The methods that could be applied in pattern analysis could be: (1) quadrant count method, (2) kernel density estimation (K-means), (3) nearest neighbor distance (G-function, F-function, K-function) [11–14]. There are three types of pattern based on the spatial association between the locations: (1) independent, (2) regular, and (3) clustered (Fig. 1).

A binomial random process with distribution is given by (Fig. 2a)

$$p[N(A)] = \frac{n!}{k!(n-k)!}\left(\frac{a(A)}{a(D)}\right)^k \left(1 - \frac{a(A)}{a(D)}\right)^{n-k}, \quad k = 0, 1, \ldots, n$$

where $a(.)$ is the unit area for specific region and $A \in D \subset R^2$. As an extension, the Poisson random process with distribution is given by (Fig. 2b)

$$p[N(A)] = \frac{\lambda a(A)}{k!}e^{-\lambda a(A)}, \quad k = 0, 1, 2, \ldots$$

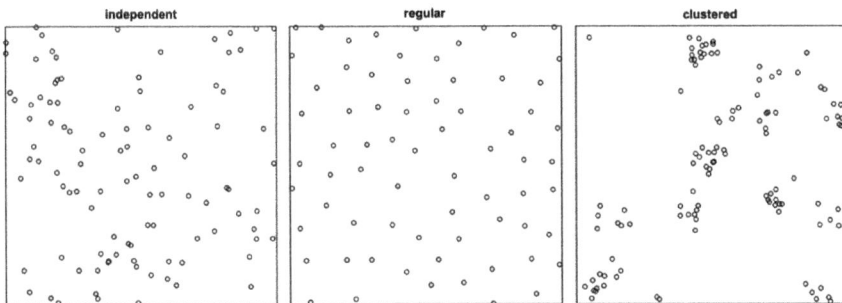

Fig. 1 Types of cluster patterns

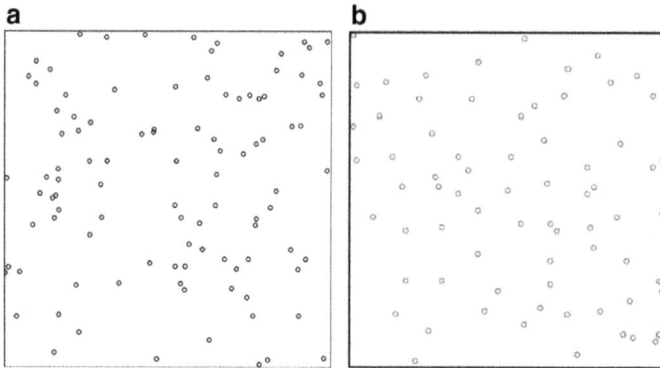

Fig. 2 **a** Realization of a binomial point process; **b** realization of a Poisson point process

The most effective way to visualize spatial pattern data is to illustrate them as a region over which the events are observed as points.

In point process procedure, each pattern is a set of point locations $\mathbf{s} = \{s_1, s_2, \ldots, s_n\}$ explaining the spatial association between different regions as in Fig. 3 where clustering regions are appearing clearly [15]. The random variable $N(A)$ is the number of events in the set $A \subset X$ introducing a random process. Based on spatial statistical measures, under Poisson point process, considering expected values of the process **first-order properties** have been defined with $\lambda(s) = \lim_{ds \to 0} \left\{ \frac{E(N(ds))}{|ds|} \right\}$ where $d(\mathbf{s})$ is a small region around the point $\mathbf{s}$, $E(.)$ is the expected value, and

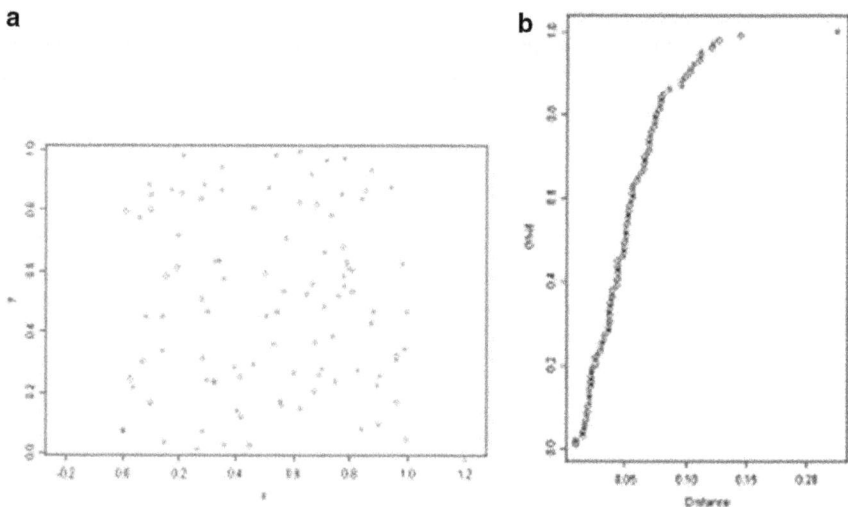

Fig. 3 **a** Simulated Poisson point data; **b** G-function of the simulated Poisson data

$N(d(\mathbf{s}))$ is the number of events in the small region [11–14, 16]. Alternatively considering covariance relation of the process **second-order properties** has been defined with $\gamma\left(s_i, s_j\right) = \lim_{ds_i,ds_j \to 0}\left\{\frac{E\left(N(ds_i)N\left(ds_j\right)\right)}{|ds_i||ds_j|}\right\}$ corresponding definitions as above. A point process is stationary if the intensity is constant over A, so $\lambda(\mathbf{s}) = \lambda$ and $\gamma(s_i, s_j) = \gamma(s_i - s_j) = \gamma(d)$ (depending only on direction and distance) [11–14, 16].

3   Measures of Point Process

An analysis of the point pattern process includes the definition of the distance between neighboring locations. If the distance is small, there is evidence of clustering. To achieve to above goal, the following distances could be introduced:

1. **Pairwise distances** $s_{ij} = \left\|x_i - x_j\right\|$ between all distinct pairs of points x_i and x_j ($i \neq j$) in the pattern
2. **Nearest neighbor distances** $s_i = \min_{j \neq i} x_{ij}$ the distance from each point x_i to its nearest neighbor
3. **Empty space distances** $d(s) = \min_i \|s - x_i\|$ the distance from a fixed reference location s in the window to the nearest data point

If a random variable S is representing the short nearest neighbor distance between locations then cumulative probability distribution of S is $P(S = s)$ estimated by the $G(s)$ formula with $\hat{G}(s) = \frac{\#(S_i \le s)}{n}$, meaning that the probability of distances less than or equal to s is estimated as $\hat{G}(s)$. For simulated data, the corresponding distribution of G-function based on distances is given in Fig. 3.

In Fig. 4, the estimation of $\hat{G}(s)$ function for the epicenters for COVID-19 virus is illustrated considering different radius region d ($d = 1, d = 30$) [11–14, 17].

Fig. 4 $\hat{G}(s)$ Function of the epicenters for COVID-19 virus (different radius region $d = 1$, and $d = 30$)

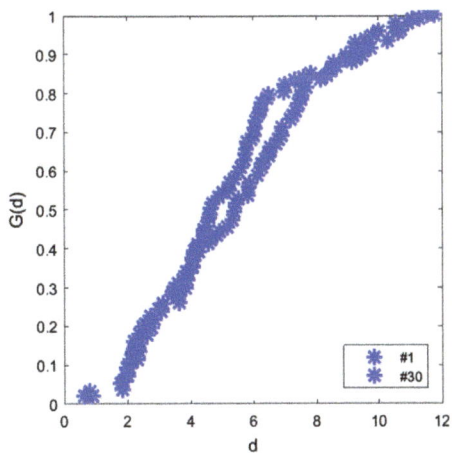

In general, *G*-function measures the distribution of distances from an arbitrary event to its nearest neighbors with

$$\widehat{G}(s) = \frac{1}{n}\sum_{i=1}^{n} l_1$$

with

$$l_i = \begin{cases} 1, & \text{if } d_i \in \{d_i : d_i \le s, \forall i\} \\ 0, & \text{otherwise} \end{cases}$$

and $d_i = \min_j\{d_{ij}, \forall j \ne i \in D\}, i = 1, 2, \ldots, n$

For a clustered pattern, observed locations should be closer to each other. *G* increases rapidly at short distance.

One of the most popular methods to estimate the first-order intensity $\lambda(s)$ is the **kernel estimator** that applies methods from kernel density estimation to obtain an estimate of λs. The general form is given by

$$\widehat{\lambda}(s) = \sum_{i=1}^{n} \frac{1}{\tau^2}k\left(\frac{s - s_i}{\tau}\right)$$

where $\kappa(\cdot)$ is a kernel function, and τ is an bandwidth satisfying standard conditions [12–14]. Reference [16] suggests various choices for $\kappa(\cdot)$ and selections for the bandwidth τ. A recommended choice for the bandwidth is $\tau = 0.68n^{-0.2}$ [11–13]. The choice of the bandwidth (τ) depends on the user. When τ is large, the kernel estimator led to a smooth estimate of the density function (i.e., small variance, large bias). If τ is small, the kernel estimator produces a rough estimate of the density function (i.e., large variance, small bias). Graphical presentation of the kernel estimator process for spatial point data is given in Fig. 5.

The most common choices for kernels formulas with the corresponding graphs (Fig. 6) are:

Fig. 5 Kernel estimator process [18, 19]

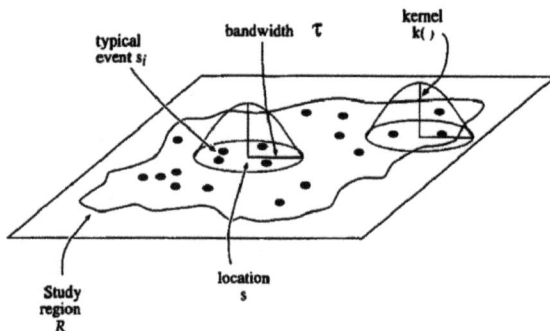

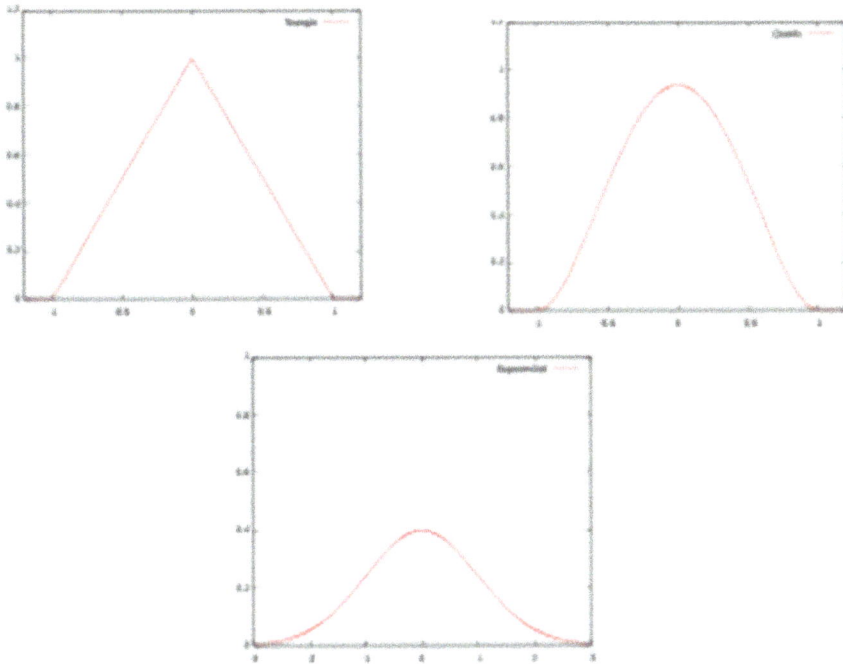

Fig. 6 Kernel forms: left to right: triangular, quartic, Gaussian [18, 19]

- Triangular: $k = 1 - \left| \frac{d_i}{\iota} \right|$
- Quartic: $k = \frac{3}{\pi} \left(1 - \frac{d_i^2}{\tau^2} \right)$
- Gaussian: $k = \frac{1}{\sqrt{2\pi}} e^{-\frac{d_i^2}{2\tau^2}}$

Based on the proposed kernel formulas, Fig. 7 illustrates the resulting spatial estimation of the point pattern for the epicenters for COVID-19 virus using Gaussian kernel form.

Considering the second-order spatial point, the appropriate measure for estimation is given by using the Ripley K-function [14] given by the form

$$K(d) = \lambda^{-1} E \left[\text{number of points distance } d \text{ of an arbitrary point} \right]$$

This is defined as the expected number of events within a distance d of an arbitrary event and describes how the spatial dependence varies through space. The K-function for a homogeneous process with no spatial dependence is πd^2. If there is clustering, there would be an excess of events at short distances $\left(K(d) > \pi d^2 \right)$. $K(t)$ can be estimated empirically from point pattern realizations. If $\lambda K(d)$ is defined as the expected number of events within distance d of an arbitrary event, then a direct estimate can be obtained for a points realization with n number of events as

Fig. 7 Gaussian kernel
estimation for the epicenters
for COVID-19 virus

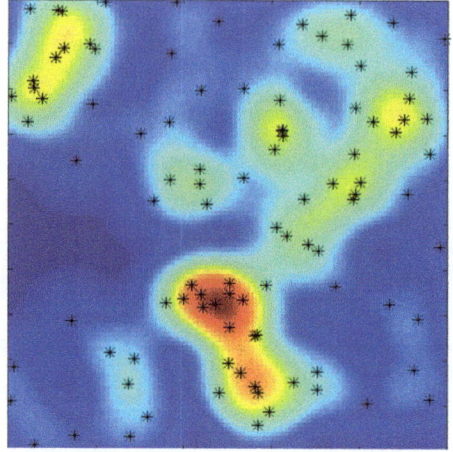

$$\lambda \, \widehat{K}(d) = \frac{1}{n} \sum_{i=1}^{n} M_i(d) = \frac{1}{n} \sum_{i=1}^{n} (\text{number of events within distane } d \text{ of the event } i)$$

To estimate $M_i(d)$ indicator values for all events i is defined as:

$$l_i = \begin{cases} 1, & \text{if } d_i \leq d \\ 0, & \text{otherwise} \end{cases}$$

where d_{ij} is the distance between event i and event j.

$$M_i(d) = \sum_{j=1}^{n} I_{ij}(d), \quad j \neq i$$

Then estimation of the K-function is given by the form:

$$\widehat{K}(d) = \frac{1}{\lambda n} \sum_{i=1}^{n} \sum_{j=1}^{n} I_{ij}(d)$$

The unbiased estimate of the K-function can be finally written by:

$$\widehat{K}(d) = \frac{1}{\lambda n} \sum_{i=1}^{n} \sum_{\substack{j=1 \\ j \neq i}}^{n} w_{ij} I_{ij}(d)$$

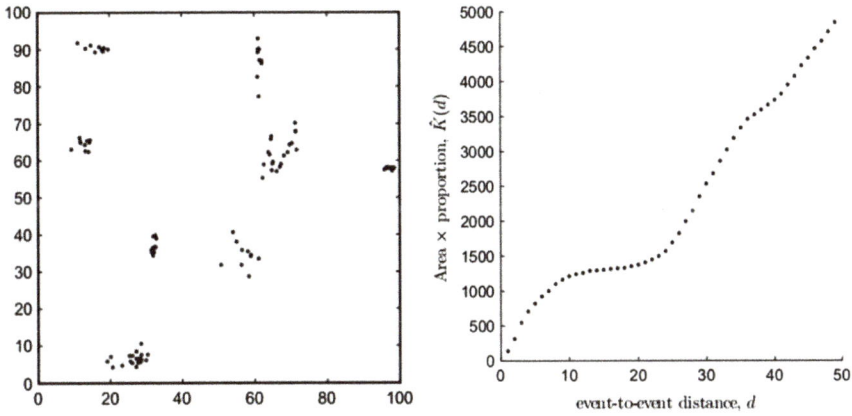

Fig. 8 Graph Ripley K-function for clustered spatial data

Fig. 9 Ripley K-function
for the epicenters of
COVID-19 virus

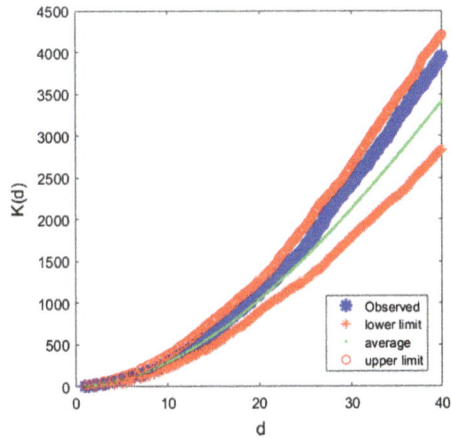

where w_{ij} is a weighting term equal to this circle's proportion of the entire area. It is clear that as d goes large, $w_{ij} \to \infty$. A valuable investigation for a point pattern is a plot of $\widehat{K}(d)$ versus πd^2. If $\widehat{K}(d) > \pi d^2$ then there is evidence of clustering (Fig. 8).

In Fig. 9, the Ripley K-function for the epicenters of COVID-19 virus is illustrated considering observed data, average data, low and upper limits, and average values.

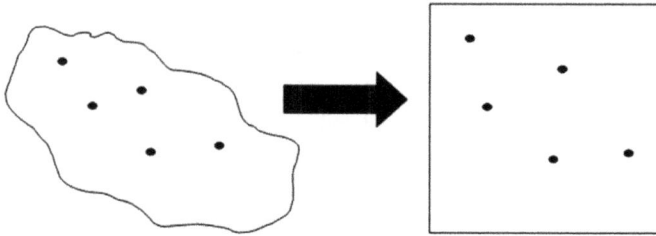

Fig. 10 Transformation from mapping to lattice point process

4 Conclusions

Spatial statistics is a powerful tool, for investigating relationships between different locations of a particular event applying specific statistical technique. In this work, we have described how spatial point methods can be applied using spatial statistical techniques.

Kernel estimation was suggested as an appropriate method for prediction of mean intensities for data. The application of the K-function to spatial background rate estimates for point spatial pattern using simulated and real data based on the position of COVID-19 virus. Investigation of distance analysis is proposed based on the choice of appropriate kernel form. As a proposed kernel was considered the Gaussian form which applied to real data considering a small region in Greece. Data have been collected by transforming a geographical map as a lattice system by marking the appearances of the positions (epicenters) from the map to the lattice system. Figure 10 illustrates the methodology.

Acknowledgements The author would like to thank Prof. Phaidon Kyriakidi for his useful support for the programming implementation of this work.

References

1. Dale MRT (1999) Spatial pattern analysis in plant ecology. Cambridge University Press
2. Baddeley A (2008) Analysing spatial point patterns in R. CSIRO, Canberra, Australia
3. Diggle PJ, Tawn JA, Moyeed RA (1998) Model-based geostatistics (with discussions). Appl Stat 47(3):299–350
4. Besag J (1974) Spatial interaction and the statistical analysis of lattice systems. J Roy Stat Soc Ser B 3:192–236
5. Besag J (1986) On the statistical analysis of dirty pictures. J Roy Stat Soc Ser B 48:259–302
6. Cross GR, Jain AK (1983) Markov random field texture models. IEEE Trans Pattn Anal Mach Intell 5(1):25–39
7. Aykroyd RG, Zimeras S (1999) Inhomogeneous prior models for image reconstruction. J Am Stat Assoc (JASA) 94(447):934–946
8. Zimeras S (1997) Statistical models in medical image analysis. Ph.D. thesis, Leeds University, Department of Statistics

9. Aykroyd RG, Haigh JGB, Zimeras S (1996) Unexpected spatial patterns in exponential family auto-models. Graph Models Image Process 58(5):452–463
10. Cliff AD, Ord JK (1981) Spatial processes: models and applications. Pion Limited
11. Cressie NA (1993) Statistics for spatial data (revised edition). Wiley
12. Diggle PJ (2003) Spatial analysis of spatial point patterns, 2nd edn. Arnold Publishers
13. Diggle PJ (1983) Spatial statistics. Wiley
14. Ripley BD (1981) Spatial statistics. Wiley, New York
15. Zimeras S, Georgiakodis F (2005) Bayesian models for medical image biology using Monte Carlo Markov Chain techniques. Math Comput Model 42:759–768
16. Upton G, Fingleton B (1985) Spatial data analysis by example. Wiley
17. Baddeley AJ, Silverman BW (1984) A cautionary example of the use of second order methods for analyzing point patterns. Biometrics 40:1089–1093
18. Gatrell AC, Bailey TC, Diggle PJ, Rowlingson BS (1996) Spatial point pattern analysis and its application in geographical epidemiology. Trans Inst Br Geogr 21(1):256–274
19. Wang F (2014) Quantitative methods and socio-economic applications in GIS (2^a edn). CRC Press

Chapter 11
Intelligent Multi-Sensor System for Remote Detection of COVID-19

G. Zaz, M. Alami Marktani, A. Elboushaki, Y. Farhane, A. Mechaqrane, M. Jorio, H. Bekkay, S. Bennani Dosse, A. Mansouri, and A. Ahaitouf

Abstract The worldwide spread of COVID-19 pandemic creates an urgent need for research and development of safe and efficient solutions for early COVID-19 detection. In this paper, an intelligent, reliable, and low-cost system detecting the main symptoms of COVID-19 disease (fever, cough, and breathing difficulties) is proposed. This system applies the principle of multi-sensor data fusion to provide a robust, precise, and complementary analysis between these symptoms to tell whether or not an individual is a carrier of COVID-19 disease. Using machine learning tools, the system is trained on infrared images to recognize the fever. The obtained thermal images are also used to control the breathing rate by monitoring temperature changes around the nasal areas on the faces. This signature is recognized through a well-trained thermal image processing model from online databases. To identify the third symptom of COVID-19 (cough), the system is associated with a network of microphones. Using specific artificial intelligence (AI) model based on mel-frequency cepstral coefficients (MFCC) convolutional neural network (CNN) architecture, it is possible to detect the cough sound. The combined use of the thermal and sound sensors allows merging data of the multi-sensor system. This approach is often the most suitable response to operational needs requiring a complete, efficient, and reactive diagnosis. The system presented in this paper is designed to be used for public hosting institutions. The objective is to contribute to slowing or even stopping the spread of COVID-19. This system can also be adapted as a useful means of early detection of many other diseases.

Keywords COVID-19 detection · SARS-CoV-2 · Body temperature · Cough · Respiratory rhythm · Artificial intelligence · Deep learning · Multi-sensor system

G. Zaz (✉) · A. Mechaqrane · M. Jorio · A. Ahaitouf
LSIGER, FST-FES, Sidi Mohammed Ben Abdellah University, Fez, Morocco
e-mail: ghita.zaz@usmba.ac.ma

M. A. Marktani · A. Elboushaki · Y. Farhane · S. B. Dosse · A. Mansouri
LSIGER, ENSA-FES, Sidi Mohammed Ben Abdellah University, Fez, Morocco

H. Bekkay
LESETI, ENSA-OUJDA, Mohamed Premier University, Oujda, Morocco

1 Introduction

Since the apparition of the SARS-CoV-2 virus, at the origin of the COVID-19 global pandemic, countries worldwide have been facing multiple challenges for its diagnosis in order to reduce its transmission rate [1, 2]. The World Organization for Health (WHO) urged government authorities to take preventive directives, measures, and recommendations to fight this pandemic and to stop its spread [3].

In the absence of widely available vaccine for different variants of the virus and approved treatment during the pandemic, the efficient solution is to implement preventive measures, such as social distancing, hand-washing, and face masks, to be taken in an attempt to limit the virus transmission as much as possible until a reliable cure is found. For the government and health services providers, it becomes a challenge to make rapid forward planning to evaluate the transmission rate of the SARS-CoV-2 without ready access to diagnostic techniques and future planning based on the sustainability of healthcare systems to cope with the outbreak [4].

To take part in the mobilization, all around the world, against this virus, Morocco launched a call for researches projects aiming to develop solutions, tools, and methods allowing not only to minimize any danger associated with this virus spread but also to stop its span [5]. Ideas about both procedural and therapeutic solutions resulting in the limitation of virus circulation and contamination were encouraged and supported in order to save as much as possible lives and to avoid saturation in the reanimation department in hospitals.

In this framework, our study deals with the development and setting up of a smart portal, useful for public and private institutions to monitor in real-time conditions the appearance of some symptoms in persons likely to be carriers of the virus, such as fever, coughing, and breathing rate. This paper therefore describes one of the multi-criteria techniques for early diagnosis of COVID-19 symptoms.

The proposed supervision portal is a non-invasive, real-time, and efficient system composed of a centralized network of communicating and portable electronic devices equipped with sensors (thermal and RGB cameras and microphones), as shown in Fig. 1. These sensors are in synergy with data mining and artificial intelligence applied to the detection of the main symptoms of the COVID-19. The portal will be placed at the entrance of high-frequented buildings such as universities, companies, factories, and hospitals.

This solution can be used in the future as a general preventive solution, beyond COVID-19, for any others infectious disease which results in respiratory symptoms and fever.

In the current study, the most common signs and symptoms of COVID-19 are overseen to formulate a practicable approach to detect and assess COVID-19's course of infection to counter outbreaks by reducing the transmission rate through early sensing and adopting appropriate measures.

The contributions are summarized as follows:

- Detection of the effective body temperature using low-cost thermal imaging solution based on specific facial landmarks and a transfer learning technic.

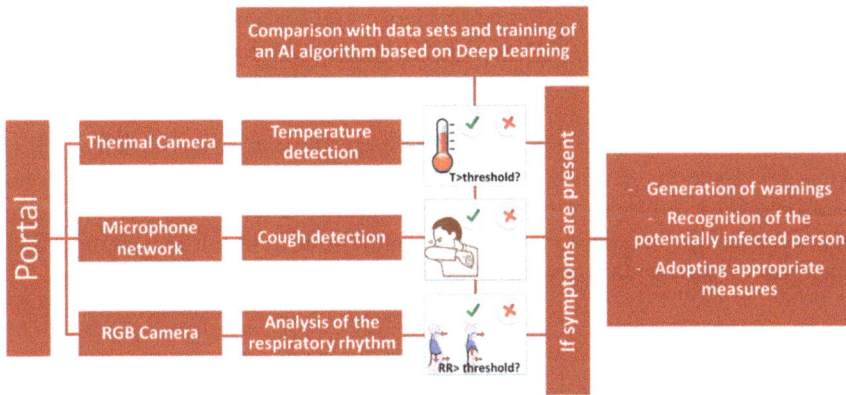

Fig. 1 Overall block diagram of the proposed monitoring system

- COVID-19' cough detection using a network of microphones and a developed deep learning algorithm.
- Analysis of breathing rate by using infrared thermography and exploiting the fact that temperature around the nostrils fluctuates during the respiratory cycles.

In the following, Sect. 2 gives an overview of monitoring systems for the main COVID symptoms (fever, cough, and breathing rate). Section 3 describes the methodology of COVID-19' detection. Experimental results are presented and discussed in Sect. 4. Section 5 summarizes the outcomes of the study.

2   Backgrounds

2.1   Body Temperature Detection

Normal body temperature may differ from a person to another, but it lies within the range of 36.5–37.5 °C. A temperature of 38 °C or higher is considered as fever [6]. One of the most common symptoms of infection by COVID-19 is fever [7], which is considered as a temporary increase in the body temperature. It is also called a high temperature, hyperthermia, or pyrexia, and it is usually a sign that the body is working to keep the healthy from the infection.

Therefore, contactless measurement of body temperature at the entrance of buildings or public places can be very useful for screening people who may be carriers of illnesses such as COVID-19. However, in order to achieve an accurate reading body temperature, finding the exact locations on the human body must be a serious task.

Contactless body temperature measurement techniques work by quantifying the intensity of infrared radiation. This approach is very attractive in the case of infectious disease since the measured body does not have to touch the sensor, which is a thermal camera. In this work, a thermal camera is used to detect the body temperature (HikVision DS-2TD1217B-3/PA model [8]). In general terms, thermal cameras (or infrared cameras) detect temperature by recognizing and capturing different levels of infrared light. Inside of the camera, there are a bunch of tiny measuring devices that capture infrared radiation, called microbolometers, corresponding to the camera pixels [9]. Once the temperature of each pixel is recorded by the corresponding microbolometer, a visual color image is created, by associating a color with the received intensity, in order to facilitate the direct reading of the temperature: Each temperature corresponds to an image color.

2.2 Cough Detection

Cough is the body's way for the clearance of the central airways of inhaled and secreted material that may be present because of many lung ailments [10]. The frequency and intensity of patient's cough have often been used by clinicians and clinical researchers in diagnosis and treatment of many diseases, such as asthma, tuberculosis, and pneumonia [11, 12]. However, such audio signals were detected by using stethoscopes or through manual auscultations at schedule visits. Recently, the automatic cough detection has received an important attention. Motivated by the success of machine learning, in particular, deep learning, several researches have shown promise performance in automatic audio analysis to diagnostic various diseases like asthma [13]. This novel approach would provide some advantages that may complement the issues with the existing biological testing approaches. Indeed, medicine needs to be increased and the artificial intelligent can provide faster and easy to access solutions.

In the COVID-19 context, the artificial intelligence (AI) coughing detection is an opportunity for earlier and rapid diagnosis. It would be an effective tool in early identification for testing, isolation, and contact tracing.

The most common approach for the AI model development dealing with cough detection is to find features of the acoustics signal that make possible to discriminate, with high accuracy, between cough and non-cough sounds, such as speech and laughter [14], and then to identify COVID-19 specific cough from non-COVID ones. Recently, many research groups have focused on recording and diagnosis of coughs. Imrane et al. [15] reported a preliminary study to detect coughs related to COVID-19 collected with smartphones applications, where 48 COVID-19 positive tested patients versus others pathology coughs on which a combination of deep models are trained. The same problem has been investigated by Brown et al. [16], giving a binary COVID-19 prediction model trained on a dataset of crowd-sourcing cough and breathing samples from public members. Carnegie Mellon University [17] and the Cambridge University [18] used web applications to upload population cough sounds along with some additional information such as their demographics and medical history. A group out of École Polytechnique Fédérale de Lausanne has created a web-based application, which distinguish COVID-19 cough from other cough categories such as normal cold and seasonal allergies. All these applications ask users to cough into the microphone on their device. Therefore, the recorded coughs would be voluntary in nature. According to the literature, there are key differences between voluntary and reflex cough events [19]. To address this issue, Han et al. [20] analyzed speech recorded from hospital patients to automatically affected patients. In Ref. [21], a novel coronavirus cough database (NoCoCoDa) is presented that has been collected through online interviews with COVID-19 positive individuals as reflex COVID-19 cough events.

2.3   Breathing Rate

Symptoms of every disease vary depending on the function of the cells that were damaged; in COVID-19, the virus infects the alveoli. Alveoli are the blood/lungs interface at which oxygen passes from the lungs to the blood and carbon dioxide passes from the blood to the lungs. Without functioning alveoli, each respiration does not remove as much carbon dioxide from the blood or provide as much oxygen to the blood as it used to. To make up for the loss of efficiency per breath, the body is forced to undergo more respirations to provide the same oxygen supply [22, 23].

Respiratory rate, reported in respirations (breaths) per minute (rpm), typically ranges from 12 to 20 rpm at rest. Each respiration has two phases: Inhalation and exhalation. During inhalation, oxygen is brought into the lungs from where it is transported throughout the body via the bloodstream, and during exhalation carbon dioxide is eliminated. Respiratory rate is a remarkably stable metric, increased respiratory rate can be considered to be one of the earliest markers of physiological distress particularly in COVID-19 positive cases [24].

3   Methodology

3.1   Body Temperature Detection

In order to measure the body temperature, one solution is to focus the camera on the tear duct point of the face, since it represents the strongest correlation between the internal and external temperature among other body parts [25, 26]. Despite the precise results of this solution, it will generally slow the process since it requires a preprocessing stage to localize the tear duct point on the face. In addition, by using this method, the temperature will be computed for each individual person with a slight stop each time to remove the wearable objects such as glasses, which is not practical especially in the crowded scenarios.

In order to overcome this problem, this paper describes another solution that is based on averaging the temperature over all the face parts. The proposed technique

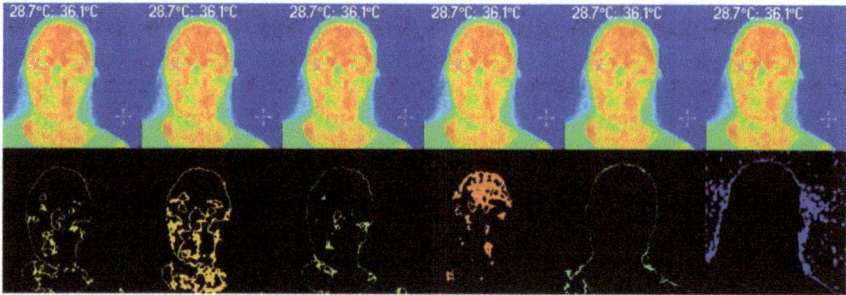

Fig. 2 Some masks made with random colors from the palette (image obtained with the HikVision DS-2TD1217B-3/PA camera)

starts by detecting the human faces that appear in the camera's field of view. For this purpose, the well-known face detection algorithm Haar cascade is applied [27, 28] on the normal RGB image. Then the obtained results are projected on the thermal image. In the subsequent step, for each detected face a color mask is assigned, starting with the color of the highest temperature in the color palette (see Fig. 2). If this color does not exist in the face image, the mask is moved to the color with less degree of temperature. This process is repeated until the maximum temperature of a face is detected. At each iteration, and to avoid false detection of temperature, the number of pixels is checked in the mask (it must be between 0 and 2000 pixels). Once the obtained mask corresponds to predefined condition, the temperature equivalent to the color of the mask is selected as the maximum temperature of the face. At the end of this process, the highest temperature is selected, and the persons corresponding to this temperature are considered as having fever.

3.2 Cough Detection

For the COVID-19 cough detection, a high accuracy classifier based on the mel-frequency cepstral coefficients (MFCC) convolutional neural network (CNN) machine learning architecture, is used.

Data collection:

Two types of dataset are used in this work: cough labeled COVID-19 and not.

- **Audio Set**: This database of audio is taken from multiple YouTube videos events. There are 871 cough events with no further descriptions other than segments being labeled as a cough [29].
- **Freesound**: This is a free access collaborative sounds database recorded by many users. There are 791 cough events [30]. The audio descriptions vary significantly.
- **Coswara and Virufy**: These databases have been developed with COVID-19 cough audio collected from a web and mobile applications. They were combined

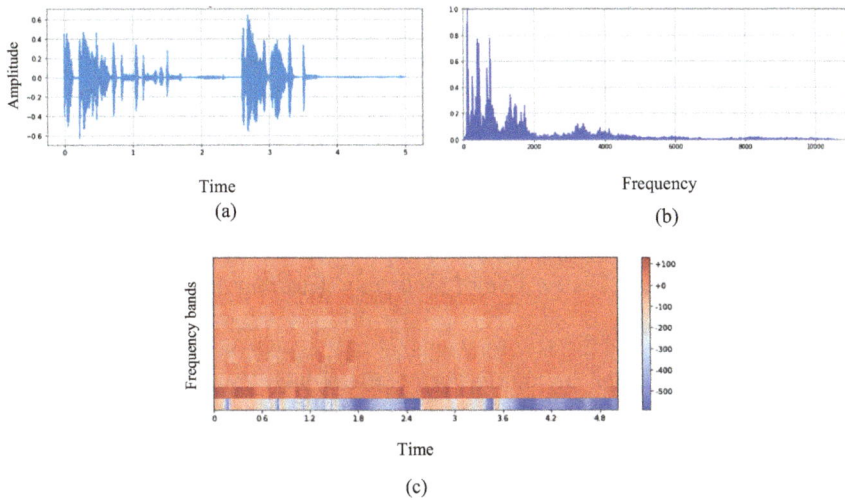

Fig. 3 a Cough signal, b cough spectrum, c cough MFCC

due to the low number of available COVID-19 audio samples (17 total audio samples) [10, 31].

Feature extraction and classifier:

After loading audio data, the raw sound waveform is resampled. Then, sound features are extracted, by using Librosa python library for audio processing. A wide array of features considered by previous researchers [30–32] but the MFCCs are among the most useful in the field of automatic sound detection [33]. The MFCC is obtained from the short-term power spectrum, based on a linear cosine transform of the log power spectrum on a nonlinear mel scale [18], and it converts the audio to a type of spectrogram, which is in a 2D image (see Fig. 3).

Inspired by the recent success of the CNN, an end-to-end CNN model is developed, that ingest MFCC images and directly predicts a binary classification label indicating the presence or not of cough and then pointing out the cough, characteristic of the COVID-19 [34].

3.3   Breathing Rate

In recent years, there has been an increasing demand for unobtrusive and contact-less but also reliable monitoring alternatives of breathing rate (BR). Therefore, new monitoring solutions based on Doppler radar and imaging sensors (visible, mid-wave infrared, and long-wave infrared imaging sensors) have been being proposed and developed [35]. Thermal imaging, also denominated infrared thermography (IRT), emerged as a promising monitoring in a wide range of medical fields [36].

This paper presents a new algorithm to remotely monitor BR by exploiting thermal images, taken from the same infrared camera, used for temperature detection. This approach permits an automatic detection and tracking of the region of interest ROI (around the nostrils and/or mouth areas) in the first frame as well as a precise estimation of the BR.

The proposed approach is based on the fact that temperature around the nostrils fluctuates during the respiratory cycle (inspiration and expiration). Whereas during inhalation cold air from the environment is inspired, during expiration warm air from the lungs is exhaled. IRT is capable of accurately detecting this nasal temperature modulation.

Object or region tracking is a fundamental task in video surveillance, human–computer interaction, and monitoring activities. A large number of methods for tracking have been proposed over the recent years. In this paper, open-source visual tracking based on the Open Computer Vision (OpenCV) library, is used. Existing algorithms are online boosting (OLB), multiple instance learning (MIL), median flow tracker (MFT), tracking-learning-detection (TLD), and kernelized correlation filters (KCF) [37].

In the proposed approach, the MFT is used because it is suitable for very smooth and predictable movements when object is visible throughout the whole sequence [38], which is the case in the video used for our system testing [39]. The first step is the extraction of different frames from the video using a MATLAB developed code. Then, the computational pipeline that has been commonly applied to studies on thermal imaging-based physiological computing is implemented. It consists of three main steps: the ROI selection, automatic ROI tracking, and spatial interpretation as shown in Fig. 4 [40].

The simple averaging of temperatures on the ROI has been used to represent the breathing signal. For each frame, the mean temperature value of the ROI was calculated.

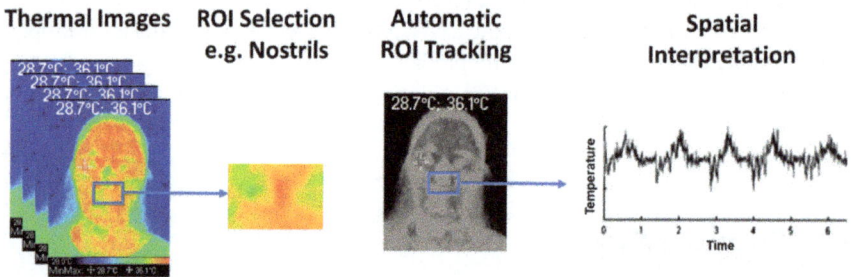

Fig. 4 Computational pipeline commonly applied to studies on thermal imaging-based physiological computing [40]

Table 1 Performance of our approach on the 15 thermal images

Temperature obtained by thermal camera	Temperature obtained by our algorithm	Error
36.3	36.27	−0.03
36.3	36.54	+0.24
36.0	36.17	+0.17
36.3	36.58	+0.28
36.5	36.68	+0.18
36.3	36.47	+0.17
36.0	36.15	+0.15
35.6	35.81	+0.21
37.8	37.92	+0.12
36.1	36.33	+0.23
37.3	37.4	+0.1
36.3	36.46	+0.16
36.0	36.23	+0.23
35.8	36.02	+0.22
Average error		±0.17

4   Results and Discussion

4.1   Body Temperature Detection

In terms of time complexity, the proposed method, using Python implementation, can successfully run in real time. Mainly, the actual implementation takes about 0.04 s to process one frame, which is considered efficient for processing thermal videos with a real-time speed between 25 and 30 frames per second (FPS).

As for the overall performance, the proposed method is evaluated on a dataset of 15 thermal images containing different persons. This has been resulted in an average error of 0.17 °C between the temperature given directly by the thermal camera and the one computed by our approach. Table 1 resumes the obtained results.

4.2   Cough Detection

Our model is trained on more than 800 samples collected from non-labeled COVID-19 dataset (Audio Set and Freesound). It can differentiate between cough sounds and non-cough ones with accuracy of 96.5%. Due to the mall size of Coswara and Virufy datasets and the lack of recognized database for this kind of cough, a continuous progressive work is launched to discriminate the COVID-19 cough. Nevertheless, the model shows good performances with the actual, even small, available database.

Indeed, the MFCC CNN architecture seems to be the higher accuracy audio classifier compared to some other based on support vector machine [41], logistic regression [41], and random forest [10].

The performance and accuracy of the model can be improved by increasing the size of the database. This is possible by collecting cough sounds (labeled and not COVID-19) from new open-source databases [42] or by developing our own database collected from clinical studies and recording.

Currently, some improvements are undertaken to increase of the classifier and feature extractor performances to improve performance of our model. A collaborative work is also initiated with the Hassan 2 Hospital in Morocco to collect audios from COVID-19 patients increasing thus the size of our database and to perform multiple clinical tests to validate the real-time performances of our model.

Finally, our model will not only detect the cough sound but also locate the person who coughs in a crowd-sourcing. The objective is to identify the coughed person and lead them to the test, isolation, and contact tracing, which can stop the spread of the virus. To do this, an image processing of motions will be used that occurs when a person coughs, such as covering the nose and mouth or sneezing into the elbow. This will be the subject of future works.

4.3 Breathing Rate

After frames extraction, the ROI is tracked on successive images. Then, the average temperature is calculated in all obtained images.

With a video rate of 25 frames per second, the BR is about 1 cycle per 2 s (30 rpm). This is slightly higher than the normal values of a healthy moving person (13–20 rpm). The accuracy of the results stills in progress. Indeed, the person on the video moves slightly his head, and therefore, the extracted ROI is not accurate on all frames. The improvement of the tracking process is crucial and currently examined closely.

The thermally expressed shape of the nostril can undergo significant deformations. These deformations result from head movements and breathing dynamics. OpenCV 3 comes with a new tracking API that contains implementations of many single object tracking algorithms (8 different trackers are available in OpenCV 3.4.1 including the MFT). Investigations to find a more accurate algorithm have already started.

A dual-mode imaging system based on visible and long-wave infrared wavelengths can be used. The addition of RGB images will provide more accurate and faster detection and tracking of face and facial tissue than in thermal images [43].

5   Conclusion

Social measures have been efficient to significantly reduce the spread on the COVID-19 pandemic, but not to stop it. In fact, several countries in the world are currently suffering from several wave of the pandemic which is more damaging. Unfortunately, scarcity, cost, and delay clinical testing are significant factors behind the spread of the virus. Motivated by the urgent need, this paper presents an intelligent, reliable, and low-cost system that can remotely detect the main symptoms of COVID-19 disease: fever, cough, and breathing difficulties. This is a multimodal approach where two modalities sound and image together work toward creating a promising solution for screening dense areas.

The first results of this solution are encouraged but it stills some efforts to be done. Indeed, collecting more data mainly from open-sources COVID-19 databases and clinical recording is in progress. This will allow analysis of a larger dataset, with more advanced AI model. It has also been shown there is age difference in temperature, cough, and breathing. The future work could focus on adapting the model to differentiate age groups.

Our work will be extended to other approaches, such as image processing for cough detection from movements and facial recognition with and without masks. These approaches will make possible to identify suspect persons and then lead them to clinical test, isolation, and contact tracing.

The relevant exploitation of the synergies between the different sources, on the one hand, and of the volume and variety of available data, on the other hand, requires the implementation of data fusion methods able to handle the specificities of multi-sensor systems.

A first prototype will be installed at the entrance of the University Sidi Mohammed Ben Abdellah (USMBA) at the faculty of Sciences and Technology of FEZ.

Finally, this solution is not meant to compete with clinical testing. Instead, it offers a suitable tool to operational needs requiring complete, timely, efficient, cost-effective, and reactive diagnosis. Most importantly, it will give a safe monitoring, suspects tracing and thus controlling the spread of the pandemic.

For those who can be interested by this project, a call for collaborative and innovative idea is launched here, all over the world.

Acknowledgements This work is financially supported by the Moroccan Ministry of National Education, Professional Formation, Higher Education and Scientific Research, the National Center for Scientific and Technical Research of Morocco, and the University Sidi Mohamed Ben Abdellah of Fez, we warmly thank them for this opportunity.

Also, we would like to thank our colleagues thank Pr. H. BEJJIT, Pr. C. Benjelloun, Pr. H. Saikouk, Pr. C. AlaouI, Pr. M. Ouazzani Jamil and Pr. A. Lakhssassi and our students Nour Meyazi, Ismail Laissaoui, Houria El Ansari, Imane Zakri, Taha Jadid, Youssra Derraz, Mehdi Samouh, Akram El Hachimi, Youssef El Kantri, Anas Mansouri, Ghita Ouazzani Taybi, Khalid Aoujdad, Bahija Tantaoui, Hamza Amraoui, Imane El Amri, Hafizah Aboubacar Attaou, Nezha Elbourkhissi, and Zakaria Mnah for their fruitful contributions and discussion in the framework of this project.

References

1. Hashmi HAS, Asif HM (2020) Early detection and assessment of covid-19. Front Med 7:311. https://doi.org/10.3389/fmed.2020.00311
2. Chakkor S, Baghouri M, Cheker Z, el Oualkadi A, el Hangouche JA, Laamech J (2020) Intelligent network for proactive detection of COVID-19 disease. In: 2020 6th IEEE congress on information science and technology (CiSt), IEEE, Agadir - Essaouira, Morocco, pp 472–478. https://doi.org/10.1109/CiSt49399.2021.9357181
3. Coronavirus disease (COVID-19) pandemic. https://www.who.int/emergencies/diseases/novel-coronavirus-2019?gclid=EAIaIQobChMIyfrYpOjB8AIVkbrtCh1jqQddEAAYASAAEgI7gvD_BwE. Last accessed 2021/05/11
4. Mehrdad S, Wang Y, Atashzar SF (2021) Perspective: wearable internet of medical things for remote tracking of symptoms, prediction of health anomalies, implementation of preventative measures, and control of virus spread during the era of COVID-19. Front Robot AI 8:610653. https://doi.org/10.3389/frobt.2021.610653
5. Scientific and Technological Research Support Program in connection with "Covid-19". https://www.cnrst.ma/index.php/fr/component/k2/item/433-programme-de-soutien-a-la-recherche-scientifique-e-t-technologique-en-lien-avec-le-covid-19. Last accessed 2021/05/11
6. Sund-Levander M, Forsberg C, Wahren LK (2002) Normal oral, rectal, tympanic and axillary body temperature in adult men and women: a systematic literature review. Scand J Caring Sci 16(2):122–128. https://doi.org/10.1046/j.1471-6712.2002.00069.x. PMID: 12000664
7. Wang C, Horby PW, Hayden FG, Gao GF (2020) A novel coronavirus outbreak of global health concern. Lancet 395(10223):15–21, 470–473
8. Temperature Screening Thermographic Turret Camera: https://www.hikvision.com/fr/products/Thermal-Products/Thermography-thermal-cameras/temperature-screening-series/ds-2td1217b-3-pa/. Last accessed 2021/05/13
9. Lee AW, Hu Q (2005) Real-time, continuous-wave terahertz imaging by use of a microbolometer focal-plane array. Opt Lett 30:2563–2565
10. Sharma N, Krishnan P, Kumar R, Ramoji S, Chetupalli SR, Ghosh PK, Ganapathy S (2020) Coswara—a database of breathing, cough, and voice sounds for COVID-19 diagnosis. Interspeech 4811–4815. https://doi.org/10.21437/Interspeech.2020-2768
11. Li S-H, Lin B-S, Tsai C-H, Yang C-T, Lin B-S (2017) Design of wearable breathing sound monitoring system for real-time wheeze detection. Sensors 17:171. https://doi.org/10.3390/s17010171
12. Oletic D, Bilas V (2016) Energy-efficient respiratory sounds sensing for personal mobile asthma monitoring. IEEE Sens J 16(23):8295–8303
13. Toop LJ, Thorpe CW, Fright R (1989) Cough sound analysis: a new tool for the diagnosis of asthma. Fam Pract 6:83–85
14. Deshpande G, Schuller BW (2020) An overview on audio, signal, speech, & language processing for COVID-19. arXiv:2005.08579v1 [cs.CY] 5 pages
15. Imran A, Posokhova I, Qureshi HN, Masood U, Riaz MS, Ali K, John CN, Hussain MI, Nabeel M (2020) AI4COVID-19: AI enabled preliminary diagnosis for COVID-19 from cough samples via an app. Inform Med Unlocked 20:100378. https://doi.org/10.1016/j.imu.2020.100378
16. Brown C, Chauhan J, Grammenos A, Han J, Hasthanasombat A, Spathis D, Xia T, Cicuta P, Mascolo C (2020) Exploring automatic diagnosis of COVID-19 from crowdsourced respiratory sound data. In: Proceedings of the 26th ACM SIGKDD international conference on knowledge discovery & data mining. Presented at the KDD'20: The 26th ACM SIGKDD conference on knowledge discovery and data mining, ACM, Virtual Event CA USA, pp 3474–3484. https://doi.org/10.1145/3394486.3412865
17. Carnegie Mellon University Home Page: https://cvd.lti.cmu.edu/. Last accessed 2021/04/27
18. COVID-19 Sounds App, University of Cambridge. https://www.covid-19-sounds.org/en/. Last accessed 2021/04/27

19. Magni C, Chellini E, Lavorini F, Fontana GA, Widdicombe J (2011) Voluntary and reflex cough: similarities and differences. Pulm Pharmacol Ther 24(3):308–311. https://doi.org/10.1016/j.pupt.2011.01.007
20. Han J, Qian K, Song M, Yang Z, Ren Z, Liu S, Liu J, Zheng H, Ji W, Koike T, Li X, Zhang Z, Yama-moto Y, Schuller B (2020) An early study on intelligent analysis of speech under COVID-19: severity, sleep quality, fatigue, and anxiety. arXiv:2005.00096 [eess.AS] 5 pages
21. Cohen-McFarlane M, IEEE Student Member, Goubran R, IEEE Fellow, and Knoefe F (2020) Novel coronavirus (2019) cough database: NoCoCoDa. IEEE ACESS 8:154087–154094. https://doi.org/10.1109/ACCESS.2020.3018028
22. The Importance of Respiratory Rate Tracking During The COVID-19 Pandemic. https://www.whoop.com/thelocker/respiratory-rate-tracking-coronavirus/. Last accessed 2021/04/27
23. Miller DJ, Capodilupo JV, Lastella M, Sargent C, Roach GD, Lee VH, et al (2020) Analyzing changes in respiratory rate to predict the risk of COVID-19 infection. PLoS ONE 15(12):e0243693. https://doi.org/10.1371/journal.pone.0243693
24. Kranthi KL, Alphonse PJA (2021) A literature review on COVID-19 disease diagnosis from respiratory sound data. AIMS Bioeng 8(2):140–153. https://doi.org/10.3934/bioeng.2021013
25. Ghahramani G, Castro G, Karvigh SA, Becerik-Gerber B (2018) Towards unsupervised learning of thermal comfort using infrared thermography. Appl Energy 211:41–49. ISSN: 0306-2619. https://doi.org/10.1016/j.apenergy.2017.11.021
26. Tan JH, Ng EYK, Rajendra Acharya U, Chee C (2009) Infrared thermography on ocular surface temperature: a review. Infrared Phys Technol 52(4):97–108. ISSN: 1350-4495. https://doi.org/10.1016/j.infrared.2009.05.002
27. Viola P, Jones M (2001) Rapid object detection using a boosted cascade of simple features
28. Viola P, Jones MJ (2004) Robust real-time face detection. Int J Comput Vision 57(2):137–154
29. Gemmeke JF, Ellis DPW, Freedman D, Jansen A, Lawrence W, Moore RC, Plakal M, Ritter M (2017) Audio set: an ontology and human-labeled dataset for audio events. In: Proceedings of IEEE ICASSP 2017. New Orleans, LA
30. Freesound—Sounds browse, n.d. https://freesound.org/browse/. Last accessed 09 May 2021
31. Virufy T (n.d.) Home|Virufy. https://virufy.org//en/ https://freesound.org/browse/, last accessed 09 May 2021
32. Matos S, Member S, Birring SS, Pavord ID, Evans DH, Member S (2006) Detection of cough signals in continuous audio recordings using hidden Markov models. IEEE Trans Biomed Eng 53(6):1078–1083
33. Drugman T, Urbain J, Dutoit T (2011) Assessment of audio features for automatic cough detection. In: Proceedings of 19th European signal processing conference, no. 32
34. Larson EC, Lee T, Liu S, Rosenfeld M, Patel SN (2011) Accurate and privacy preserving cough sensing using a low-cost microphone. In: Proceedings of 13th international conference on ubiquitous computing, p 375
35. Pereira C, Yu X, Czaplik M, Rossaint R, Blazek V, Leonhardt S (2015) Remote monitoring of breathing dynamics using infrared thermography. Biomed Opt Express 6:4378–4394
36. Ruminski J, Kwasniewska A (2017) Evaluation of respiration rate using thermal imaging in mobile conditions. In: Ng EY, Etehadtavakol M (eds) Application of infrared to biomedical sciences. Series in bioengineering. Springer, Singapore, pp 311–346. https://doi.org/10.1007/978-981-10-3147-2_18
37. Lehtola V, Huttunen H, Christophe F, Mikkonen T (2017) Evaluation of visual tracking algorithms for embedded devices. In: Scandinavian conference on image analysis, vol 10269. Lecture notes in computer science. Springer, pp 88–97. https://doi.org/10.1007/978-3-319-59126-1_8
38. TrackerMedianFlow Class Reference on Open-Source Computer Vision Homepage, https://docs.opencv.org/3.4/d7/d86/classcv_1_1TrackerMedianFlow.html. Last accessed 2021/04/28
39. Video used for test: Thermal Body Temp Monitoring Solution—Dahua. https://youtu.be/VJy2869i_K8. Last accessed 2021/04/27
40. Cho Y, Bianchi-Berthouze N (2019) Physiological and affective computing through thermal imaging: a survey. arXiv e-prints

41. Chatrzarrin H, Arcelus A, Goubran R, Knoefel F (2011) Feature extraction for the differentia-tion of dry and wet cough sounds. In: IEEE international symposium on medical measurements and applications. IEEE
42. Hershey S, Chaudhuri S, Ellis DPW, Gemmeke JF, Jansen A, Moore RC, Plakal M, Platt D, Saurous RA, Seybold B, Slaney M, Weiss RJ, Wilson K (2016) CNN architectures for large-scale audio classification
43. Hu MH, Zhai GT, Li D, Fan YZ, Chen XH, Yang XK (2017) Synergetic use of thermal and visible imaging techniques for contactless and unobtrusive breathing measurement. J Biomed Opt 22(3):36006. https://doi.org/10.1117/1.JBO.22.3.036006. PMID: 28264083

Chapter 12
A Comparative Study of Deep Learning Models for COVID-19 Diagnosis Based on X-Ray Images

Shah Siddiqui[ID]**, Elias Hossain**[ID]**, Rezowan Ferdous**[ID]**, Murshedul Arifeen**[ID]**, Wahidur Rahman**[ID]**, Shamsul Masum**[ID]**, Adrian Hopgood**[ID]**, Alice Good**[ID]**, and Alexander Gegov**[ID]

Abstract Background: The rise of COVID-19 has caused immeasurable loss to public health globally. The world has faced a severe shortage of the gold standard testing kit known as reverse transcription-polymerase chain reaction (RT-PCR). The accuracy of RT-PCR is not 100%, and it takes a few hours to deliver the test results. An additional testing solution to RT-PCR would be beneficial. Deep learning's superiority in image processing is characterised as the most effective COVID-19 diagnosis based on images. The small number of COVID-19 X-ray images in existing deep learning methods for COVID-19 diagnosis may degrade the performance of deep learning methods for new sets of images. Our priority for this research is to test and compare different deep learning algorithms on a dataset consisting of many COVID-19 X-ray images. **Methods**: We have merged the publicly available image data into two groups (COVID and Normal). Our dataset contains 579 COVID-19 cases and 1773 Normal cases of X-ray images. We have used 145 COVID-19 cases and 150 Normal cases to test the deep learning models. Deep learning models based on CNN, VGG16 and 19, and InceptionV3 have been considered for prediction. The performance of these models is compared based on measurements of accuracy, sensitivity, and specificity. In the deep learning models, the SoftMax activation function is used along with the Adam optimiser and categorical cross-entropy loss. A customised hybrid CNN model found in literature is considered and compared to explore how the inclusion of many COVID-19 X-ray images could impact the model's performance.

S. Siddiqui · S. Masum · A. Hopgood (✉) · A. Good · A. Gegov
The University of Portsmouth (UoP), School of Computing, Faculty of Technology, Lion Terrace, Portsmouth, Portsmouth PO1 3HE, UK
e-mail: adrian.hopgood@port.ac.uk

S. Siddiqui
e-mail: shah.siddiqui@timerni.com

S. Siddiqui · E. Hossain · R. Ferdous · M. Arifeen · W. Rahman
Time research and innovation (Tri), 189 Foundry Lane, Southampton, Southampton SO15 3JZ, United Kingdom

336/7, TV Road East Rampura, Khilgaon Dhaka 1219, Bangladesh

Results: The accuracy of the considered deep learning models using InceptionV3, VGG16, and VGG19 algorithms achieved 50%, 90%, and 83%, respectively, in predicting the X-ray images of COVID-19. We have shown that number of COVID-19 X-ray images does have a significant impact on the model's performance. A customised hybrid CNN model found in the literature failed to perform well on a dataset consisting of a large number of COVID-19 X-ray images. The customised hybrid CNN model reached an accuracy of 71% on many COVID-19 X-ray images. In contrast, it achieved 98% accuracy on a small number of COVID-19 X-ray images. It is also observed from the experiments that the VGG16 performs well with an increased number of images. **Conclusions**: A maximised number of COVID-19 X-ray images should be considered in building a deep learning model. The deep learning model with VGG16 performs the best in predicting from the X-ray images.

Keywords Coronavirus (COVID-19) · RT-PCR · Machine Learning (ML) · Deep Learning (DL) · X-ray images

1 Introduction

The COVID-19 pandemic is caused by the novel coronavirus known as severe acute respiratory syndrome coronavirus (SARS-COV-2) found in Wuhan city, China, at the end of 2019 [21]. SARS-COV-2 and other viruses from the corona family known as MERS-COV 2 are responsible for causing respiratory disease in humans. The death and the transmission rate by this virus are very high, and it can survive a few hours to few days in the environment. The primary symptoms of COVID-19 are fever, cough, headache, muscle pain, and shortness of breath [13, 17, 24]. It spread so fast from Wuhan, China, to the whole world that it was declared as a global pandemic [18, 21]. Around 2.9 million people had died due to covid infection by 8 April 2021 [6]. On the contrary, some countries are now facing the third wave, and the virus is changing its variant and spreading so fast that it is hard to imagine what will happen shortly [5]. Therefore, the early detection of a corona virus-infected person is of great importance to slow down the spread and death [11, 17]. The gold standard diagnostic technique for COVID-19 is the reverse transcription-polymerase chain reaction (RT-PCR) [7]. However, RT-PCR is not fast enough for the diagnosis of COVID-19 because RT-PCR takes about 4–6 hours to provide an outcome which is time-consuming. Also, the shortage of RT-PCR kits creates another challenge [11, 12]. Furthermore, RT-PCR's implementation and standardisation were hampered in many countries due to the cost, availability, and technology. Therefore, many lower- and middle-income African, Asian, and Latin American countries have failed to implement RT-PCR testing at the beginning of the pandemic. It was also confirmed that in China, the CT chest findings' sensitivities were as high as 98% compared to 71% for RT-PCR [1, 8].

Many deep learning models are currently being used experimentally to detect COVID-19 from X-ray and CT imaging [6, 17, 26]. In our previous study, "Deep

Learning models for the diagnosis and screening of COVID-19: A systematic review (accepted)", in Google Scholar and PubMed, we found a total number of 188 titles in September 2020, whereas a recent search during 12 April 2021 with the exact keywords have found 646 titles. Through our analysis, we have observed that X-ray mainly was used for the theoretical implementation of deep learning as CT images pose several challenges in detecting COVID-infected region using deep learning algorithms. The challenges imposed by CT images are less applicability, slower image acquisition, high cost, and limited sensitivity [24]. On the contrary, chest X-ray-based ML models for COVID-19 diagnosis have become the popular early detection technique recent days because of its various salient features like it can be used in emergencies using portable device [25], cheaper, faster, and reliable method [24]. An automated, faster, and reliable COVID-19 diagnosis method algorithm with better performance can save the RT-PCR kits [20]. Therefore, there is increasing interest in deep learning models in many publications in the last several months.

We have also observed the cost and availability between the RT-PCR and X-ray to diagnose COVID-19. X-ray imaging is painless and cheap compared to RT-PCR [15, 30]. In the UK, we have noticed that the lowest cost of an X-ray is £, but the maximum cost is £ compared to the PCR test which is £ [29]. However, RT-PCR kits are widely available in high-income countries, whereas in Bangladesh, an X-ray cost is from 450 BDT to 1200 BDT, but RT-PCR is 3000 BDT [10, 16]. In India, X-ray images' cost is a minimum of RS183, and the maximum is RS 1370. On the other hand, RT-PCR has a RS 980 to RS 1800 [15, 30]. X-ray images in South Africa cost approximately R2500. RT-PCR, on the other hand, has a minimum price of R1150 [9, 27]. Furthermore, many low- and middle-income countries (LMICs) struggle to cope with the shortages of the RT-PCR testing kits and the technology, and we need an alternative testing solution to RT-PCR [14].

Researchers are trying to investigate the power of AI or ML techniques in diagnosing COVID-19 based on medical imaging and are restructuring the system to alleviate this problem. Therefore, deep learning with X-ray images is gaining popularity to use available data and technology [3]. But the main concern with most of these ML models is that they are trained and tested with smaller size dataset, and practical implementation is limited. Therefore, considering all the above scenarios, we have decided to progress our research to inspect some of the published literature and models based on the chest X-ray image.

In healthcare services like ML-based automated diagnosis systems, performance depends on the reasonable amount of dataset used during the training phase. Generally, a large dataset enhances the classification performance of the predictive model. Still, a small dataset leads to an overfitting problem [2]. More specifically, CNN-based ML models require a large dataset to work correctly [4]. Since most of the models' design for X-ray-based COVID-19 diagnosis includes the CNN model, it is a significant issue to analyse their performance in various dataset sizes.

To overcome this issue related to the dataset, we have collected the dataset used in various papers by emailing the corresponding authors. Combining all the datasets, we have prepared a dataset of 2352 images containing 1773 standard images and 579 COVID images. In particular, we have taken a hybrid deep learning model proposed in [17] and reimplemented it with the paper's dataset for training and testing. The proposed model used DarkNet to detect and classify COVID-19 patients as binary classification and multiclass classification using CXR images. The dataset used for this model includes 125 COVID-19 cases, 500 normal and 500 pneumonia cases. COVID sample dataset is very low in number for training the deep model. We also tested the model based on our prepared dataset. We have also considered VGG16, VGG19, and InceptionV3 algorithms and used our training and testing dataset. Furthermore, we have compared these models' performance in terms of performance metrics like accuracy, sensitivity, and specificity.

This study is divided into four main sections; in Sect. 2, we have described our methodology's detailed procedure, including data preparation, preprocessing, model training, and testing. In Sect. 3, we discussed the results found from the experiment and compared the results. Finally, Sect. 4 concludes this paper.

2 Methodology

2.1 Dataset Preparation

We have considered all the papers included in our previous systematic review work. We have emailed all the authors to request their dataset and received seven datasets in response. Three of the datasets are used here (of the others, one was CT images and three were corrupted). The datasets for this study are represented in the below table. The dataset contains 2352 images, consisting of 1773 normal images and 579 COVID images. The training dataset consists of 2057 images, which contains 1623 normal images and 434 COVID images. In contrast, the test dataset consists of 295 images containing 150 normal images and 145 COVID images (Table 1).

Table 1 Dataset considered for this study

Ref	COVID	Non-COVID	Pneumonia	MERS	SARS	Streptococcus	Varicella	Augmented
Toğaçar et al. [26]	271	65	98	×	×	×	×	×
Ozturk et al. [17]	125	500	500	×	×	×	×	×
Islam et al. [11]	183	208	1525	×	×	×	×	912
Pereira et al. [19]	0	1000	11	10	11	12	10	×

2.2 Data Preprocessing

Data preprocessing has been carried out in several steps. Firstly, the image sizes were reduced and image augmentation was performed. Since the images in the training sample were of different sizes, they had to be resized before being used as inputs to the algorithm. Square images were resized to 256×256 pixels in resolution. Rectangular images were resized to 256 pixels on their shortest line, and then the image's middle 256×256 square was cropped. Image data training augmentation was used to have 224×224 images.

2.3 Models

The models that are considered for comparison are VGG16 and VGG19 [22]. VGG16 and 19 are the variants of the VGG machine learning model, which focuses on the CNN's depth feature. This model consists of the input layer, hidden layer, convolutional layer, fully connected layer. Besides VGG, we also considered the InceptionV3 [23] model, whose underlying architecture is also CNN. This version improves several features like label smoothing. Finally, the customised CNN model was proposed in [17] to detect COVID-19 patients from chest X-ray images. We have reimplemented this model and compared it with the models mentioned above.

2.4 Model Training and Testing

The environment for implementing our models comprises Intel Core i9-10885H (8 Core, 16MB Cache, 2.40–5.30 GHz, 45W, vPro), 32GB RAM, NVIDIA Quadro RTX 5000 w/16GB GDDR6. We used the train and test dataset mentioned in Sect. 2.1 for training and testing our models. The loss function used was categorical cross-entropy loss with Adam optimiser. The target size of the image and batch size was 224×224 and 32, respectively. Finally, the epoch size was taken as 50 for each of the models.

2.5 Performance Metrics

We have considered three metrics to evaluate and compare the models that we have considered: accuracy, sensitivity, and specificity. They are defined as follows in terms of true positive (TP), true negative (TN), false positive (FP), and false negative (FN).

$$\text{Accuracy} = \frac{TP + TN}{TP + TN + FP + FN}$$

The accuracy of the model is the proportion of the total dataset that is classified correctly.

$$\text{Sensitivity} = \frac{TP}{TP + FN}$$

The sensitivity defines the model's ability to generate a correct positive result for people with COVID-19.

$$\text{Specificity} = \frac{TN}{TN + FP}$$

The specificity defines the model's ability to generate a correct negative result for people who do not have COVID-19.

3 Results and Discussion

We have considered various datasets initially to inspect the models. First, we have run all the models with the smaller dataset to see the performance and actual data. For this review, we present only the results of the final datasets to understand the performance.

Table 2 compares the models' image quantity, accuracy, sensitivity, and specificity. We have compared and evaluated the models using these metrics, where VGG16 performed well in all three categories compared to all other models. VGG16 gained 90% accuracy compared to 83% for VGG19, 71% for Customised CNN [2] and 50% InceptionV3. VGG16 achieved 90.90% in the sensitivity parameters, whereas VGG19 achieved 76.92%, InceptionV3 and Customised CNN [2] achieved 69.93% and 70.71% respectively. VGG16 achieved 89.04% for specificity parameters, whereas VGG19 achieved 88.43%, Customised CNN [2] achieved 70.32%, and InceptionV3 achieved 28.57%.

Table 3 compares the models' performance using our dataset and the actual dataset of Ozturk et al. [17], showing how the performance varies with the image quantity. The Customised CNN achieved 98% accuracy, where it contains only 125 COVID images. With our dataset of 579 COVID images, the Customised CNN accuracy was

Table 2 Performance comparison

Models	Image quantity	Accuracy (%)	Sensitivity (%)	Specificity (%)
InceptionV3	2352	50	69.93	28.57
VGG16	2352	90	90.90	89.04
VGG19	2352	83	76.92	88.43
Customised CNN [17]	2352	71	70.71	70.32

Table 3 Comparison based on image quantity

Models	Image quantity	Accuracy (%)	Sensitivity (%)	Specificity (%)
Customised CNN [17]	1127	98.08	95.13	95.30
VGG16	1127	98	62.50	100
Customised CNN [17]	2352	71	70.71	70.32
VGG16	2352	90	90.90	89.04

71%. The model with the VGG16 algorithm shows the same performance with our dataset as the Ozturk et al. dataset. The model with the VGG16 algorithm failed to match the sensitivity and specificity on Ozturk et al. compared with the 125 COVID images. However, the model with the VGG16 algorithm gained the correct balance between sensitivity and specificity with our dataset, which contains 579 COVID images.

Figure 1a, b shows the accuracy and loss curve of VGG16, respectively, where the training and test data almost overlap each other, showing the satisfactory performance of this model. Figure 1c, d shows the accuracy and loss graph of VGG19, respectively, where the training and test data are separated from each other, showing worse performance compared with VGG16. However, InceptionV3 performed very poorly compared with both VGG16 and VGG19. Figure 1e, f shows the accuracy and loss graph of InceptionV3, where the training and test data are widely separated.

Figure 2a, b, and c shows the confusion matrix for VGG16, VGG19, and InceptionV3. VGG16 wrongly predicted 13 COVID cases and 16 normal cases. In contrast, VGG19 wrongly predicted 33 COVID cases and 17 normal cases. Finally, InceptionV3 wrongly predicted 43 COVID cases and 106 normal cases. Therefore, the performance of VGG16 is the best of the three.

We have also compared the customised CNN [17] model with our dataset. Figure 3a shows the confusion matrix of their dataset, which can be compared with our dataset in Fig. 3b. In the first run (a), 6 COVID cases and 1 normal case were wrongly predicted. In contrast, in the second run (b), 46 COVID cases and 41 normal cases were wrongly predicted. Therefore, this CNN model is not performing adequately when we ran an extensive dataset. We present some of the alternative models and their performance in this section. Table 4 represents the selected models with different datasets. We have observed that the proposed model by Toğaçar et al. [26] is a MobileNetV2 and SqueezeNet-based COVID-19 diagnosis method based on X-ray images. The experimental phase of the proposed model shows a dataset that includes only 295 images.

Rajaraman et al. [21] proposed an iteratively pruned deep learning model for COVID-19 diagnosis. Though the dataset contains 6761 normal, 5412 pneumonia, and 2538 bacterial images, the COVID cases were only 268. Islam et al. [11] used the dataset of 613 COVID-19 cases, 1525 pneumonia, and 1525 normal cases of X-ray images for training and testing their proposed combined CNN and LSTM-based deep learning model COVID-19 diagnosis. Although the dataset for that study con-

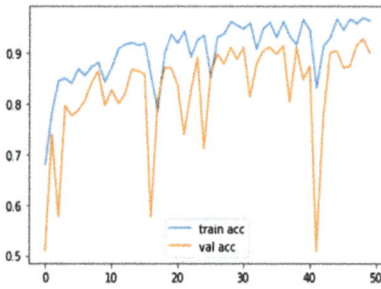

(a) VGG16 Accuracy

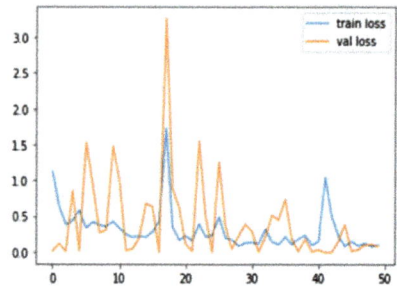

(b) VGG16 Loss

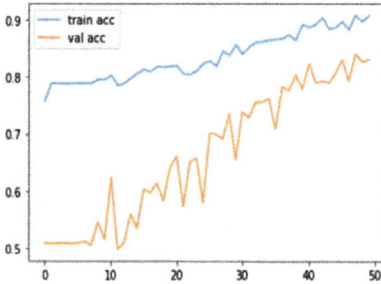

(c) VGG19 Accuracy

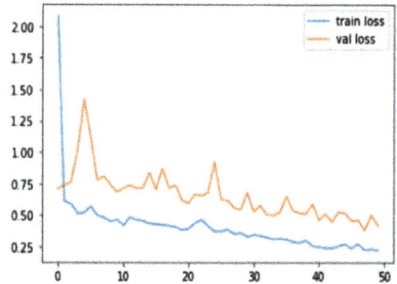

(d) VGG19 Loss

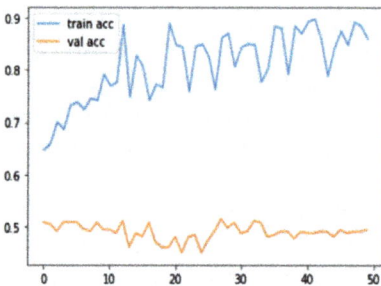

(e) InceptionV3 Accuracy

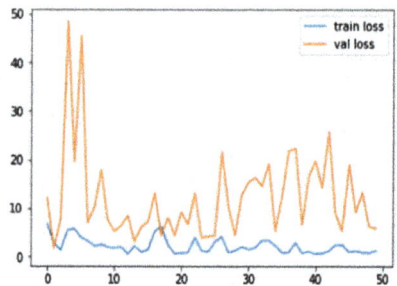

(f) InceptionV3 Loss

Fig. 1 Training loss, validation loss, training accuracy, and validation accuracy for VGG16, VGG19, and InceptionV3

tains more COVID images than our dataset, we could not confirm their performance with any of our experiments as we did not receive either their model or dataset to make a comparison. Ucar and Korkmaz [28] considered two large datasets for their experimental purpose but unfortunately, among 5949 images of one dataset, only 76 COVID cases were present. The other dataset comprises only 45 COVID cases. Panwar et al. [18] used a small dataset of 337 total images and 192 COVID X-ray cases. However, the authors performed image augmentation to increase the number

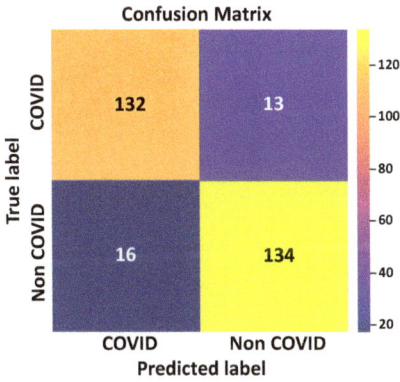

(a) VGG16 Confusion matrix

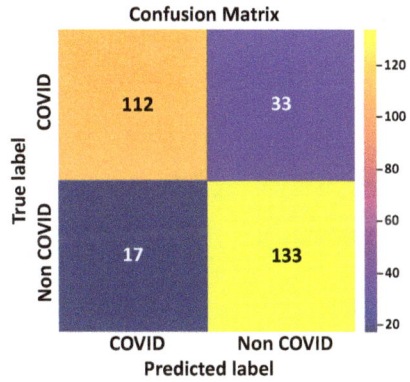

(b) VGG19 Confusion matrix

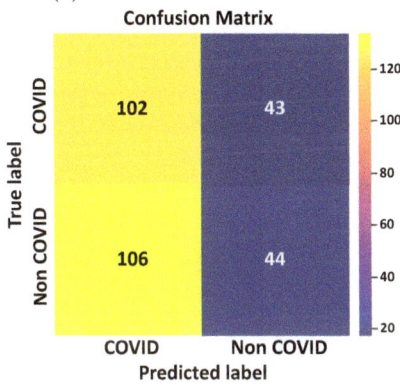

(c) InceptionV3 Confusion matrix

Fig. 2 Confusion matrices of the experimental models

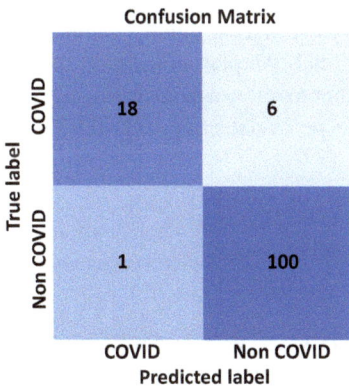

(a) Confusion matrix Ozturk et al. (2020)

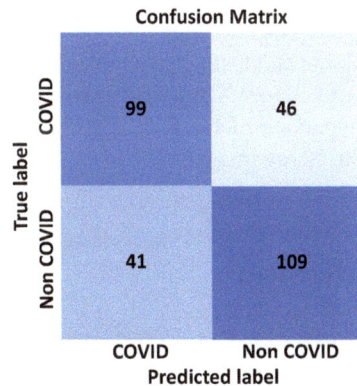

(b) Confusion matrix for our dataset

Fig. 3 Confusion matrices of customised CNN models

Table 4 Selected models with different datasets

Model name	Image quantity	COVID images	Accuracy (%)	Sensitivity (%)	Specificity (%)
VDSNet [26]	348	295	70.80	64	62
Pruned [21]	14,979	268	99	99	99
CNN+LSTM [11]	3363	613	99.20	99.30	99.20
COVIDiagnosis-Net [28]	5949	76	98.3	98.2	99.1
Panwar, Gupta [18]	529	192	97	97.62	78.57

of images. All the above studies have used limited datasets for their models, so their training and testing ML models may not perform well for a larger COVID dataset.

We have identified some high accuracy, sensitivity, and specificity models with the corresponding results compared with other papers. We can see that CNN+LSTM, pruned deep learning model, and COVIDiagnosis-Net achieved 99% results. However, the number of samples of COVID images considered for training and testing is small in size. If those models are trained and tested with a larger dataset, they might also show poor performance.

4 Conclusions and Future Work

This study has prepared a large dataset containing COVID-19 chest X-ray images and non-COVID-19 X-ray images, which has been achieved by combining publicly available data found in the existing literature. Comparative analysis of algorithms suggests that the model with VGG16 algorithms performs best and outperforms others in predicting COVID-19 from X-ray images. This study indicates that the number of COVID-19 X-ray images in the dataset plays a vital role in the model's performance. The maximum number of COVID-19 X-ray images should be considered for training and testing a reliable deep learning model. We plan to increase the number of COVID-19 X-ray images in our dataset further to allow a deeper investigation and comparison of the deep learning methods' scope in detecting COVID-19 patients from X-ray images.

References

1. Ai T, Yang Z, Hou H, Zhan C, Chen C, Lv W, Tao Q, Sun Z, Xia L (2020) Correlation of chest ct and rt-pcr testing for coronavirus disease 2019 (covid-19) in China: a report of 1014 cases. Radiology 296(2):E32–E40
2. Althnian A, AlSaeed D, Al-Baity H, Samha A, Dris AB, Alzakari N, Abou Elwafa A, Kurdi H (2021) Impact of dataset size on classification performance: an empirical evaluation in the medical domain. Appl Sci 11(2):796

3. Bar Y, Diamant I, Wolf L, Greenspan H (2015) Deep learning with non-medical training used for chest pathology identification. In: Medical imaging 2015: computer-aided diagnosis. vol 9414, p 94140V. International Society for Optics and Photonics

4. Barbedo JGA (2018) Impact of dataset size and variety on the effectiveness of deep learning and transfer learning for plant disease classification. Comput Electron Agric 153:46–53

5. BBC: Coronavirus: cases of new variant appear worldwide (2020). https://www.bbc.com/news/world-europe-55452262. Accessed 09 May 2021

6. Elflein J(2021) Coronavirus deaths worldwide by country|Statista (2021). https://www.statista.com/statistics/1093256/novel-coronavirus-2019ncov-deaths-worldwide-by-country/. Accessed 09 May 2021

7. Fan DP, Zhou T, Ji GP, Zhou Y, Chen G, Fu H, Shen J, Shao L (2020) Inf-net: Automatic covid-19 lung infection segmentation from ct images. IEEE Trans Med Imaging 39(8):2626–2637

8. Fields BK, Demirjian NL, Gholamrezanezhad A (2020) Coronavirus disease 2019 (covid-19) diagnostic technologies: a country-based retrospective analysis of screening and containment procedures during the first wave of the pandemic. Clin imaging 67:219–225

9. Govùk: tuberculosis testing in South Africa—GOV.UK. https://www.gov.uk/government/publications/tuberculosis-test-for-a-uk-visa-clinics-in-south-africa/tuberculosis-testing-in-south-africa. Accessed 09 May 2021

10. Icddrb: icddr,b lab services: test details (2021). http://labservices.icddrb.org/test-details/529?, Covid-19 Real Time RT PCR. Accessed 09 May 2021

11. Islam MZ, Islam MM, Asraf A (2020) A combined deep CNN-LSTM network for the detection of novel coronavirus (covid-19) using x-ray images. Inform Med Unlocked 20:100412

12. Jawerth N (2020) How is the covid-19 virus detected using real time RT-PCR. International Atomic Energy Agency. Vienna International Centre, PO Box 100

13. Khanna RC, Cicinelli MV, Gilbert SS, Honavar SG, Murthy GV (2020) Covid-19 pandemic: lessons learned and future directions. Indian J Ophthalmol 68(5):703

14. Peplow M (2020) Special to C&EN: developing nations face COVID-19 diagnostic challenges. C&EN Global Enterprise, vol 98, no 27, pp 25–27. https://doi.org/10.1021/cen-09827-feature2

15. Medifee: X-Ray Cost (2021). https://www.medifee.com/tests/x-ray-cost/. Accessed 09 May 2021

16. NHFB: National heart foundation of bangladesh. http://www.nhf.org.bd/hospital_charge.php?id=6. Accessed 09 May 2021

17. Ozturk T, Talo M, Yildirim EA, Baloglu UB, Yildirim O, Acharya UR (2020) Automated detection of covid-19 cases using deep neural networks with x-ray images. Comput Biol Med 121:103792

18. Panwar H, Gupta P, Siddiqui MK, Morales-Menendez R, Singh V (2020) Application of deep learning for fast detection of covid-19 in x-rays using ncovnet. Chaos, Solitons & Fractals 138:109944

19. Pereira RM, Bertolini D, Teixeira LO, Silla CN, Costa YM (2020) Covid-19 identification in chest x-ray images on flat and hierarchical classification scenarios. Comput Methods Programs Biomed 194:105532

20. Rahman S, Bahar T (2020) Covid-19: the new threat. Int J Infect 7(1)

21. Rajaraman S, Siegelman J, Alderson PO, Folio LS, Folio LR, Antani SK (2020) Iteratively pruned deep learning ensembles for covid-19 detection in chest x-rays. IEEE Access 8:115041–115050

22. Simonyan K, Zisserman A (2014) Very deep convolutional networks for large-scale image recognition. arXiv preprint arXiv:1409.1556

23. Szegedy C, Vanhoucke V, Ioffe S, Shlens J, Wojna Z (2016) Rethinking the inception architecture for computer vision. In: Proceedings of the IEEE conference on computer vision and pattern recognition, pp 2818–2826

24. Tahir AM, Chowdhury ME, Khandakar A, Rahman T, Qiblawey Y, Khurshid U, Kiranyaz S, Ibtehaz N, Rahman MS, Al-Madeed S et al (2021) Covid-19 infection localization and severity grading from chest x-ray images. arXiv preprint arXiv:2103.07985

25. Tartaglione E, Barbano CA, Berzovini C, Calandri M, Grangetto M (2020) Unveiling covid-19 from chest x-ray with deep learning: a hurdles race with small data. Int J Environ Res Public Health 17(18):6933

26. Toğaçar M, Ergen B, Cömert Z (2020) Covid-19 detection using deep learning models to exploit social mimic optimization and structured chest x-ray images using fuzzy color and stacking approaches. Comput Biol Med 121:103805

27. Traveldoctor: Corona Testing for Travellers—Travel Doctor. https://traveldoctor.co.za/home/corona-testing-for-travellers/. Accessed 09 May 2021

28. Ucar F, Korkmaz D (2020) Covidiagnosis-net: deep bayes-squeezenet based diagnosis of the coronavirus disease 2019 (covid-19) from x-ray images. Med Hypotheses 140:109761

29. UK PH (2019) How much does a private MRI scan cost in the UK? https://www.privatehealth.co.uk/conditions-and-treatments/mri-scan/costs/. Accessed 09 May 2021

30. Waghmare A (2020) Covid-19 Tests Price in India State-wise: how much do Coronavirus tests cost in India? A state-wise breakup. https://indianexpress.com/article/india/covid-19-test-prices-rates-india-6896237/. Accessed 09 May 2021

Chapter 13
Fuzzy Cognitive Maps Applied in Determining the Contagion Risk Level of SARS-COV-2 Based on Validated Knowledge in the Scientific Community

Márcio Mendonça, Rodrigo H. C. Palácios, Ivan R. Chrun, Acácio Fuziy, Douglas F. da Silva, and Augusto A. Foggiato

Abstract Due to the current pandemic that is causing psychological problems, sequelae, in some cases, irreparable damage, and, mainly, leading people around the planet to death; this work aims to create an intelligent application from a validated table, presented by Texas Medical Association. Specifically, the application of fuzzy cognitive map can facilitate the contagion risk level's inference of SARS-CoV-2 from information on human behavior of everyday life. As a possible contribution of this investigation, in addition to the listed and classified risks, the individual's behavior should mitigate or increase his contagion risk level. The results are presented and normalized on a scale from 0 to 10.

Keywords Fuzzy cognitive maps · Diagnosis · Simplified comportamental dynamic fuzzy cognitive maps · COVID-19 contamination risk level

1 Introduction

The current COVID-19 pandemic is due to the severe acute respiratory syndrome coronavirus 2 (SARS-CoV-2), which belongs to the Coronaviridae family, of the B-line of beta-coronaviruses [1]. The viral structure consists of a single-stranded

M. Mendonça (✉) · R. H. C. Palácios
Federal University of Technology—Parana (UTFPR), Cornélio Procópio, Brazil
e-mail: mendonca@utfpr.edu.br

R. H. C. Palácios
e-mail: rodrigopalacios@utfpr.edu.br

I. R. Chrun
Technical-Professional Innovation and Engineering University (FEITEP), Maringá, Brazil

A. Fuziy · A. A. Foggiato
Foggiato Research Institute, Jacarezinho, Brazil

D. F. da Silva
State University of Northern Paraná (UENP), Jacarezinho, Brazil

© The Author(s), under exclusive license to Springer Nature Singapore Pte Ltd. 2022   175
R. J. Howlett et al. (eds.), *Smart and Sustainable Technology for Resilient Cities and Communities*, Advances in Sustainability Science and Technology,
https://doi.org/10.1007/978-981-16-9101-0_13

ribonucleic acid (RNA), with four main structural proteins encoded by the coronaviral genome in the envelope, the nucleocapsid protein (N) being the spike protein (S), a small membrane protein (SM), and the glycoprotein membrane (M) with an additional glycoprotein membrane (HE), which allow entry and replication in the host cell [1, 2].

COVID-19 has the capacity to be transmitted mainly from person to person through direct contact, or through respiratory droplets, which are released when an infected person coughs or sneezes [3]. Thus, during health activities and procedures, such as intubation and airways aspiration, allow the exhalation and inhalation of aerosols, enabling the infection to occur between individuals [4].

As it is an emerging acute respiratory infection, and for the time being, there is no specific drug recommended to prevent or treat the disease, and it lacks effective methods to control and treat the infection [5]. The first prevention approaches were social isolation measures [6]. In addition, public health agencies have also adopted methods of quarantine, social distance, and community containment measures [7].

It is not within the paper's scope to discuss the virus structure and behavior itself. However, it can be found in the literature several papers addressing those; it can be cited in topics of transmission [8], incubation [9, 10], behavior inside humans [11, 12], virus survival rates and contamination [13–15], symptoms in patients [16–21], preventive measures [5–7], and precautions and hygiene for professionals and general population [22–28].

To better understand the COVID-19 and its effects in the population and the economy, several papers applied intelligent systems to analyze the current situation. For instance, in [29], it is presented a novel software utility for 2-D ANOVA without replication, an intuitionistic fuzzy two factor ANOVA, an extension of the classical ANOVA. In order to analyze imprecise numbers, this tool is based on intuitionistic fuzzy sets (IFSs) using a software implementation of index matrices (IMs) to calculate the results. Thus, this utility is applied to find the dependencies of the COVID-19 case notification rate per 100,000 people in European countries, up to June 24, 2020, aiming to investigate the effect of "density" and "climate zone" factors. At last, the results are compared between the proposed utility software and the classical ANOVA.

In [30], the author's proposed method uses a new swarm intelligence (SI) method, called marine predators' algorithm (MPA), improved by the use of the moth-flame optimization (MFO). It is used as a multi-level thresholding (MLT) method for image segmentation and medical image segmentation, such as COVID-19 CT images. The proposed method was extensive compared to other several techniques, such as GWO, SSA, CS, among others. The results presented showed that the proposed method outperforms the other methods in terms of structural similarity index (SSIM), peak signal-to-noise ratio (PSNR), and fitness value.

The work of [31] proposes an integrated methodology based on machine learning for evaluation and benchmarking various classifiers for COVID-19 diagnosis. Aiming as an assist tool to help the decision-makers in the medical and health organization to decide which the best classifiers system should be used for COVID-19 diagnosis by evaluating different classifiers models, using chest X-ray data. It is compared 12 diagnosis models based on 12 well-known algorithms in the literature, e.g., neural network, naive Bayes, logistic regression, and others; through an integrated MCDM method with TOPSIS and entropy, where the and the integration of TOPSIS and entropy methods. Where the former is used for benchmarking and ranking purpose, while the latter to calculate the weights of criteria.

In this context, the main objective of this research is to quantify the level of COVID-19 contagion risk through an already valid table in the literature and the behavior of individuals. Sciophyte objectives can be seen in the virus contagion biological basis. And finally, formalize an adaptation of Dynamic-FCM, already validated tool by the scientific community, as it has been published in IEEE journals and conferences on several occasions, for the proposed application. The motivation for this research is social. After validating the results of the tool, it will be to develop a software that, through a form, provides its information and obtains the level of contagion risk of the virus, possibly even an APP for cell phones. The APP development will be possible due to the low computational complexity of the fuzzy cognitive maps, which will be presented on the next section.

This paper is divided as follows. Section 2 justifies, formalizes, and mentions some application areas, an intelligent computational tool applied to estimate the risk level. Section 3 presents an adaptation of extension D-FCM, proposed in this work that sCD-FCM. Section 4 presents examples of individuals and discusses results. Finally, Sect. 5 concludes and addresses future work.

2   Fuzzy Cognitive Maps' Background

Proposed in 1986 by Kosko [32], fuzzy cognitive maps (FCMs) can be seen as a class of artificial neural networks (ANN), which represent knowledge in a symbolic way and variable states reports, based on exits and entrances events, using a cause-and-effect approach.

The FCMs, when compared to artificial neural networks, have several important advantages such as the relative ease of representing knowledge structures and the simplicity of the inference that is calculated by numeric matrix operations [33, 34]. In short, FCM combines aspects, such as the robustness of fuzzy logic and neural networks [35].

FCMs aim at modeling and simulating dynamic systems. They exhibit numerous advantages, such as a model of transparency, simplicity, and adaptability to a given domain, among others. FCMs have been applied to numerous industrial areas, such as the work of Mendonça et al. [36]. The proposed extension modifies the simulation model of a classic Kosko FCM [32], due to the canonical FCM not dealing with time, some of which can be found in [37]. One of the difficulties of the work was to determine the aspects, variables, or concepts of the FCM. The construction of these models can be done in two ways, based on the knowledge of specialists in the area (as was the case with this research) and based on historical data [38], or even an approach that uses both complementary methods [39]. Finally, the FCM can be

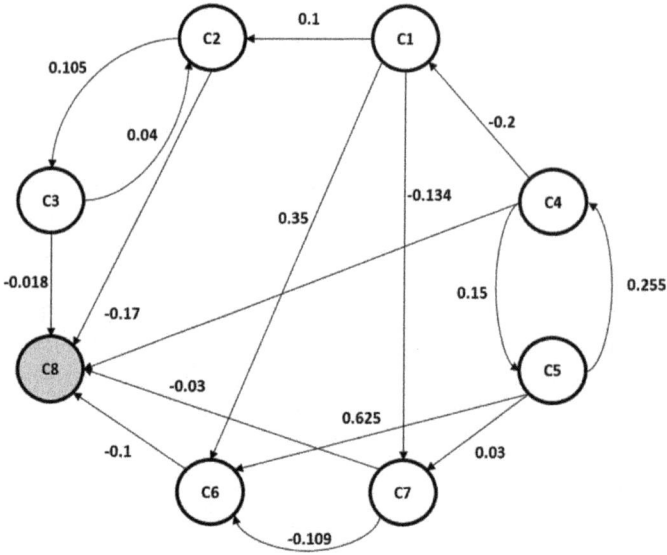

Fig. 1 Example of a cyclical FCM

cyclic or acyclic. The former has one or more cycles between concepts and their causal relationships as shown in Fig. 1.

It can be seen in this figure that there are several cycles between the connections of causal relationships and their respective concepts. For example, the cycle formed between the concepts C6, C7, and C8. It is noteworthy that the proposed model of this research uses an acyclic cognitive model.

The fuzzy cognitive maps (FCM) consist of an approach developed by the knowledge of experts for the construction of scenarios, often used to register several mental models with the variables of the problem concepts (circles in the graph), and the arcs are the relations of cause and effect and can be determined in two ways, manually and automated; the former is used for this work (it can result in large and complex models that are difficult to analyze because they occur indirect effects, loops feedback and time intervals [32]). The main reason for its use is to infer decisions by applying human reasoning methods in uncertain environments. Several fields of research are being investigated using FCM, for example, industrial, logistics, medical, among others [39].

There are usually two types of FCM, manual FCMs and automated FCMs [40]. Manual FCMs are produced manually by specialists (the development methodology applied in this research), and automated ones are produced by historical data. In addition, there are different types of adjustment functions and are used in the evolution of the values of the FCM concepts (FCM inference) [38]. Another possible construction of the FCM is the coexistence of two methods, which use historical knowledge and expert knowledge [39].

There are several applications of FCMs in the literature, such as Virtual Worlds [41], Mendonça and collaborators [42], social systems [43], decision making on fast access roads [44], modeling and decision making in corporate environments and electronic commerce [45], spot detection in images generated by a stereo camera system [46], autonomous navigation [47], swarm robotics [48], agriculture [49], among others. Some of its application areas can also be seen in [37].

Despite the aforementioned advantages of FCM, the classic version of Kosko has a drawback. The canonical version does not deal with time. Thus, the community proposed to diversify FCM extensions or even cognitive models inspired by Kosko's original proposal. One can cite the works of [50] which uses concepts: functions of pertinence, causal relations: base of rules. Two other important extensions found in the literature are DCN [51] which uses concepts that assume a set of values, causal relationships represented by dynamic systems and TAFCM [52] which employs the addition of concepts of temporal automata, alteration of the cognitive map [53], among others.

However, despite the temporal disadvantage of the canonical version, the FCM has a low computational complexity and allows, for example, to be embedded on a low-cost controller, such as the work [54].

The formal representation of the FCM adopted in this work is in the tuple format (C, W, S, f): C is the set of concepts used to build the FCM [55], adopting values ranging from -1 to 1. Following (i) and (ii) in coherence with Eqs. (1) and (2) [56]. A formalism computes similar tuple-like structure that can be used to model an FCM similar to the work of Mendonça and collaborators [54].

(i) $C = \{C1, C2, \ldots\}$: set of concepts n FCM.
(ii) $W : (,)$ is the weight (causal relationship) that links the concepts of input and output.

Among the various FCM inference relationships found in the literature, Eqs. 1 and 2 represent two of the most used ones. In these equations, f_c is the concept activation function; wi is the concept value j, representing the causal relationship between i and j; and λ is the learning rate.

$$f(x_i) = f_c\left(\sum_{j=1}^{n} W_j * x_j\right) \tag{1}$$

$$f_c = \frac{1}{1 + e^{-\lambda x}} \tag{2}$$

3 sCD-FCM Development

In this simplified comportamental dynamic fuzzy cognitive map (sCD-FCM), an adaptation of D-FCM to include the necessary human behavior in the development of this research. More details of the D-FCM and its applications can be found in the works of Arruda and collaborators artificial life [57–61]. In these papers, the D-FCM uses a state machine managed by an architecture inspired by Brooks' subsumption, in which a robot's sub-behaviors are activated by rules or events and change its model's structure.

The main changes are: The state machine is replaced by a form in which, according to the answers, the model can be changed. As, for example, the concept related to attending the gym can have zero weight, and consequently, the cognitive model should vary according to different behaviors of people. In this model, the adjustment of the concepts will be made according to the individual's behavior in a similar way to the classic FCM, whereas the other concepts will have the weights pre-defined in Table 1, based on the Texas Medical Association COVID-19 risk chart [62]. One observation is that weights will vary within their range [0–1], in which the closer to zero, the better the individual's behavior in the investigated concept. In summary, the intensity of the weights will be in accordance with the normalized values in the table and the individual's behavior will adjust the weight of the concept within its range. A priori, three concepts about behavior will be included in the cognitive model which are, social distance, washing your hands with alcohol gel, water, and soap or both when it touches a public surface, and the use of PPE.

An algorithm with the development stages of the sCD-FCM can be presented in Table 2.

The data entered in the FCM are as follows: the concepts are based on the Texas table, and the subject must answer about the framework presented in the form. The weight is given as shown in Fig. 2, with the addition of three weights as already mentioned. The individual will give a grade from zero to ten, according to the three items mentioned about his behavior. Posteriori, this score will be normalized, since the sCD-FCM weights must vary from 0 to 1 for a better computational processing. After completing the form, the system will calculate the individual's risk level according to his behavior and items framing in Table 1. The following

Table 1 sCD-FCM development stages

Step	Description
1	Abstract the cognitive model with the cause-and-effect relationships in accordance with the table variables
2	Identify and verify each individual analyzed in their context. In this step, the cause-and-effect relationships will be normalized and according to the values established in Table 1
3	Behavioral relations will be assigned values from 0 to 1
4	Execute the computational code according to data provided by the users which are normalized by a simulated worst-case experiment, considered to be the maximum level of risk
5	Validate the results obtained by the sCD-FCM with the instantiation of at least 3 cases of different individuals

Table 2 COVID-19 contagious risk level by activity

Risk Level	Activity	Group
0.1	Opening mail	Low-risk activities
0.2	Getting takeout	
0.2	Pumping gasoline	
0.2	Participating in a tennis game	
0.2	Camping	
0.3	Going to the supermarket	Moderated-low risk activities
0.3	Walking, running, or bike riding with others	
0.3	Playing golf	
0.4	Spending two nights at a hotel	
0.4	Waiting in a doctor's office	
0.4	Spending time at the library or museum	
0.4	Eating outside at restaurants	
0.4	Walking in a busy sidewalk downtown	
0.4	Going to the playground for an hour	
0.5	Dinning at someone else's house	Moderated-risk activities
0.5	Going to a backyard barbecue	
0.5	Spending time at the beach	
0.5	Going to the mall	
0.6	Taking the kids to camp, school, or day care	
0.6	Working, for a week, in an office building	
0.6	Going to the public pool	
0.6	Visiting a friend, or a senior relative, house	
0.7	Getting a haircut at a hair salon or barbershop	Moderated-high-risk activities
0.7	Eating inside at restaurants	
0.7	Going to a wedding or funeral	
0.7	Taking a plane	
0.7	Playing a game of basketball	
0.7	Playing a game of football	
0.7	Greeting a friend, hug or handshake	
0.8	Going to buffet restaurants	High-risk activities
0.8	Going to the gym	
0.8	Spending time at an amusement park	
0.8	Watching a movie at the theater	
0.9	Going to a music concert	
0.9	Attending sports matches at the stadium	
0.9	Attending to 500+ worshipers religious services	
0.9	Drinking or eating at a bar	

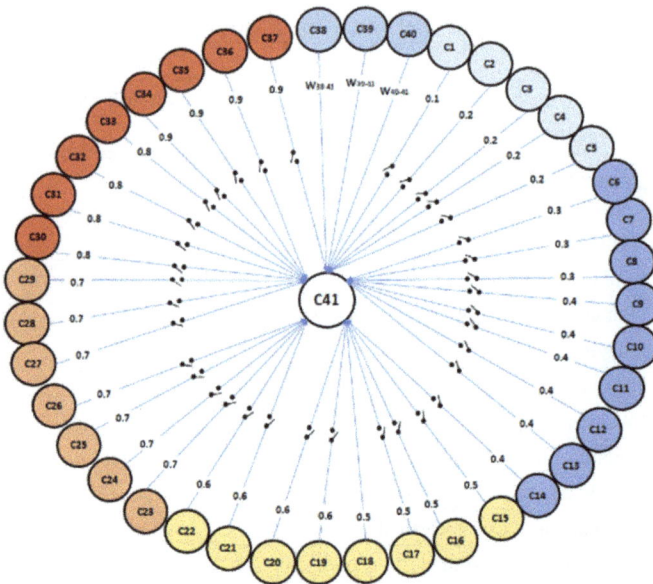

Fig. 2 Concepts C1 to C32 are related to certain activities (e.g., going to the mall, playing tennis) and C33 to C35 as preventive measures (e.g., using PPE). The model infers this data through weights and selection concepts that dynamically change its structure and evaluate the data to a normalized value, which represents a contagion risk level

criterion is applied in order to have a plausible result: the worst case will be hypothetically an individual who fits all items in the table with the highest score on their behavior, emphasizing that the score attributed to each item of behavior is inversely proportional to the individual's behavior.

In order to normalize the data from 0 to 100%, the value obtained in the hypothetical worst situation, in which the individual fits all the items in the TMA chart [62], and has a bad behavior, assigning maximum score in the three modeled items, which will be used as the denominator of all analyzes. Equation 3 presents the standardization methodology adopted for the risk level of each individual.

$$\text{Risk Level} = \frac{\text{sCD - FCM current evaluation}}{\text{Worst case}} \quad (3)$$

4 Results

As mentioned above, the worst-case scenario, in which the individual fits all items in the table and has bad behavior, will be used to normalize the risk levels. Figure 3 shows its result.

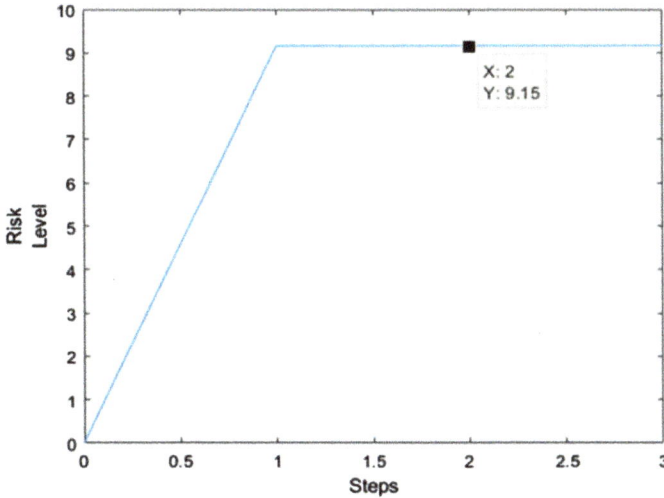

Fig. 3 Risk evaluation of the worst-case scenario

According to the development presented, to instantiate this tool is presented three real cases, where the subjects answered the form.

Three individuals were analyzed, a one younger, less than 30 years, an elder one, around 76 years old, and a half age one, of 50 years old. It is considered the following criteria for the risk level, very low in 0 to 20%, low 21 to 40%, medium 41 to 60%, high 61 to 80%, and very high 81 to 100%.

Figure 4 presents the risk level of the individual of approximately 30 years of age, who falls under items 1, 2, 3, 6, 10, 12, 15, 22, 24, in the table. Regarding the behavior of wearing a mask 0.2, social distance 0.3 and washing hands 0.4. The risk level of this individual will be 34.32%, a low-risk level.

Figure 5 presents the 76 years old risk level. This individual falls under items 1, 2, 3, 6, 10, 12, 13, 16, 23, and 24. Regarding the behavior, keeping the mask hygienize 0.4; social distance 0.7 and washing your hands when you return home 0.1.

The level of risk for the elderly, evaluated by the proposed tool, is 24.04% also considered low, however tending toward the very low-risk range.

The middle-aged individual, almost 50 years old, falls under items 1, 2, 3, 6, 7, 10, 12, 13, 15, 16, 22, 23, 24, and 31. Regarding the behavior of wearing a mask 0.3, social distance 0.3, and handwashing 0.1. Despite attending a high-risk environment gym, this individual compensated with low levels of behavior and obtained a 23.6% risk level, a low-risk tending to very low risk. Its results can be seen in Fig. 6.

With this initial sample, the age group was not so relevant. The highest index was for the 30-year-old; however, the middle-aged individual had a lower level than the elderly. Emphasizing that the elderly belongs to the risk group due to age, have high blood pressure and is pre-diabetic.

It is noteworthy that even the middle-aged individual had partaken in a high-risk activity (item 34—going to the gym), the sCD-FCM results were lower risk than

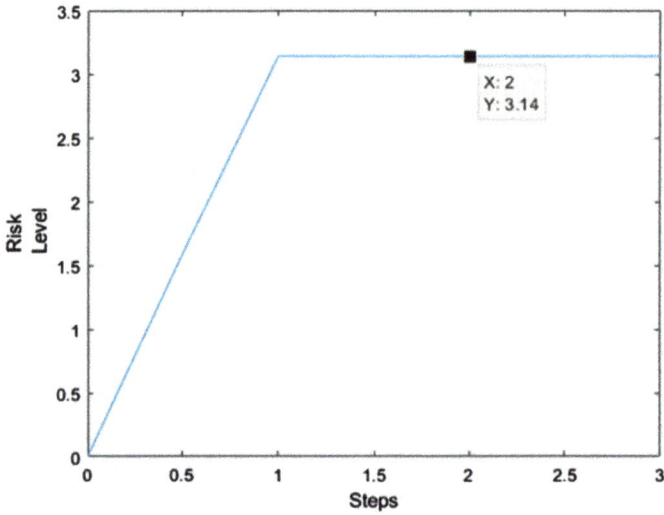

Fig. 4 sCD-FCM risk evaluation: 30-year-old individual

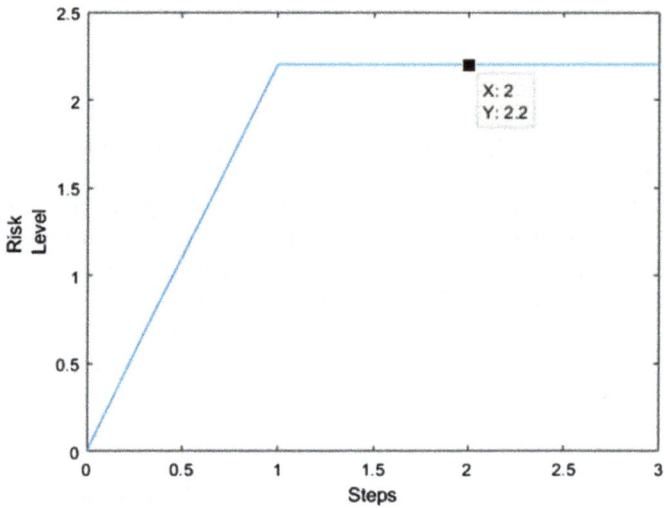

Fig. 5 sCD-FCM risk evaluation: 76-year-old individual

the younger individual, due having more preventive actions toward the behavior of wearing mask, social distance, and hand sanitizing.

These results, although initial, should certainly help raise the population's awareness of the risk of contagion from COVID-19. An observation that can be made about the convergence of the sCD-FCM is that due to its acyclic graph, it converged in just one step and reached a fixed point.

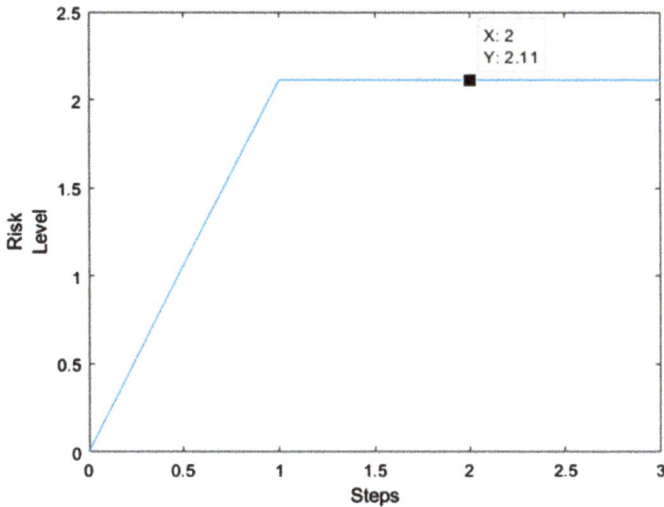

Fig. 6 sCD-FCM risk evaluation: 50-year-old individual

5   Conclusion

The analysis of the risk levels of SARS-CoV-2 disease with fuzzy cognitive map, will help people to raise awareness and prevent the possible places and environments that the users should pay more attention.

The sCD-FCM also presented the response behavior regarding the importance of using masks, social distancing, and cleaning hands, which are within a control range of every individual. Most of the risk activities presented can be avoided; however, some are a constant activity in peoples lives, e.g., using public transportation to go to work. Independently of the activity risk, these three behaviors have a great importance on reducing the contamination risk level and should be adopted. As can be seen in the presented results.

In this way, it is expected to have contributed to this pandemic with an intelligent computational tool that quantifies a possible risk of virus contamination.

Future work aims to investigate and possibly expand the risk analysis with factors of ventilation, occupation, and other types of behavior. An application, a priori for cell phones, is being developed in order to raise awareness and make life easier for users. And finally, the risk analysis of healthcare professionals, due to greater inherent exposure, and behaviors. A more accurate analysis considering the exposure risk of the Texas table and its frequency to exposure.

References

1. Velavan TP, Meyer CG (2020) The COVID-19 epidemic. Tropical medicine and international health. Blackwell Publishing Ltd1 mar. 2020. Available at: https://onlinelibrary.wiley.com/doi/full/10.1111/tmi.13383. Accessed on: 31 ago. 2020

2. Zu Z et al (2020) Coronavirus disease 2019 (COVID-19): a perspective from China radiology NLM (Medline), 1 ago. 2020. Available at: https://doi.org/10.1148/radiol.2020200490. Accessed on: 31 ago. 2020

3. Rothan HA, Byrareddy SN (2020) The epidemiology and pathogenesis of coronavirus disease (COVID-19) outbreak. J Autoimmun 109:102433

4. Gandhi RT, Lynch JB, Del Rio C (2020) Mild or moderate covid-19. New England J Med

5. Luo H et al (2020) Can Chinese medicine be used for prevention of corona virus disease 2019 (COVID-19)? A review of historical classics, research evidence and current prevention programs. Chin J Integr Med 26(4):243–250

6. Schuchmann AZ et al (2020) Vertical social isolation X Horizontal social isolation: the health and social dilemmas in coping with the COVID-19 pandemic. Braz J Health Rev 3(2):3556–3576

7. Wilder-Smith AC, Chiew J, Lee VJ (2020) Can we contain the COVID-19 outbreak with the same measures as for SARS? The Lancet Infectious Diseases Lancet Publishing Group, 1 may 2020. Available at: https://pmc/articles/PMC7102636/?report=abstract. Accessed on: 31 ago. 2020

8. Uddin M et al (2020) SARS-CoV-2/COVID-19: Viral genomics, epidemiology, vaccines, and therapeutic interventions. Viruses 12(5):526

9. Pereira MD et al (2020) Epidemiological, clinical, and therapeutic aspects of COVID-19. J Health Biol Sci 8(1):1

10. Zhai P et al (2020) The epidemiology, diagnosis, and treatment of COVID-19. Int J Antimicrob Agents 55(5):105955

11. Jin Y et al (2020) Virology, epidemiology, pathogenesis, and control of COVID-19. Viruses 12(4):372

12. Sun P et al (2020) Understanding of COVID-19 based on current evidence. J Med Virol 92(6):548–551

13. PAHO (2020) Fact sheet COVID-19—PAHO and WHO office in Brazil—PAHO/WHO Pan American Health Organization. Available at: https://www.paho.org/en/covid19\#superficies. Accessed on: 31 ago 2020

14. FIOCRUZ (2020) How long does the coronavirus remain active on different surfaces? Available at: https://portal.fiocruz.br/pergunta/Quanto-tempo-o-coronavirus-permanece-ativo-em-diferentes-superficies. Accessed on: 31 ago 2020

15. Kampf G et al (2020) Persistence of coronaviruses on inanimate surfaces and their inactivation with biocidal agents. Journal of Hospital Infection, WB Saunders Ltd1 mar. 2020. Available at: https://doi.org/10.1016/j.jhin.2020.01.022. Accessed on: 31 ago 2020

16. Lauer SA et al (2020) The incubation period of coronavirus disease 2019 (CoVID-19) from publicly reported confirmed cases: estimation and application. Ann Intern Med 172(9):577–582

17. Quan LL et al (2020) COVID-19 patients' clinical characteristics, discharge rate, and fatality rate of meta-analysis. Journal of Medical Virology, John Wiley and Sons Inc., 1 jun 2020. Available at: https://onlinelibrary.wiley.com/doi/full/10.1002/jmv.25757. Accessed on: 31 ago 2020

18. Long QX et al (2020) Clinical and immunological assessment of asymptomatic SARS-CoV-2 infections. Nat Med 26(8):1200–1204

19. Rodriguez-Morales AJ et al (2020) Clinical, laboratory and imaging features of COVID-19: a systematic review and meta-analysis. Travel Med Infect Dis 34:101623

20. Yang W et al (2020) Clinical characteristics and imaging manifestations of the 2019 novel coronavirus disease (COVID-19): a multi-center study in Wenzhou city, Zhejiang. China. J Infect 80(4):388–393

21. Lai J et al (2020) Factors associated with mental health outcomes among health care workers exposed to coronavirus disease 2019. JAMA network open 3(3)
22. Mcintosh K (2020) Coronavirus disease 2019 (COVID-19). Available at: https://www.cmim. org/PDF/_covid/Coronavirus/_disease2019/_COVID-19/_UpToDate2.pdf. Accessed on: 31 ago 2020
23. Murthy S, Gomersall CD, Fowler RA (2020) Care for critically ill patients with COVID-19JAMA—Journal of the American Medical Association, American Medical Association21 apr 2020. Available at: http://www.remapcap.org. Accessed on: 31 ago 2020
24. Poggio C et al (2020) Copper-alloy surfaces and cleaning regimens against the spread of SARS-CoV-2 in dentistry and orthopedics. From fomites to anti-infective nanocoatings. Materials (Basel, Switzerland) 13(15):3244
25. Shahbaz M et al (2020) Food safety and COVID-19: Precautionary measures to limit the spread of Coronavirus at food service and retail sector, Journal of Pure and Applied Microbiology16 abr. 2020. Available at: https://microbiologyjournal.org/food-safety-and-covid-19-precautio nary-measures-to-limit-the-spread-of-coronavirus-at-food-service-and-retail-sector. Accessed on: 31 ago 2020
26. Adams JG, Walls RM (2020) Supporting the health care workforce during the COVID-19 global epidemic, JAMA—Journal of the American Medical Association, American Medical Association, 21 apr 2020. Available at: https://pubmed.ncbi.nlm.nih.gov/32163102/. Accessed on: 31 ago 2020
27. Neto ARS, Bortoluzzi BB, Freitas DRJ (2020) Personal protective equipment to prevent infection by Sars-Cov-2. JMPHC, Journal of Management and Primary Health Care 12:1–7
28. Avancini (2020) Cam disinfectants for use in the sanitary context of covid-19. Infect Health Prev Mag
29. Traneva V, Mavrov D, Tranev S (2020) Fuzzy two-factor analysis of COVID-19 cases in Europe. In: 2020 IEEE 10th international conference on intelligent systems (IS)
30. Elaziz MA et al (2020) An improved marine predators algorithm with fuzzy entropy for multi-level thresholding: real world example of COVID-19 CT image segmentation. IEEE Access 8:125306–125330
31. Mohammed MA et al (2020) Benchmarking methodology for selection of optimal COVID-19 diagnostic model based on entropy and TOPSIS methods. IEEE Access 8:99115–99131
32. Kosko B (1986) Fuzzy cognitive maps. Int J Man Mach Stud 24(1):65–75
33. Ndousse TD, Okuda T (1996) Computational intelligence for distributed fault management in networks using fuzzy cognitive maps. In: Proceedings of ICC/SUPERCOMM '96—international conference on communications, pp 1558–1562
34. Parsopoulos KE et al (2003) A first study of fuzzy cognitive maps learning using particle swarm optimization. In: Proceedings of the IEEE 2003 congress on evolutionary computation (IEEE CEC 2003), Canberra, Australia, pp 1440–1447
35. Aguilar J (2004) Dynamic random fuzzy cognitive maps. Comput Sist 7(4):260–270
36. Mendonça M, Angélico BA, Arruda LVR, Neves F Jr (2013) A subsumption architecture to develop dynamic cognitive network-based models with autonomous navigation application. J Control Autom Electr Syst 1:3–14
37. Papageorgiou EI (2014) Fuzzy cognitive maps for applied sciences and engineering. Springer, Heidelberg
38. Yesil E, Ozturk C, Dodurka MF, Sakalli A (2013) Fuzzy cognitive maps learning using artificial bee colony optimization. In: 2013 IEEE international conference on fuzzy systems (FUZZ-IEEE), Hyderabad
39. Mazzuto G, Ciarapica FE, Stylios C, Georgopoulos VC (2018) Fuzzy cognitive maps designing through large dataset and experts' knowledge balancing. In: 2018 IEEE international conference on fuzzy systems (FUZZ-IEEE), Rio de Janeiro
40. Ghazanfari M, Alizadeh S (2008) Learning FCM with simulated annealing
41. Dickerson JA, Kosko B (1996) Virtual worlds as fuzzy dynamical system. In: Sheu B (ed) Technology for multimedia, 1st edn., IEEE Press, Hoboken, pp 1–35

42. Mendonça M, da Silva ES, Chrun IR, Arruda LVR (2016) Hybrid dynamic fuzzy cognitive maps and hierarchical fuzzy logic controllers for autonomous mobile navigation. In: 2016 IEEE international conference on fuzzy systems (FUZZ-IEEE)

43. Perusich K (1996) Fuzzy cognitive maps for policy analysis. IEEE Purdue University South Bend

44. Lee KC, Lee S (2003) A cognitive map simulation approach to adjusting the design factors of the electronic commerce web sites. Expert Syst Appl 24(1):1–11

45. Pajares G, De La Cruz JM (2006) Fuzzy cognitive maps for stereovision matching. Pattern Recogn 39(11):2101–2114

46. Papageorgiou E, Stylios C, Groumpos P (2007) Novel for supporting medical decision making of different data types based on fuzzy cognitive map framework. In: Proceedings of the 29th annual international conference of the Ieee Embs Cité Internationale, Lyon, France August 23–26

47. Mendonça M, Arruda LVR, Neves F (2011) Autonomous navigation system using event driven-fuzzy cognitive maps. Springer Science+Business Media

48. Mendonça M, Chrun IR, Neves-Jr F, Arruda LVR (2017) A cooperative architecture for swarm robotic based on dynamic fuzzy cognitive maps. Eng Appl Artif Intell 59:122–132

49. Makrinos A, Papageorgiou E, Stylios C, Gemtos T (2007) Introducing fuzzy cognitive maps for decision making in precision agriculture. Precision agriculture 2007—papers presented at the 6th European conference on precision agriculture

50. Carvalho JP, Tome JAB (2001) Rule based fuzzy cognitive maps-expressing time in qualitative system dynamics. In: 10th IEEE international conference on fuzzy systems

51. Miao Y, Liu Z-Q, Siew CK, Miao CY (2001) Dynamical cognitive network—an extension of fuzzy cognitive map. IEEE Trans Fuzzy Syst 9(5):760–770

52. Acampora G, Loia V (2009) A dynamical cognitive multi-agent system for enhancing ambient intelligence scenarios, fuzzy systems, 2009. In: IEEE International Conference on FUZZ-IEEE 2009, Jeju Island, pp 770–777

53. de Souza LB, Soares PP, Barros RVPD, Mendonça M, Papageorgiou E (2017) Dynamic fuzzy cognitive maps and fuzzy logic controllers applied in industrial mixer. Int J Adv Syst Meas 10(3):222–233

54. Mendonça M, Chrun IR, Neves-Jr F, Arruda LVR (2017) A cooperative architecture for swarm robotic based on dynamic fuzzy cognitive maps. Eng Appl Artif Intell 59(March):122–132

55. Stach W, Kurgan L, Pedrycz W, Reformat M (2005) Evolutionary development of fuzzy cognitive maps. IEEE international conference on fuzzy systems, pp 619–624

56. Nápoles G, Bello R, Vanhoof K (2013) Learning stability features on sigmoid fuzzy cognitive maps through a swarm intelligence approach. In: CIARP 2013: progress in pattern recognition, image analysis, computer vision, and applications. Lecture notes in computer science, vol 8258

57. Arruda LVR, Mendonça M, Neves-Jr F, Chrun IR, Papageorgiou E (2018) Artificial life environment modeled by dynamic fuzzy cognitive maps. IEEE Trans Cogn Dev Syst 10(1):88–101

58. Mendonça M, da Silva ES, Chrun IR, Arruda LVR (2016) Hybrid dynamic fuzzy cognitive maps and hierarchical fuzzy logic controllers for autonomous mobile navigation. In: 2016 IEEE international conference on fuzzy systems (FUZZ-IEEE), Vancouver, BC, pp 2516–2521

59. Soares PP, de Souza LB, Mendonça MR, Palacios HC, de Almeida JPLS (2018) Group of robots inspired by swarm robotics exploring unknown environments. In: 2018 IEEE international conference on fuzzy systems (FUZZ-IEEE)

60. Mendonça M, Kondo HS, Botoni de Souza L, Palácios RHC, de Almeida JPLS (2019) Semi-unknown environments exploration inspired by swarm robotics using fuzzy cognitive maps. In: 2019 IEEE international conference on fuzzy systems (FUZZ-IEEE), New Orleans, LA, USA

61. Mendonça M, Palacios RHC, Papageorgiou E, de Souza LB (2020) Multi-robot exploration using dynamic fuzzy cognitive maps and ant colony optimization. In: 2020 IEEE international conference on fuzzy systems (FUZZ-IEEE), Glasgow, United Kingdom
62. Texas Medical Association: TMA, what's more risky, going to a bar or opening the mail? Available at: https://www.texmed.org/TexasMedicineDetail.aspx?id=53977. Accessed on: 20 july 2020

Part III
Changes in Teaching and Learning Practices in Response to a Pandemic

Chapter 14
Education After COVID-19

Manuel Mazzara, Petr Zhdanov, Mohammad Reza Bahrami, Hamna Aslam, Iouri Kotorov, Muwaffaq Imam, Hamza Salem, Joseph Alexander Brown, and Ruslan Pletnev

Abstract The year 2020 has brought life-changing events for many and affected numerous professional sectors. Education has been one of those fields heavily impacted, and institutions have almost worldwide switched to forms of online education, which has become a common practice. With a fourth industrial revolution happening in front of our eyes, some elements of the existing education system are showing themselves as out- dated. However, despite the realization that online teaching is here to stay, frontal classes are a millennia-old practice that cannot be entirely replaced without neglecting human nature. Instead, old and new can coexist, and humanity and machines can cooperate for societal development. In this paper, we present the past, present, and future of education, what we have learned by the experience of teaching online, and how we see and are getting ready for future developments in the field.

M. Mazzara (✉) · P. Zhdanov · M. R. Bahrami · H. Aslam · M. Imam · H. Salem · J. A. Brown · R. Pletnev
Innopolis University, Innopolis, Russia
e-mail: m.mazzara@innopolis.ru

P. Zhdanov
e-mail: pe.zhdanov@innopolis.ru

M. R. Bahrami
e-mail: mo.bahrami@innopolis.ru

H. Aslam
e-mail: h.aslam@innopolis.ru

H. Salem
e-mail: h.salem@innopolis.university

J. A. Brown
e-mail: j.brown@innopolis.ru

R. Pletnev
e-mail: r.pletnev@innopolis.university

I. Kotorov
North Karelia University of Applied Sciences, Joensuu, Finland
e-mail: iouri.kotorov@karelia.fi

© The Author(s), under exclusive license to Springer Nature Singapore Pte Ltd. 2022   193
R. J. Howlett et al. (eds.), *Smart and Sustainable Technology for Resilient Cities and Communities*, Advances in Sustainability Science and Technology,
https://doi.org/10.1007/978-981-16-9101-0_14

1 Introduction

For more than a century, the commonly used systems of education and, especially, higher education systems have been strongly influenced by the models and principles formulated yet during the first industrial revolution [1]. The modern education systems were essentially formed in the eighteenth century in Europe and the mid-nineteenth century in North America [2] with their dominant features being:

(1) top-down management, (2) outcomes designed to meet societal needs, (3) age-based classrooms, and (4) focus on producing results. Even though this view of the system can be debated,[1] such a retrospective view can still be a valid starting point to reflect and collect insights on how the current educational systems are outdated [3]. It is worth pointing out that outdated does not mean that every element does not apply to the current realities; instead, the elements require to be reconsidered with a focus on what should be kept or updated and what should go.

With the considerable impact of the regime of the global pandemic on education, the year 2020 might have initiated the changes to higher education that are here to stay and can potentially revolutionize education. The almost worldwide switch to the different forms of online education alone is enough to speculate that the future of education seems radically different from what it has been for decades. The precise effect of the global pandemic is a much-debated topic. Based on the attempts to demonstrate that the modern pre-pandemic education systems were developed upon the philosophy and principles of the past centuries that are outdated, this paper highlights the importance of the changes that the global pandemic has recently brought. In the aftermath, certain adjustments will be needed and necessary for the educational establishments to stay effective at educating and attractive for students.

In this paper, we attempt to discuss the past, the present, and the future of education, especially its aspects that are likely to change even after the global pandemic ends. After this introduction, the paper is structured as follows: Sect. 2 gives a brief review of the philosophy that influenced the modern educational systems; Sect. 3 describes the emergent themes of the first industrial revolution that are most likely influencing even modern education; Sect. 4 makes some consideration about the way the administrations of educational establishments often thought right before the global pandemic; Sect. 5 provides vital lessons that the global pandemic taught the educational establishments; Sect. 7 discusses the way online education will become part of regular academic life; Sect. 8 just scratches the surface of the relations between artificial intelligence and the future of education while Sect. 9 draws conclusive remarks on the analysis made in this paper.

---

[1] https://www.washingtonpost.com/news/answer-sheet/wp/2015/10/10/american-schools-are-modeled-after-factories-and-treat-students-like-widgets-right-wrong/.

2   Philosophy of Educational System

In any successful education system, the most critical issue is the choice of values, coherent goals, and philosophies. The importance of these issues is that any educational system seeks to educate an ideal human being in some way, and without a value system and goals of education, such a human being is not possible.

The ideal human being, the product of communist educational philosophy, is educated for the general purpose of this philosophy, which is a classless society, and its values and anti-values are institutionalized in him based on this school and through education. We claim to nurture him. Lack of knowledge or basically ignorance of the importance of the philosophical school that should be followed causes the disintegration and confusion of the entire educational system from elementary to university. Therefore, knowing the philosophers of education should be one of the main tasks of experts in any educational system.

The rest of this section talks about the rules of some outstanding philosophers of antiquity [4, 5], and the Middle Ages in the philosophy of education in their era [6].

2.1   Antiquity

We start from the philosophy of Socrates in education. It should be noted that since Socrates did not write down any book/notes; therefore, Socrates' ideas and theories about education are coming from Plato's writings (Plato was Socrates' student). Socrates did not teach using methods as we have them now, like books, a school, or a particular place. According to Plato's writings, Socrates moved around Athens with his students and discussed things such as law, justice, and politics since he believed that everything is open to question and education is not a process [4].

Plato was the founder of a university and implemented the Spartan method of education in his method. To Plato, education is an essential key to dispense people from a primal state of ignorance [5].

According to the Aristotelian thought, education should carry on during the whole life and accomplish via a combination of habits and logic. Aristotle founded his school Lyceum in Athens, where one could study as long as he is willing [4].

2.2   Middle Ages

John Comenius brought a new philosophy into education that called Pansophism based on worldwide understanding and peace which uses children's feeling to make learning fun.

While John Locke was developing the principles of empiricism in education, which states that individuals may learn by external virtues, Jean-Jacques Rousseau, who supported John Comenius's philosophy of education, expressed the willingness of the majority to make decisions. Johann Heinrich Pestalozzi implemented Rousseau's beliefs in practice and hereby is called the first psychologist of applied education. Pestalozzi believed that teachers needed to be taught to develop the education of children.

Johann Friedrich Herbart, which is known as the "father of scientific pedagogy" introduced the system of philosophy in education in where use logic and metaphysics and aesthetics as guiding elements [7].

2.3 First Industrial Revolution

In the early 1800s, thanks to the industrial revolution and the need for factories to have educated workers, schools were established in the territory of factories where workers could study. These schools were called factory schools and have their advantages and disadvantages (discussion of these are beyond the scope of this work), and the learning method was called Lobby training [7].

In the mid-nineteenth century, a new learning method, correspondent learning, was developed and spread rapidly. Many tutors consider correspondence education as a forerunner of remote education that used various communication technologies such as telephone, radio, or television for teaching.

3 The Past of Education

Before discussing the state of education in the modern post-fourth industrial revolution era, in general, and in the post-COVID-19 era, it would be useful to review the challenges that the first industrial revolution introduced to the education of mankind. Those challenges and responses have arguably laid the fundamentals for the education systems that nowadays are used across the world. It is certain, though, that all four revolutions, each in its own way, have transformed the economic and political life, especially in Europe and the USA, having thus an effect on many other domains, including education.

Education has long been known to have a double purpose of assisting people in becoming educated and assisting society in becoming good [8]. Starting from the late eighteenth century and the beginning of the first industrial revolution, the double purpose became even more apparent. The industrial revolution brought steam. The steam brought manufacturing, and the manufacturing brought manufacturing towns with the moderate presence of industrial and commercial middle classes and a large working class. Mass production heavily relied on children required through regularity, self-discipline, obedience, and trained effort [9]. Therefore, after the first industrial revolution, the concept of education was largely focused on the utility of it for future adult work, with schools principally teaching reading, writing, and religion, often even avoiding arithmetic as too difficult for the teacher [10]. However, as Smith [11] pointed out in his magnum opus "the Wealth of Nations," reading, writing, and arithmetic should be the essential part of education that needs to be administered by the government. The role of government in the spread of mass schooling in the nineteenth century can hardly be underestimated as the educational establishments were extensively financially supported by the government that might have thus created dependencies as well as limited the scope of teaching to the current interests of the government. However, the spread of schooling itself could seem to be uneven depending on local relationships between elites and government. The elites or manufacturers saw the benefits of education for their factories and the society and had more control over government expenditure; the education was more likely to be accessible [12]. Such an approach to education clearly demonstrates the top-down way of setting educational agenda that is supposed to reach certain manufacturing-related or societal results.

As for the higher education establishments, it can be argued that starting from the beginning of the nineteenth century and until the mid-twentieth century, they had inconsistent success at its core activity of education, as they also could often be found in significant dependency from their supervisory authority that could adjust the educational systems according to its political and philosophical aims [13]. Also, the period since the first industrial revolution was characterized by the replacement of the universities as such with the professional institutions and the appearance of the French model of colleges with the military discipline and strict control of the curriculum, conformity of views, and even personal habits [14]. However, higher education in the nineteenth century has produced ideological despotism and developed a top-down and outcome-oriented approach to teaching but saw the emergence of the German model that encouraged students to think liberally and explore the world into account the laws of science [14]. However, such academic freedom was successfully restricted with the Carlsbad Decrees in 1819 and then restored only later in the mid-nineteenth century.

Although the prevalent view of education on the onset of economic growth, especially in Europe, after the first industrial revolution may seem to be heavily reliant on the government and its funds and thus was accountable for what it taught, the beginning of the student-oriented perspective can already be found at those long past times. The German model might be what initiated the change of the focus from teacher-centered or "what should be taught" perspective to learner-centered or "how do students learn" perspective that humankind might need the most in the post-COVID-19 era.

4 The Fourth Industrial Revolution: Old Versus New

Even though it could be argued that the state of education in the post-fourth industrial revolution era is still largely based on the fundamental principles developed during the first industrial revolution, a brief overview of the changes introduced in the twenty-first century might help to better understand the requirements for the education of the future, especially in the post-COVID-19 era. With the term "the fourth industrial revolution" being first defined by Klaus Schwab, founder and executive chairman of the World Economic Forum, in 2015,[2] it can be claimed that the fourth industrial revolution itself has started to impact the human society even earlier. The development of cyber-physical systems, artificial intelligence, big data, and not seen before connectivity, have allowed full automation of manufacturing and industrial practices using machine-to-machine communication (M2M) and the Internet of things (IoT) without the need for human intervention. Based on such developments, there are assumptions that in the future, low-skilled jobs will be replaced mainly by machines, middle-skilled jobs will face a reduction in numbers, and only highly skilled and knowledgeable professionals will be in high demand [15]. Even though this tendency is often articulated as the future prevalence of the Science, Technology, Engineering, and Mathematics (STEM) education, there are arguments that the whole person education should be still prioritized. The concept of Science, Technology, Engineering, the Arts, and Mathematics (STEAM) is proposed where the education is based on STEM, but the additional study of "the arts" is pursued in order to develop creativity, critical thinking, and other soft skills [16]. Both perspectives clearly dictate that to adapt to the quickly changing demands of the future economy and the future society; people will have to embrace lifelong learning adjusting to the new digital reality. Thus, the dramatic change in the production processes that happened in the twenty-first century has certainly already and likely will even more so significantly affect the knowledge production, education process, and the learning experience.

In today's digitally interconnected world, education and, specifically, higher education become more accessible than ever, in other words, from anywhere and anytime. In light of this, our education system with the higher education establishments largely attracting only local talents is certainly outdated. The growing number of universities starts to use the massive open online course platforms (MOOC) to provide some of their courses on demand and mostly for free or low-cost. Using MOOCs, students from any part of the world can pause, rewind, fast-forward, play in double speed, or, if necessary, skip the lessons. Such initiatives urge the appearance of new business, financial, and revenue models traditionally associated with the tuition fees and academic labor and provide new opportunities for the innovations in learning models [17] such as encouraging lifelong learning and active online engagement. Adapting to the need of future generations, education may be more likely transformed into learner-centered instead of teacher-centered.

With the more than two-century-long history of the influence on the education systems, the industrial revolutions are yet to revolutionize the way people are taught again. Even though the advantages of the new approach are apparent, especially from the students' perspective, particularly in terms of convenience, there are also several

---

[2] https://www.foreignaffairs.com/articles/2015-12-12/fourth-industrial-revolution.

Table 1 Education: old versus new

Educational Factor	Old	New
Objectives	Skills-based	Whole person
Focus	Teacher-centered	Learner-centered
Learner Experience	Passive	Active
Target Age	School-age	Lifelong learning
Access	Physical	Anytime, anywhere

challenges associated with the changes, discussed in the following sections. Table 1 summarizes the distinctive traits of the old and the new models of education. The cell in red emphasizes the distinctive characteristic that emerged in the twenty-first century.

5   Lessons Learned from COVID-19

The year 2020 has been a remarkable one for the world in general and, in our case, specifically for education. Many certainties vanished in a matter of days or weeks and will not be back. We have been forced to reorganize the educational process entirely literally in a matter of one week. This implied improving the ICT infrastructure, purchasing or extending licenses for specific platforms such as Microsoft Team[3] or Zoom,[4] making sure that connectivity was suitable and identifying adequate methods of instructions. Some challenges have been institutional, some individual. Even something as simple as setting up an effective home office with good connectivity has been, at times, challenging. The reluctance of teachers to indulge in the new format can also be a challenge. However, we, as an IT university, did not experience this much. Changes come as a hurricane, and it is not leaving. Online education, in the blended format, is here to stay. This is discussed in Sect. 7.

The experience of the year 2020 showed some distinctive emerging traits that we believe can be generalized to any educational organization worldwide:

- **New methods of teaching**: Education as online streaming existed as an integrative format for some organizations or was a prerogative of some commercial online

---

[3] https://www.microsoft.com/en-us/microsoft-365/microsoft-teams/group-chat-software.

[4] https://zoom.us/.

platforms. Traditional academia was mostly reluctant to move and adopt it. The year 2020 is a game-changer.

The recorded lecture and flipped classroom are one way to accommodate; however, video alone lack social presence. There is not much opportunity for interaction between students and professor or among students [18]. Therefore, the instructional approaches mostly shifting to online modality should be considered in light of different factors. Teachers will be more as moderators not as instructors [19].

- **Evaluation and assessment by virtual means**: It is undoubtedly challenging to adapt courses and programs to the online format, but the most challenging thing is about assessing students' results and proctoring.
- **Assessments will be done regularly throughout the semester and will not solely depend on final examinations**: The traditional final examination will lose its importance and will be replaced by continuous assessment. Many universities moved before to this format; the year 2020 is an accelerator of this process.
- **Educational programs will include digital content**: This has already happened for some time, 2020 is a year of non-return. In this paper, we also advocate the importance of traditional content, such as books (See Sect. 9).
- **Greater use of Open Educational Resources**: Closed access resources turned out to be a blockage for online education. Some of the advantages of online education are affordability and accessibility (see Sect. 7), and to achieve this Open Educational Resources are fundamental.
- **Professional development for teaching staff on digital education**: Our organization, Innopolis University, is a young and dynamic IT university which experienced marginal issues in the switch, being the teaching staff mostly IT specialists with long experience as software users. However, we noticed that the changes have been problematic for some of our colleagues from less IT-related departments, and we have observed significant problems in other universities and faculties. The path of complete digitization for established teachers is long ahead.
- **Distinctive features of the educational experience must compensate for the decrease of campus experience**: As we will discuss in Sect. 9, education is not only about content delivery, but campus experience, human networking, sports, and social activities, where young adults learn a 360° perspective on life and profession. Online education, even in a blended format, cannot offer a comparable experience in this sense. We certainly hope that part of these offline activities can be eventually restored. However, it is necessary to rethink education to offer specific features that can compensate for this emerging gap.
- **Greater emphasis will be placed on collaborative projects**: As a consequence of loss of campus interaction and networking, collaborative projects, even if executed remotely, can reinstate back some feeling of the community belong in and horizontal learning.
- **ICT infrastructure is critical**: while in the pre-COVID classic delivery mode, teaching could be potentially delivered with a shortage of ICT infrastructure, now this is a *"condition sine qua non."* ICT was before a support infrastructure, and every teacher was able occasionally to deliver a functioning class without a

projector, a laptop, or an Internet connection. This is not possible now, and often even a slightly sub-optimal bandwidth can make things frustrating and difficult to follow. Universities that want to win the race have to put the development of ICT infrastructure on top of the list. Before considered a distraction by many lecturers, laptops in the class room are now the primary tools of operations. Of course, this is also bringing problems, as discussed in Sect. 7.

- **Paradigm shift in teachers training programs**: Teachers' training programs are mostly designed for classroom-based in-person teaching. With online education as a norm, teachers have to be trained accordingly. Online education presents more challenges in terms of ensuring students' attention. Furthermore, classroom interactions allow teachers to have some idea of students' mental state, and teachers can provide support accordingly. Online education makes it difficult to know students' issues (especially with their cameras and microphone off most of the time). It might become mandatory for teachers to have basic mental health education to assess students' behavior and provide support accordingly.
- **New ways of establishing discipline and roles**: With education being learner-centric, classroom disciplinary restrictions are more relaxed as compared to the situation some decades ago. Students have a say in educational reforms, and any significant alteration in the educational process takes place with the teacher and student consensus. With online education, classroom boundaries need to be revisited in consultation with teachers and students. By boundaries, we refer to teacher–student agreements that ensure discipline required to maintain quality of education. Some of the points that require consideration involve camera and microphone on/off issues and lecture delivery modes such as recorded or live sessions.

6   Strategic Planning, or Lack of It

Before the year 2020, e-learning was growing by 15.4% yearly [20].

During the year 2020, educational institutions had to provide most of their services online, including lectures, and different assessments via several platforms for over 60% of students around the world [21]. The time for decision-making was very limited, and organizations had to act without action plans had been developed in advance. Having now about a year of experience, the community is starting to develop a correct understanding of e-learning and develop new strategies for coping with new problems. Perhaps, by using proven strategic planning tools and trying to minimize risks, educational institutions could avoid most of the mistakes made in the past.

Tools such as PDCA (plan, do, check, act [22]) or LEAP (learning, evaluation, and planning) [23], or any other similar approach, would have led to a better decision-making process. Within an organization, special working groups should be settled to make prompt and correct decisions and plan the list of actions that have then to be implemented. These groups should represent the interests of all stakeholders of an educational institution, with the main focus on students and professors.

Based on tools as PDCA or LEAP, a list of quick actions may have looked as follow:

1. Establish a group of heads, professors, and students that will have the responsibility to develop the education response to the COVID-19 pandemic;
2. Schedule regular communication between the stakeholders;
3. Define the principles which will guide stakeholders;
4. Focus on new or prioritize old curriculum goals under new delivery methods;
5. Define clear expectations for professors and students;
6. Define learning support activities under new circumstances and guide self-study and learning process;
7. Generate full support for students under new delivery methods;
8. Support communication and collaboration between students and encourage mutual learning;
9. Create a mechanism to support and educate professor and students for new delivery methods;
10. Create a road map for a period of 2 years;
11. Create metrics and evaluation tools;
12. Provide continues evaluation on the process minimum two and maximum four times per year;
13. Report to all stakeholders results and ask for continued support.

It is worth noting that this list is not meant to be a general strategy for the entire educational institution, but it is just focusing on how to quickly transform teaching and adjust the curriculum that was developed for an on-campus delivery method.

7 Online Education is Here to Stay: The 4-Move Checkmate

The 4-Move Checkmate (or Scholar's Mate) is a common checkmate pattern among beginners. Almost all chess players have fallen for or delivered this checkmate at some point in their lives. We use it here as a metaphor to explain the four steps happening to make routine something that was seen with reluctance just a few years back. Online education is not a temporary patch to cope with an emergency but a move to a blended and modern form of instruction. This change will happen in four steps. At the moment of writing, our organization (Innopolis University) is between the first and second steps. It is interesting to note that before 2020, any attempt to provide online education, if not under exceptional circumstance, would be opposed. Opposition can be seen as the step zero of any evolutionary change. We identify the following steps:

- We had no choice than teaching online.
- Online teaching will be encouraged.
- Students will demand loudly online teaching.

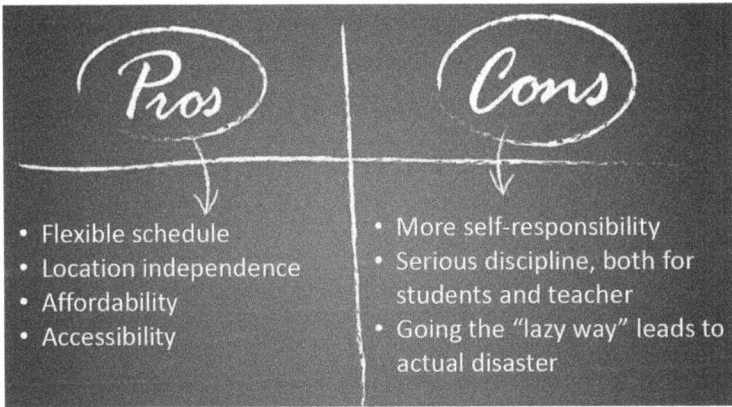

Fig. 1 Pros and cons of online education

- Online education will be part of routine.

In the context of the fourth industrial revolution, societal and economic needs are changing. Environmental pressure should also be taken into account. Distance learning and smart work will increase in relevance. Laptops in the class- room were opposed by numerous lecturers just a few months ago. Now, without a laptop, the learning process cannot happen. The mindset of students and lecturers has to change and adapt to the fluid changing situation. As every process of change, it can be seen with reluctance. The opposition will arise, emphasizing only the cons; promoters will see only the pros. Both pros and cons are on the table. A successful organization will be able to exploit the pros and limit the cons. As discussed further in Sect. 9, the key of success (in the authors' opinion) stays in the ability to keep the community alive and compact both on the virtual and physical levels. The feeling of belonging should not be lost. Figure 1 summarizes the advantage and disadvantages of new format.

On the positive side, from our experience, we identify:

- **Flexible Schedule**: Online education comes with video recording. This allows attendance of the classes at any time bringing advantages to students who are working in need of financial support. The institution may not need to run evening courses anymore. Clearly, this becomes a problem to overcome for those orga- nizations having this as a core business. They should focus on added value, for example, the community.
- **Location Independence**: Students (and lecturers) can be anywhere and avoid long commuting. This can have an impact also on the environment. It is important not to lose the community and keep a sense of belonging. Ideally, a core group of students should still be on-campus.
- **Affordability**: Cost-cutting in transportation and living costs can become the norm. Students who cannot afford to live far from home may have a chance to participate to a remote study program.

- **Accessibility**: Flexible schedule, location independence, and affordability converge in general terms into higher accessibility of the programs for disadvantaged categories of students (low-income family, disabled, chronically ill, etc.). A particular critical aspect here is supporting *digital accessibility of educational resources*. For example, helping blind students with screen readers to access and use online contents.

On the negative side, we see the following:

- **More self-responsibility**: The burden of study now relies more on the proactivity of the students. Teachers have limited means for direct support, especially in remote areas with limited bandwidth and access to electricity.
- **Serious discipline**: Discipline, as to come endogenously since the exogenous part, is reduced to the very minimum. Success is mostly determined by self-motivation.
- **Going the "lazy way"**: It seems convenient to do things "online" sitting from home, not taking the bus or train, no need to dress-up or mentally switch. Also, once at the laptop, many things can be attempted to be done simultaneously (attendance of multiple classes or meetings, for example). In the absence of self-responsibility and discipline from both teachers and students, the online format can turn out to be very ineffective, to say the least.
- **Violation of "oikos"**: the ancient Greek word *oikos* (ancient Greek:, plural:; English prefix: eco- for ecology and economics) refers to three related but distinct concepts: *the family, the family's property, and the house* opposed to the term *"polis,"* indicating the city and the political activities. This separation is particularly important for psychological, sociological, and economical reason. Blurring such separation, starting from losing the use of different words for indicating two different places, may lead to complex problems in the self-management and management of the society.

8   Education and the Turing Line: The Role of AI

At the beginning of the twentieth century, the future was imagined as a place dominated by machines and where humans and robots would peacefully co-operate. In Fig. 2, the learning process is depicted like a radio broadcasting generated by a book-eating machine. It is not very different from a modern podcast, often used for pedagogical reasons. In future decades dystopias presented the relationships between humans and machines in a much less harmonic way, possibly starting with the Russian novel by Yevgeny Zamyatin "We" in 1924 [24] and notoriously followed by other major novels and Hollywood movies. All these human artistic expressions present the coexistence of humans and machines as problematic, in the best case, and oppressive in the worst scenario.

The year 2020 is ultimately different from everything that was depicted before; still, machines and algorithms are dominant in our lives. It is now the right time to ask whether the future of education will be an emerging dystopic scenario, or

Fig. 2 Françoise Foliot—La radio à l'école by Jean-Marc Côté, CC BY-SA 4.0

more positively, an opportunity to embrace enhanced learning approaches, making the "factory model" of education history. It should be clear that educators also have a role in this process, not only in politics or major IT corporations.

In "Computing Machinery and Intelligence" (1950), Turing draws "*a fairly sharp line between the physical and intellectual capacities of a man.*" [25]. However, the techno-social environment in which we live today makes this line slightly more fuzzy than it was at the time Turing's paper was written.

There is a possibility that humans may tend to behave more "machine-like" as long as they are more and more surrounded, and therefore constrained, or competed against, by machines. At the same time, machines themselves are becoming more and more "human-like" as it is clearly visible in current technology. Think, for example, of anthropomorphic robots [26] or personal digital assistants [27]. With this tendency in mind, the sharp line of turning does not appear that neat anymore and less, and less will be when humans are equipped with portable embedded devices. A question naturally arises. What traits of humanity and machines can be blended for the benefits of the new generation when it comes to education?

Frontal classes are a millennia-old practice, cannot be (and should not be) fully replaced. The idea that education can be delivered fully online, always and in all the cases, under all circumstances seems not to take into account the very nature of learning and mental development, especially for kids. Human beings are thinking as well as feeling organisms [28] and evolutionary and psychological research has shown, for example, that touch is an important part of bond-building and emotion communication. The connection between loneliness and physical contact is explored in [29]. Taking these peculiarities into account, artificial intelligence can provide valuable support for all human activities, and education is not last. You can imagine personalized learning schemes, even more, likely to function when education is provided remotely. Intelligent tutoring systems (ITS) are part of this development [30]. Education for children with disabilities has been a challenge ever since, and new technological development may help in eliminating the divide [31]. An overview of how artificial intelligence in education is applied these days can be found in [32].

9   Conclusions: The Importance of Community and ICT

Education is not only about content delivery, and even not principally about it. What creates the unique university experience is human networking, sports, and social activities, all the involvements where young adults learn a 360 degree perspective on life and profession guided by senior colleagues and faculty. Online education, even in a blended format, cannot offer this experience. The significant risk is that students can escape "reality" and hide in a "cocoon." After decades of exaggerated emphasis on "soft skill" where introverts were pushed to their limits and extroverts could more easily thrive, we now push young generations to the opposite extreme and make introverts thrive. Even if one could see a sort of "divine justice" in such change, it is not by moving from one extreme to another that we balance a situation. In some sense, injustice cannot compensate for another injustice. We need to have a balanced approach. Work is now necessary to find a way to overcome such a possible degeneration and avoid the "*student in a learning cocoon*" approach. Even in the post-2020 era, the community still counts.

The future will see a sharp separation between those educational institutions able to catch up with the pace and those left behind, both in terms of new pedagogical methods and supporting ICT infrastructure. In the pre-COVID classic delivery mode, teaching could be delivered with a modest ICT infrastructure: projectors and laptops were often sufficient, and an Internet connection was not always necessary. Every teacher was able to deliver functioning classes without the need for any particular device. Now even a slightly sub-optimal bandwidth can make things impossible.

References

1. Coluccia D (2012) The first industrial revolution (c1760–c1870). Palgrave Macmillan, London, pp 41–51
2. Carl J (2009) Industrialization and public education: social cohesion and social stratification. Springer, Netherlands, Dordrecht, pp 503–518
3. Marshall HH (1990) Beyond the workplace metaphor: the classroom as a learning setting. Theory Into Practice 29(2):94–101
4. Barnes J et al (1984) Complete works of Aristotle, volume 1: the revised Oxford translation, volume 192. Princeton University Press
5. Cooper JM, Hutchinson DS et al (1997) Plato: complete works. Hackett Publishing
6. The history of the development of learning: antiquity—middle ages—modern times (Rus). Accessed: 7 Feb 2020
7. Clark DR. History of learning & training

8. Kemmis S, Edwards-Groves C (2018) Understanding education. History, politics and practice
9. Gordon P, Lawton D (2019) A history of western educational ideas. Routledge
10. Boyd W (1966) The history of western education (revised by ej king). Adam & Charles Black, London
11. Smith A (1937) The wealth of nations [1776]
12. Andersson J, Berger T (2019) Elites and the expansion of education in nineteenth-century Sweden. Econ Hist Rev 72(3):897–924
13. Gerbod P (2004) Relations with authority. In a history of the university in Europe
14. Rüegg W (2004) A history of the university in Europe: Volume 3, universities in the nineteenth and early twentieth centuries (1800–1945), volume 3. Cambridge University Press
15. Jung J (2020) The fourth industrial revolution, knowledge production and higher education in South Korea. J High Educ Policy Manag 42(2):134–156
16. Ilori MO, Ajagunna I (2020) Re-imagining the future of education in the era of the fourth industrial revolution. Worldwide Hospitality and Tourism Themes
17. Peters MA (2017) Technological unemployment: educating for the fourth industrial revolution
18. Sun A, Chen X (2016) Online education and its effective practice: a research review. J Inform Technol Educ 157–190
19. Cahapay MB (2020) Rethinking education in the new normal post-covid-19 era: a curriculum studies perspective. Aquademia 4(2):ep20018
20. Stub ST (2020) Countries face an online education learning curve: the coronavirus pandemic has pushed education systems online, testing countries' abilities to provide quality learning for all. 2020. shorturl.at/qDEFV, note = Accessed: 27 April 2020
21. Unesco's support: Educational response to covid-19. https://en.unesco.org/covid19/education response/support. Accessed: 19 May 2020
22. Ren M, Ling N, Wei X, Fan S (2015) The application of pdca cycle management in project management. In: 2015 international conference on computer science and applications (CSA), pp 268–272
23. Leap: A manual for learning evaluation and planning in community learning and development. Published by the Scottish Government, November, 2007. https://www.scdc.org.uk/what/leap. ISBN: 978-0-7559-5517-6
24. Zamyatin Y (2010) We: introduction by Will Self. Random House
25. Turing AM (1950) Computing machinery and intelligence. Mind 59(236):433–460
26. Fink J (2012) Anthropomorphism and human likeness in the design of robots and human-robot interaction. In: Ge SS, Khatib O, Cabibihan J-J, Simmons R, Williams M-A (eds) Social robotics. Springer, Berlin, pp 199–208
27. Milhorat P, Schlögl S, Chollet G, Boudy J, Esposito A, Pelosi G (2014) Building the next generation of personal digital assistants. In: 2014 1st international conference on advanced technologies for signal and image processing (ATSIP), pp 458–463
28. LeDoux JE (2012) Evolution of human emotion: a view through fear. Prog Brain Res 195:431–442
29. Tejada H et al (2020) Physical contact and loneliness: being touched reduces perceptions of loneliness. Adapt Hum Behav Physiol
30. Kokku R, Sundararajan S, Dey P, Sindhgatta R, Nitta S, Sengupta B (2018) Augmenting classrooms with AI for personalized education. In: 2018 IEEE international conference on acoustics, speech and signal processing (ICASSP), pp 6976–6980
31. Rajagopal A, Vedamanickam N (2019) New approach to human AI interaction to address digital divide AI divide: Creating an interactive AI platform to connect teachers students. In: 2019 IEEE international conference on electrical, computer and communication technologies (ICECCT), pp 1–6, 2019
32. Chen L, Chen P, Lin Z (2020) Artificial intelligence in education: a review. IEEE Access 8:75264–75278

Chapter 15
Equipping European Higher Education Teachers for Successful and Sustainable e-Learning with Home Remote Work

Inés López-Baldominos, Vera Pospelova, and Luis Fernández-Sanz

Abstract COVID-19 consequences in the shape of restrictions and lockdowns caused a sudden need of transition from traditional higher education (HE) teaching scenario to e-learning with teachers working remotely from home. HE entered in emergency mode and was hardly capable of keeping education service. Sustainability of HE during possible future crisis depends on a good analysis of what has happened during the contingency period. A combined study based on literature review and a specific survey to HE teachers in Europe has collected consistent results to suggest which should be the contents for equipping teachers for the new teaching paradigm. Conclusions show that ensuring HE education service sustainability requires giving teachers more support and training in several areas: distance learning pedagogy, technical troubleshooting, cybersecurity, data privacy, IPR and inclusion through digital accessibility and cultural and gender considerations in e-learning. These are essential factors to enable HE teachers to strength their digital education readiness through e-learning.

Keywords e-learning · COVID-19 · Higher education · Sustainability · Teacher support

1 Introduction

Relevant challenges for the HE community have emerged worldwide during the COVID-19 pandemic. Teachers have had to move urgently and unexpectedly from face-to-face university courses to online teaching. This was the only option to keep

I. López-Baldominos · V. Pospelova · L. Fernández-Sanz (✉)
Universidad de Alcalá, Alcalá de Henares, Spain
e-mail: luis.fernandez.sanz@uah.es

I. López-Baldominos
e-mail: ines.lopezb@edu.uah.es

V. Pospelova
e-mail: vera.pospelova@uah.es

running education during lockdown as all institutions closed their doors, replacing face-to-face activities with online courses as contingency measure. The decision to close temporarily HE centres was prompted by the principle that large gatherings of persons constitute a serious risk to safeguarding public health during a pandemic. The impact on teachers is the focus of our analysis of the evident challenge of working in the continuity of the teaching activity. In summary, the most evident impact on teachers is the expectation, if not the demand, of the continuity of teaching activity using a virtual modality. Far of being a trivial and easy reflection, this problem needs an exhaustive analysis to ensure the sustainability of HE in future critical situations.

A survey with 93 responses from 35 countries [1] through 17 European Universities Alliances (representing 114 institutions) has identified some of the aspects most strongly impacted by the COVID-19 crisis. These are: staff teleworking (8.8), organization of exams (8.8), online teaching (8.8), online assessment (8.4) and digital infrastructure (7.0) have attracted high scores (10 is most impacted) where even work–life balance (6) has been mentioned.

Living this situation has added to the stresses and workloads experienced by university faculty and staff [2]. They were already struggling to balance teaching, research and service obligations, apart from the limitation of work–life balance [3, 4]. Teachers of all backgrounds and ages have had to prepare and deliver their classes in the distance from home. This entails all types of practical and technical challenges, frequently without proper technical support so they mostly had to troubleshoot by themselves [5]. Known risks such as those connected to cybersecurity, data privacy and confidentiality and the proper policies of IPR have gained an enormous importance after the massive increment of the digitalization and remote work in the activities of teachers. However, the most significant challenge for university teachers has been the lack of the pedagogical content knowledge (PCK) [6] which is key for the transition to e-learning [7–9]. This deficit was clearly shown in many cases during the sudden shock of adaptation to lockdown situations. As [10] has revealed, teachers with better PCK background reported better communication with students and more effective online sessions.

The complexity of the instructional situation was aggravated by the shortcomings in planning and organization so teachers faced hundreds of incoming "tips and tricks" without even scarce contextual knowledge: they needed to decide which method would work best in their specific situation in situation of uncertainty. The sources of information ranged from documented reports provided by reputed organisms [11] to mere advise available in ad-hoc Internet blogs.

Other problems previously considered as minor ones have gained importance in the context of the crisis. They have been highlighted by a work focused on the challenge of the digital/technological connection in university teaching as consequence of COVID-19 [12]. It mentions some fundamental factors of the process of teaching inspired in the study of [13]:

- Digital accessibility: teachers and universities need to ensure access for all people who want to attend education, irrespectively of their special needs.

- Teaching scheme: ease the process of learning in the distance context and not a mere integration of traditional model in the new technology context.
- Organizational-cultural consideration: reviewing activities style to analyse the possible impact of cultural background of learners or some gender differences in learner's performance.

All surveys tend to show that almost all (98%) teachers already used communication media [10] when analysing ICT specific aspects of the transition to e-learning. However, in opposition to expectations, the acquisition of digital skills is still under expected levels even in early career teachers (considered by default as "digital natives"). Probably, adoption of real digital transformation of HE centres has been rather slow. This also hinders the motivation and progress of teachers towards an effective digital autonomy. When this situation is combined with telework, the activity is impacted by the absence of interaction/communication with co-workers, the scarcity of technical infrastructure (frequently shared by several members of the family) and the limited, or totally, absent technical support. Of course, we should also mention the reconciliation of teleworking with family dedication and time/schedule management. When looking at the technical part, the standard challenges of employees that work remotely during COVID-19 crisis they include: the issues with Internet bandwidth, the risks and challenges of migration of organization data to personal devices, the exploitation by hackers of the COVID-19 situation with poorly protected environments and greater security exposure as new or inexperienced remote-working employees are frequently in long online sessions.

Considering all the aspects explained above, our research has been focused on finding out and analysing the specific teachers' needs for an effective e-learning. Our approach follows two parallel activities: (a) compiling and analysing 35 existing contributions in literature and reports and (b) the design and conduction of a specific survey to higher education teachers with 112 responses from teachers and managers from Universities of nine European countries. The analysis of results allowed us to find out that both sources of information are very consistent in their conclusions and that the survey is representative of the reality of perception by HE teachers. This consistency enables us to present a solid identification of the most relevant needs for HE teachers for an effective distance learning in COVID-19 times. The structure of content is organized as follows:

- Section 2 describes the specific online survey to HE teachers and managers and the sample that participated in the data collection process.
- Section 3 discusses issues related to pedagogy when moving to a digital environment where the traditional teaching paradigm does not work well.
- Section 4 explores challenges related to the new technical environment of university teachers working at distance, from home, to continue the teaching activity online. Aspects like digital skills of teachers to cope not only with the common office and communication tools but also with the self-management of equipment by configuring devices, network connection and troubleshooting all types of problems with scarce or no remote technical support.

- Section 5 addresses the very relevant aspect of cybersecurity when working from home and other associated risks in data privacy and protection of both teachers and students or the correct management of Intellectual Property Rights (IPR).
- Section 6 will explore the customization of methods to consider inclusion and equality in e-learning by addressing digital accessibility, possible personalization of teaching to work with cultural and gender differences.
- Section 7 compiles a set of other problems and perceptions of teachers and students during their experience with the e-learning process during the crisis.
- Finally, Sect. 8 presents conclusions and future lines of work.

2   Data Collection Process

As commented, we designed a specific survey addressed to teachers of universities from several EU countries to collect their opinion and their needs for e-learning on the different areas of interest identified in our preliminary literature review. We collected 112 responses from university professors and managers from nine European countries, where Spain was the most represented with 25% of the responses. All relevant ages are present in the sample as it included people from 25 to 64 years old: 39% are between 35 and 45 years old and 33% between 45 and 55. Gender representation is very balanced, with a slightly higher participation of women (53%).

The survey's questions were divided into five categories:

- Profile, including information on age, gender, and country of origin.
- E-learning tools and users' level of competences, the goal is to know the preferred tools and if docents have the proper skills to correctly manage e-learning.
- Difficulties in using IT for distance learning, to evaluate the nature of the difficulties encountered during the period of teaching from home.
- Digital pedagogy and innovation, to evaluate de difficulties in moving teaching methods to digital environments and other aspects as accessibility and personalization to address cultural and gender differences.
- Needs of training, to know the opinion of each user regarding the importance of receiving training in the different topics covered.

As the online survey is the main data collection instrument, we consider several mechanisms to increase its reliability and validity. Although questionnaires are generally considered high in reliability as they ask a uniform set of questions. Before disseminating the survey, we made an extensive analysis of questions to ensure the proper writing avoiding problems like ambiguity, confusing phrasing, clear and single questions, etc. Moreover, we organized a pilot phase with a reduced set of representatives of different countries to get their feedback and fix minor details which may hinder or confuse future respondents. Only after these steps, we disseminated the link to the survey. As expressed above in the description of sample, the conclusions could be enough representative given the variety of locations, ages and profiles as well as gender balance.

Obviously, this study has limitations because the sample from nine EU countries with different sizes, and typology is relevant, but responses from other places could be very interesting due to differences in the impact of COVID and the characteristics of each national educational system. The survey did not differentiate too much between teachers from more technical fields and those without such background, something that could be relevant to extract conclusions on their needs of training and preparation. However, the homogeneity of conclusions suggests that the problems and perceptions expressed by the sample of respondents are rather universal and may not differ too much in case of segmentation.

3 Pedagogy in Digital Environments

Just moving traditional content to online platforms is not the correct implementation of an e-learning program; it requires a new planning and design of processes. The new teaching model should facilitate the process of learning rather than the integration of the traditional learning model into the technological context [8, 9]. In these studies, pedagogical content knowledge (PCK), as defined by [6, 7] is considered as a key factor to achieve effectivity and sustainability in e-learning. They state that improving PCK on teachers is essential for communication with learners and for effective online sessions.

Regarding e-learning tools, the results show that Moodle was the platform most preferred by the surveyed teachers (73%). This is consistent with the statistics of Moodle with more than 170 thousand installations and more than 250 million of users (https://stats.moodle.org/). However, other popular LMS such as Blackboard were also widely mentioned (34%). Teams is the most used videoconference tool (61%), but again, other popular tools were mentioned, e.g. Zoom (36%), Skype (33%) or Blackboard Collaborate (25%). The new digital context requires the use of additional tools and resources to reach quality and effectivity, and 53% declare that they are frequently using some additional tools and resources as online quizzes (including live quizzes) or decision-based games as innovative and motivational methods.

The adaptation of learning methods has posed problems for educators. 27% of them considered the transition from traditional leaning methods to online methods as hard or very hard. Even more, 21% declare to have had relevant problems. Teachers are aware of the need of improvement. 97% of them emphasize the need for more training in e-learning methods and good practices and 96% in innovative education.

4   Digital Skills of Teachers Working from Home: Technical Troubleshooting with Limited Support

Although frequently ignored, teleworking in general requires a minimum level of digital skills. When dealing with e-learning, teachers need digital skills for the generation of materials, communication with students or colleagues, organization of digital information, use and management of cloud systems, etc. In general, it is believed that the presence of technology in daily life should have led to more solid digital skills of users. However, the reality as reflected by the analysis of more than 150,000 yearly tests during 6 years until 2019 has shown that the success rate has decreased over the years and that digital skills in Europe are not improving naturally over time [14]. This happens in all modules of tests, but it is especially relevant in the case of cybersecurity and in the one of the essential computer skills modules.

The relevant factors for the teleworkability were exposed by a recent study on telework [15], and digital skills were identified as key for technical feasibility. The study has identified the occupations related to higher education provision in a high level of teleworkability. Despite this, several contributions mentioned in the study have shown that teachers in the different education levels are far from having the recommended level of digital skills for their daily activities. This is consistent with the results of the ECDL study mentioned above [14]. Some European HE organizations like the European University Association (EUA) have declared, "Teachers need training so that they are prepared for learning environments where non-specialists are exposed to digital skills". One specific study found out that the implementation of learning virtual environments in university teaching is hindered by low skills on Moodle use, low digital skills and limited participation in training programs with differences by sex or age [16]. The same situation has been confirmed in different countries, not only in Europe [17, 18]. It seems very clear that solid basic digital skills are a prerequisite for the effective transition of teachers to e-learning. All training programs focused on promotion of sustainable transition to e-learning should devote a good part of effort to digital skills.

Of course, teachers' teleworking connected to e-learning does not only need the most common digital skills. They need to maintain their IT equipment well configured and operative to connect to university's system and to other cloud platforms. The problem is achieving it with a very limited support from the IT services department. This unit is normally dimensioned and organized to support activity within physical venue of the institution. It is not usually prepared for massive demand of assistance from users, who are using heterogeneous configurations and devices from many and varied providers, while they also have very limited access to remote systems for troubleshooting. Most of the reports on telework have highlighted the need of training teleworkers in equipment configuration and installation, network connectivity, best practices for use of system from remote locations and technical troubleshooting. This would be essential although training sometimes is ranked as less important barrier when compared to other IT issues [19]. This is not rare, large organizations (as universities tend to be large) have reported that technical issues were very important in planned programs for implementing telework [20]: the home configuration was non-standard and not supported by technical staff in most cases. At the same time, users were largely expected to fend for themselves for installation and troubleshooting. Therefore, everybody can imagine what happens during an unexpected and sudden crisis as the one created by COVID-19. Obviously, users need specific training for solving most of the issues, as it is not possible that an average technical department could be capable to assist personally all of them. Of course, this is not an excuse for not pursuing a proper IT planning for telework. However, the reality shows that this type of training really works. The study in [21] and data from Manchester Health Authority Report showed that users with at least basic digital skills (measured with the ECDL certification) reduced the request of assistance to help desk service: e.g. from 44% before training to only 10% after it.

The reality that shows the survey is that teachers suffered difficulties related to different aspects of IT during the performance of their learning activities. Issues when configuring a teaching online session or during the session are the ones that have more impact in HE educators: 71% declare having suffered these problems. However, numbers are relevant enough in other aspects: configuring an online service tool caused problems for 59% of respondents; configuring a secure network (57%); managing files and information (54%); configuring a PC (53%); using office applications (51%) or configuring other devices, as scanners or printers, at home (43%).

When it comes to receiving training for this area, the survey covered the interest on three different topics: digital basic skills, IT administration basic skills and best practices for working from home. Educators agree in attending training on these topics: 85%, 81% and 95%, of them, respectively.

5   Cybersecurity, Data Privacy and IPR: Persistent and Enhanced Challenges

Experts in technology and cybersecurity have been warning employees and general users about the increasing threat of cyber-attacks and unauthorized access to data in the last years. Attacks have dramatically increased in impact and number during the pandemic. All organisation are targets of attacks including public institutions and private businesses without forgetting phishing attacks to the general population [22, 23]. The growth in the number of employees working from home is behind this increase, especially given that many of these workers are not used to work outside the office [24]. Potential risks grow when there is a mix usage of devices by employees. They are used for working in tasks, but also in social and personal facets, without the protection of firewalls, proxies, BNS filtering and VPNs. So, the list of potentially risky activities include practices like opening e-mail attachments, having greater data access or more administrator rights than required or than the ones which can be safely managed, or downloading sensitive information onto local drives, forwarding work emails to personal accounts or sharing confidential documents. In the end, these activities may leave the door open to personal and confidential information or enable jumping over security barriers for malicious actions.

Another aspect on the information and contents is the one of Intellectual Property Rights (IPR). This has been relatively vague and imprecise in the e-learning world. However, it has become crucial for legal reasons. Educational materials and resources are expensive to create, and IPR information is vital for digital libraries and repositories. However, IPR management has been frequently poor in many universities due to several reasons: the main one is the complexity of considering all the possible aspects in large organizations [25]. Not all the universities have established clear policies on intellectual property, clarifying the role of teachers and sometimes colliding with students' creations. Some studies have confirmed lack of knowledge of teachers on essential aspects of IPR and its effects on materials when using third party resources or in allocation of IPR licenses to contents [26]. In the end, it becomes exclusively dependent on the interest and skills of these individuals rather than a more general orientation of the university. IPR should be part of any training action to support teachers when moving to e-learning (and it should also have occurred in traditional teaching environments).

Awareness on this wide topic of cybersecurity, privacy and IPR is very high among HE stakeholders. In the survey, 95% of teachers agree with the importance of data protection and self-privacy for effective e-learning. When asked about security concerns increase, 83% agree with the need of getting extra training on IT security and privacy basic skills.

6   e-Learning Customization for Equity and Inclusion: Accessibility, Gender and Cultural Inclusion

There is no doubt that the inclusion of all groups and collectives is necessary, as none should be left behind. Now, digital accessibility is more than just a moral obligation. Directive 2016/2102 of the European Union [27] compels the public sector to ensure digital accessibility of websites, intranets, extranets, published documents, multimedia files and mobile applications. Directive 2019/882 [28] extends this obligation of digital accessibility compliance to websites and mobile applications of essential activities, such as passenger transport services or e-commerce services, whether the seller is a public or private operator. Although legislation is very clear and specific, the lack of knowledge, awareness and motivation to put it into practice is still hindering the practical application to the reality. Erasmus + project WAMDIA highlighted these needs of training based on the results of 27 personal interviews as well as on a survey with 525 responses from people in 16 European countries [29]. Findings showed that the knowledge declared by participants was not consistent with the actual concept of digital accessibility. Only 55% of respondents selected the correct option when asked about the definition as opposed to 76% who indicated knowing about accessibility. Available training is scarce despite accessibility is a mandatory aspect and does not involve a big difficulty for those with basic digital skills. The WAMDIA project organized up to three editions of the training course involving 34 secondary and higher education teachers. Results were very clear; all the participants considered their previous digital skills as enough to learn how to develop accessible documents and content and the difficulty of the course was "normal" for 79% of participants. In summary, digital skills and possible difficulty of the training courses are not an obstacle to implement digital accessibility, but teachers reported a lack of support from universities and institutions.

Nevertheless, inclusion also needs that teachers may adapt and improve e-learning materials and environments according to the different learning styles of online students and their preferred way for being engaged. For example, gender could be factor to keep in mind [30]. Other researchers have examined gender differences in communication patterns in a study with 303 males and 252 females in a Web-based introductory information systems course [31]. When students must face online courses, women communicated more, perceived the environment to have greater social presence, were more satisfied with the course, found the course to be of greater value and had marginally better performance than men had. The results of this study suggest that e-learning environments that allow peer-to-peer communication and connectedness can help female students. Other studies have observed similar results. For example, a sample of 1185 students who had been doing online courses at Universidad de Granada in Spain during 2008–2010 lead the authors [32] to observe that female students are more satisfied than male students with the e-learning subjects; female students assign more importance to the planning of learning, as well as to being able to contact the teacher in various ways. Another study [33] showed that female respondents rated significantly higher the content organization and structure, the appropriateness and variety of resources, the appropriateness of vocabulary and terminology and concrete illustration of abstract concepts as key aspects of usability of e-learning.

Cultural differences are also a relevant factor to e-learning effectiveness. Cultural values have a major impact on learning. E-learning may even make stronger the impact of communication and interaction style in different cultural groups. It is frequent to debate between providing a universal approach to a course as if all learners were similar or to adapt an e-learning program to the specific cultural values of the participants. Geert Hofstede was the author of one of the most comprehensive studies of how behaviour in the workplace are influenced by culture: the so-called Hofstede model [34] uses six dimensions of national culture. It is probably the main reference for assessing the impact of a country's culture in employment and corporate life. The study of [35] showed how learnability of groups with different cultural background (characterized through the Hofstede indicators) is dependent on teacher's style, e.g. enhancing spontaneity, creativity, and individual responsiveness in high-context culture and encouraging cooperation, collaboration, and communication across members in low-context culture participants. Other studies have confirmed the relationship between cultural indicators and style for e-learning design and instruction (e.g. [36]). It seems that cultural background should be present when training teachers to enhance the effectiveness of distance learning. One final and additional consideration of cultural background is based on its influence in tele-working: this aspect impacts both the activities of teachers and students during the whole educational process based on e-learning [37].

In our survey, 82% of the teachers also believe in the interest of inclusiveness and the importance of assuring digital accessibility for the quality of an e-learning program. Moreover, 97% agrees with the main motivation for accessibility: it is essential to promote digital accessibility as a solidarity activism for ensuring that every citizen (including those with special needs) has the right of accessing digital information and services. One favourable fact is that Moodle, the most used platform, meets accessibility standards (https://docs.moodle.org/310/en/Accessibility), so the goal of offering accessible contents is then easier and mostly dependent on the attitude and work of teachers. 54% says that it is important to adapt e-learning courses' design and communication processes considering gender and cultural differences. Teachers also rate as important to be trained in these areas: 94% of respondents agree with the relevance of training in inclusive education and 82% in adding a gender and cultural perspective to e-learning.

7   Other Aspects of e-Learning with Home Telework

Other authors have researched how the COVID-19 situation has affected both teachers and students during the process of adaptation from traditional education context to e-learning. One of the issues that cannot be ignored is technical problems, troubleshooting and well-being during digital interaction. As stated in [38], 79% of 106 teachers had Internet problems and suffered anxiety about having them (74%), also experienced anxiety for not being able to clear up doubts online (68%). 59.6% have reported vision problems, tiredness, etc. Also, 1162 students participated in the study and 75.9% reported problems in understanding online content, over 50% experienced problems in communication with the teacher, and 77.3% have had vision problems, fatigue, etc.

Other contributions have focused on quality and effectiveness of e-learning during the pandemic. The work in [39] reported 53.9% of teachers who consider that online is less effective, 76.3% that there is a loss of interaction and 71.7% experienced difficulty in maintaining attention in class. Moreover, 90.8% suffered critical technical problems directly affecting their classes. Only 32.9% complain about having low digital skills, 63.1% about less job satisfaction in online teaching, 78.9% reported more student absenteeism and 75% less interest and motivation. In this study, only 19% of 407 students believe that online learning is more effective if compared to face to face. Other barriers are lack of interaction (64%) and poor quality and technical problems affecting classes (84.2%). Just 40.8% complain about having low digital skills, 68.9% report difficulty in understanding, and 59.3% experienced less interest and motivation.

Not all the researchers show pessimistic data. Teachers in [40] said that technology overload has not affected teachers' performance, even more they were able to work faster, anytime, anywhere. However, they confirmed that their teaching ability is affected by three factors:

- Technocomplexity: complexity of ICT forces them to learn new and changing skills.
- Technical insecurity: fear of being replaced by technological solutions.
- Technical uncertainty: discomfort due to frequent technology updates, changes in functionality, etc.

This change of context of HE education entails the risk of leaving the most disadvantaged behind. Bring Your Own Device and m-learning policies help to promote inclusion [41], as there are more resources available for students and these are closer to their context [42]. BYOD can be used as an assistive educational technology for the motivation and cognitive development of students. However, it cannot be done without support. Implementing a BYOD strategy at a university without pedagogical support is detrimental to the learning process as stated in [43].

8 Conclusions

When talking about digital education and e-learning readiness, there is a good number of relevant factors to be considered to ensure that university teachers are equipped with the skills and knowledge for effective design and teaching. The fast transition, which had to be adopted in European universities to keep activity during COVID-19 lockdown and restrictions in 2020, revealed that not all the teachers and not all the universities were equally prepared for that. Thus, sustainability of the HE education requires a determining action to equip teachers with the skills they need for effectively working in distance learning from their homes. Possibly, the large set of aspects which need support and training, from basic IT skills to new pedagogic methods and consideration of inclusion and equity, was a complex mix which only in few cases they had previously and completely been covered. Specific aspects had been more clearly known by teachers, but in all of them, we have found pieces of evidence of the need of more support and training. The survey has shown that teachers are demanding guidelines and training in the field to support their online teaching. This training needs to cover their basic IT skills as well as pedagogic methods also adding a perspective of inclusiveness, as this is considered very important for almost everyone and which helps to keep us growing as society. Basic digital skills are a prerequisite for an effective transition to e-learning and to self-solve basic IT incidents frequently associated to remote teaching. This should be a basic aspect in training programs on remote e-learning to face crisis like the one of COVID-19, as stated in research works [14, 15, 20]. Clearly, a modular curriculum, which covers the list of topics listed in this paper, is an essential condition to guarantee a quality and sustainable massive transition to e-learning as a reaction to COVID-19 situation.

We are already working on the training design and contents to provide a solution for all those many teachers and universities struggling to keep higher education working even in limit situations as the ones lived during COVID-19.

References

1. European Universities Initiative (2020) Survey on the impact of COVID-19 on European Universities. European Commission
2. Rapanta C, Botturi L, Goodyear P, Guàrdia L, Koole M (2020) Online university teaching during and after the covid-19 crisis: refocusing teacher presence and learning activity. Postdigital Sci Educ 2:923–945. https://doi.org/10.1007/s42438-020-00155-y
3. Houston D, Meyer LH, Paewai S (2006) Academic staff workloads and job satisfaction: expectations and values in academe. Null 28:17–30. https://doi.org/10.1080/136008005002 83734
4. Veletsianos G, Houlden S (2020) Radical Flexibility and Relationality as Responses to Education in Times of Crisis. Postdigital Sci Educ 2:849–862. https://doi.org/10.1007/s42438-020-00196-3
5. Hodges C, Moore S, Lockee B, Trust T, Bond A (2020) The difference between emergency remote teaching and online learning. Educause Rev 27
6. Shulman L (2011) Knowledge and teaching: foundations of the new reform. Harv Educ Rev 57:1–23. https://doi.org/10.17763/haer.57.1.j463w79r56455411
7. Angeli C, Valanides N (2005) Preservice elementary teachers as information and communication technology designers: an instructional systems design model based on an expanded view of pedagogical content knowledge. J Comput Assist Learn 21:292–302. https://doi.org/10.1111/j.1365-2729.2005.00135.x
8. Baldwin SJ, Ching Y-H, Friesen N (2018) Online course design and development among college and university instructors: an analysis using grounded theory. Online Learn 22(2):2018. https://doi.org/10.24059/olj.v22i2.1212
9. Kali Y, Goodyear P, Markauskaite L (2011) Researching design practices and design cognition: contexts, experiences and pedagogical knowledge-in-pieces. Null 36:129–149. https://doi.org/10.1080/17439884.2011.553621
10. König J, Jäger-Biela DJ, Glutsch N (2020) Adapting to online teaching during COVID-19 school closure: teacher education and teacher competence effects among early career teachers in Germany. Null 43:608–622. https://doi.org/10.1080/02619768.2020.1809650
11. Reimers F, Schleicher A, Saavedra J, Tuominen S (2020) Supporting the continuation of teaching and learning during the COVID-19 pandemic. Oecd 1:1–38
12. Nuere S, de Miguel L (2020) The digital/technological connection with COVID-19: an unprecedented challenge in university teaching. Technol Knowl Learn. https://doi.org/10.1007/s10758-020-09454-6
13. Sangrà Morer A (2006) Educación a distancia, educación presencial y usos de la tecnología: una tríada para el progreso educativo. Edutec-e. https://doi.org/10.21556/edutec.2002.15.541
14. López Baldominos I, Fernández Sanz L, Pospelova V (2020) Análisis de las competencias digitales básicas en Europa y en España. JENUI 5:77–84
15. Sostero M, Milasi S, Hurley J, Fernandez-Macias E, Bisello M (2020) Teleworkability and the COVID-19 crisis: a new digital divide?
16. Espinosa HR (2016) Desarrollo de habilidades digitales docentes para implementar ambientes virtuales de aprendizaje en la docencia universitaria. Sophia 12:261–270
17. Grünwald N, Pfaffenberger K, Melnikova J, Zaščerinska J, Ahrens A (2016) A study on digital teaching competence of university teachers from Lithuania and Latvia within the peesa project. Andragogy 7:109–123

18. Kühn M, Grünwald N, Pfaffenberger K, Zascerinska J, Ahrens A (2017) A study on digital teaching competence of trainers from South Africa within the PEESA project, pp 116–124
19. Office of Government wide Policy General Services Administration (2002) Final report on technology barriers to home-based telework. Whashington, DC
20. Williams R, Procter R, Dalziel P (2008) A case study of a small group teleworking pilot in a large organisation. Retrieved 25/04/2010, from http://citeseerx.ist.psu.edu/viewdoc/summary
21. Van Deursen A, Van Dijk J (2012) CTRL ALT DELETE. Lost productivity due to IT problems and inadequate computer skills in the workplace. Universiteit Twente, Enschede
22. Borkovich DJ, Skovira RJ (2020) Working from home: cybersecurity in the age of COVID-19. Issues Inform Syst 21
23. Kamal AHA, Yen CCY, Ping MH, Zahra F (2020) Cybersecurity Issues and Challenges during Covid-19 Pandemic
24. Evangelakos G (2020) Keeping critical assets safe when teleworking is the new norm. Netw Secur 2020:11–14. https://doi.org/10.1016/S1353-4858(20)30067-2
25. Casey J, Proven J, Dripps D (2006) After the Deluge: navigating IPR policy in teaching and learning materials
26. Nunes Gimenez AM, Machado Bonacelli MB, Carneiro AM (2012) The challenges of teaching and training in intellectual property. J Technol Manag Innov 7:176–188
27. European Parliament, Council of the European Union (2016) Directive (EU) 2016/2102 of the European parliament and of the council of 26 October 2016 on the accessibility of the websites and mobile applications of public sector bodies
28. European Parliament, Council of the European Union (2019) Directive (EU) 2019/882 of the European parliament and of the council of 17 April 2019 on the accessibility requirements for products and services
29. WAMDIA Erasmus+Project (2018) Information from stakeholders on digital accessibility collected through a survey and interviews
30. Garland D, Martin B (2005) Do gender and learning style play a role in how online courses should be designed. J Interact Online Learn 4:67–81
31. Johnson RD (2011) Gender differences in e-learning: communication, social presence, and learning outcomes. J Organ End User Comput 23:79–94. https://doi.org/10.4018/joeuc.201 1010105
32. González-Gómez F, Guardiola J, Martín Rodriguez Ó, Montero Alonso MÁ Gender differences in e-learning satisfaction. FEG Working Paper Series 01/11, Faculty of Economics and Business (University of Granada)
33. Zaharias P (2008) Cross-cultural differences in perceptions of e-learning usability: an empirical investigation. Int J Technol Human Interact (IJTHI) 4:1–26. https://doi.org/10.4018/jthi.200807 0101
34. Hofstede GH, Hofstede GJ, Minkov M (2005) Cultures and organizations: software of the mind. Mcgraw-hill, New York
35. Swierczek FW, Bechter C (2010) Cultural features of e-learning. In: Sampson D, Spector JM, Ifenthaler D, Isaias P, Kinshuk (eds) Learning and instruction in the digital age. Springer, Boston, pp 291–308
36. Downey S, Wentling RM, Wentling T, Wadsworth A (2005) The relationship between national culture and the usability of an e-learning system. Null 8:47–64. https://doi.org/10.1080/136 7886042000338245
37. Wojčák E, Baráth M National culture and application of telework in Europe. Eur J Bus Sci Technol 3:65–74
38. Subedi S, Nayaju S, Subedi S, Shah SK, Shah JM (2020) Impact of E-learning during COVID-19 pandemic among nursing students and teachers of Nepal. Int J Sci Healthc Res 68–76
39. Nambiar D (2020) The impact of online learning during COVID-19: students' and teachers' perspective. Int J Indian Psychol 8:783–793
40. Christian M, Purwanto E, Wibowo S (2020) Technostress creators on teaching performance of private universities in Jakarta during Covid-19 pandemic. Tech Rep Kansai Univ 62:2799–2809

41. Ossiannilsson E (2018) Increasing access, social inclusion, and quality through mobile learning. IJAPUC 104(2):29–44
42. Erazo J, Jiménez Peralta J, Varela Patiño C Bring you own device—BYOD como tecnología educativa asistida para la motivación y el desarrollo cognitivo de los estudiantes como parte de la inclusión educativa. In: Memorias del tercer Congreso Internacional de Ciencias Pedagógicas: Por una educación inclusiva: con todos y para el bien de todos, pp 1248–1256
43. Cereceda Fernández-Oruña J, Sáncehz Jiménez F, Herrera Sánchez D, Martínez Moreno F, Rubio García M, Gil Pérez V, Santiago Orozco AM, Gómez Martín MÁ (2019) Estudio sobre la cibercriminalidad en España

Chapter 16
Fully Online Project-Based Learning of Software Development During the COVID-19 Pandemic

Atsuo Hazeyama⊙**, Kiichi Furukawa, and Yuki Yamada**

Abstract The COVID-19 pandemic impelled educational institutions worldwide to deliver online distributed education. We have been developing project-based learning (PBL) style software engineering education since 1997. Because of the COVID-19 pandemic, our PBL course was applied fully online during the 2020 academic year. Our PBL method uses GitHub to manage artifacts and provide feedback to student groups, i.e., the teaching staff perform the inspection process and acceptance testing. To deliver the course during the 2020 academic year, we changed the lecture method from a face-to-face approach to one that uses an online meeting system. All the groups successfully completed their project. In the final assignment, we asked the students to describe any "difficulties they encountered during their remote learning activities and their solutions to these problems." As the results indicate, none of the students experienced any major difficulties. The teaching assistants and instructors did not encounter any major issues in fully online remote PBL because the progress of all groups could be ascertained during a progress meeting held each week, as well as through the inspection process conducted during the upstream phase and the acceptance testing, and thus, any problems could be addressed promptly. GitHub played an important role in this process. Our process and software engineering environments, which are a combination of GitHub and an online meeting system, are suitable for a fully online remote PBL.

Keywords Fully online remote project-based learning · Software engineering education · GitHub

A. Hazeyama (✉)
Tokyo Gakugei University, Koganei 184-8501, Tokyo, Japan
e-mail: hazeyama@u-gakugei.ac.jp

K. Furukawa · Y. Yamada
Graduate School of Education, Tokyo Gakugei University, Koganei 184-8501, Tokyo, Japan

223

1 Introduction

Educational institutions worldwide were compelled to implement online distributed education during the COVID-19 pandemic [1–8], and our university was among those that conducted online remote lectures during the 2020 academic year.

We have been developing project-based learning (PBL) style software engineering education since 1997 [9, 10]. Because of its characteristics, this learning style relies on the teaching staff giving feedback to the students. The results of software inspection of the artifacts of the upstream phase, acceptance testing for the developed system, and progress checks during lecture time are used as feedback in our PBL approach for software engineering education. The student groups manage all such artifacts in a repository on GitHub, which is also used for the inspection process and acceptance testing.

The PBL course examined in this study was delivered fully online during the 2020 academic year because of the COVID-19 pandemic. This paper reports an overview of the course, the preparations for fully online remote operations, and the results of its implementation. In our study, we attempted to answer the following questions. What difficulties do students with less experience in PBL style software development encounter in a fully online environment? How are these difficulties manifested?

The rest of this paper is organized as follows. Section 2 describes the state of the art of software engineering education during the COVID-19 pandemic. Section 3 provides an overview of our PBL software development course before the COVID-19 pandemic. Section 4 presents the PBL course we delivered in the 2020 academic year. Section 5 describes the results of this practical implementation. In Sect. 6, our fully online remote PBL course is evaluated. Finally, Sect. 7 summarizes this paper.

2 Related Work

Several studies have been conducted in the context of software engineering education during the current pandemic [11–18].

Barr et al. described the manner in which intensive online lectures were conducted [11]. Their results showed that the students preferred watching pre-recorded videos over participating in live online and even face-to-face classes, and the authors concluded that flipped learning is an effective means of teaching in such a situation.

In addition, Schmiedmayer et al. distributed lectures in the form of live-streamed pre-recorded videos, during which the teacher and teaching assistants allowed time for questions [12].

Kanij and Grundy presented guidelines for taking full advantage of online learning in which they included the means of making announcements to students, the learning materials, and the assessment method [13].

Mues and Howar claimed that, when the participants in fully online remote PBL do not know one another, it is important to establish communication [14] and described their approach.

Plewnia et al. reported the experience of using PBL for teaching software project laboratories in collaboration with an industry partner during the COVID-19 pandemic [15]. They found that the greatest disadvantage in this situation was the lack of direct contact between the students and the industry partners.

Bringula et al. investigated the challenges of implementing a programming project course during the COVID-19 pandemic and proposed their solutions to these challenges [16]. They identified 13 challenges; however, they did not evaluate their solutions based on feedback from the students.

Motogna et al. presented an empirical study that was aimed to improve the understanding of the manner in which the assessment of student learning changes in response to the transition from in class to online courses [17].

Finally, Yamada et al. noted the difficulties related to building a software engineering environment for Web application development using the students' own laptop computers in a fully online remote environment [18]. They developed a script for automatically building the software engineering environment.

Few studies have examined the transition to a fully online remote project-based software engineering education during the COVID-19 pandemic. In particular, to the best of our knowledge, no studies have been conducted on whether students having less experience with project-based software development find it difficult to learn in a fully online remote environment, and if so, what types of difficulties they experience.

3 Overview of Our PBL of Software Development (Before the COVID-19 Pandemic)

The PBL software development course is offered to third-year undergraduate students in the Department of Informatics Education of Tokyo Gakugei University. The quota of the department is 15 students. Therefore, the PBL course is provided at a small scale. The course consists of 15 weekly 90 min lectures. The task the students are set is Web application development using Java. In the semester preceding that in which the PBL software development course is offered, we deliver an introductory course on software engineering.

Among other aspects of PBL, we specify the software development process, artifact, and approach to grading in the information provided to the students about the operation of the course and explain them during the first lecture. Each lecture consists of announcements from the instructor or teaching assistants (TAs), explanations of the usage of tools (e.g., GitHub), and group meetings and progress checks. The TAs are master's students who previously passed this course. Certain activities may need to be conducted outside lecture hours.

Each group consists of three–five students. We perform a questionnaire survey for organizing the groups and determine the organization of each group based on the results of both the questionnaire survey and the students' grades in the introductory software engineering course.

The development process is based on the waterfall model. The types of artifacts are requirements specification, user interface design document, class diagram, database design document, sequence diagram, source code, unit/system testing report, development plan, group progress report (each week), and project completion report. A sequence diagram and source code are created, and unit testing is conducted for each function by each student.

The group progress report is presented in turn by each group member. Progress checks during the lecture time, the inspection of artifacts created during the upstream phase, and acceptance testing of the application developed by each group are conducted by the teaching staff to provide feedback to the student groups.

As the development environment, the students use their own laptop computer. We use GitHub as the source code and document repository. We use version control in the documents, the "Issues" function (the formal location of our text communications, including discussions and bug reporting, among other exchanges), and the "Pull Request" function for the review process for artifacts. The groups are allowed to use a variety of tools.

4   Our PBL in the 2020 Academic year

All face-to-face learning activities were prohibited in the PBL course of the 2020 academic year.

The task in this year's PBL software development course was based on a request from a professor at our university. It consisted of software development for an actual client, constituting a Web application for home economics education, where the students select items from the food menus stored in the system, and the system calculates the adequate level of nutrition and provides advice to the students.

We used Microsoft Teams as the infrastructure of online remote lectures and applied the following functions provided by the application:

- Creation of a "team" (private)
- Creation of a "channel" as a place of communication per group under a "team"
- Text chat
- Voice communication
- Video communication
- Screen sharing
- Creation of folders and file uploading.

We also implemented a function to transfer the content of the Issues function in GitHub to Teams. GitHub was used in the same manner as prior to the pandemic.

Table 1   Some quantitative data regarding the process of groups in the 2019 academic year and the 2020 academic year

Items	Group A in 2019	Group B in 2019	Group B in 2020	Group B in 2020
Lines of code	4591	7088	61,036	9073
Number of issues	117	77	203	187
Number of pull requests	228	207	250	254

We used Teams and GitHub for different purposes. Live lectures were delivered using the voice communication and screen-sharing functions of Teams. Discussions between the developers and the members of the Faculty of Home Economics, who were the clients, were held using Teams. By contrast, communications between the teaching staff and developers regarding the creation of artifacts based on text were typically exchanged on GitHub.

The progress meeting during the lecture hours required a longer time in the remote online environment than in a normal learning environment (10 min per group during the previous years versus 15 min per group during the 2020 academic year).

5   Result

Nine students, organized into two groups, participated in the course during the 2020 academic year. The PBL was conducted from October 23, 2020, to February 5, 2021. Both groups completed their project, presented their processes, and demonstrated their system on February 5. All the students completed the course.

Table 1 shows partial data on the developed system and the development process during the 2019 and 2020 academic years. As the table indicates, in 2020, the student groups produced more lines of code, Issues, and Pull Requests than in 2019. Because the project for 2020 was the development of a system required by actual users, it was more complicated than those for past years. Table 1 shows that the groups worked actively in a fully online remote environment. Figure 1 shows the partial communication history of a team in Microsoft Teams, and Fig. 2 shows an example screenshot of the Issues function provided by GitHub.

6   Evaluation

We clarified the difficulties encountered by the students with less experience in conducting PBL style software development in a fully online remote environment.

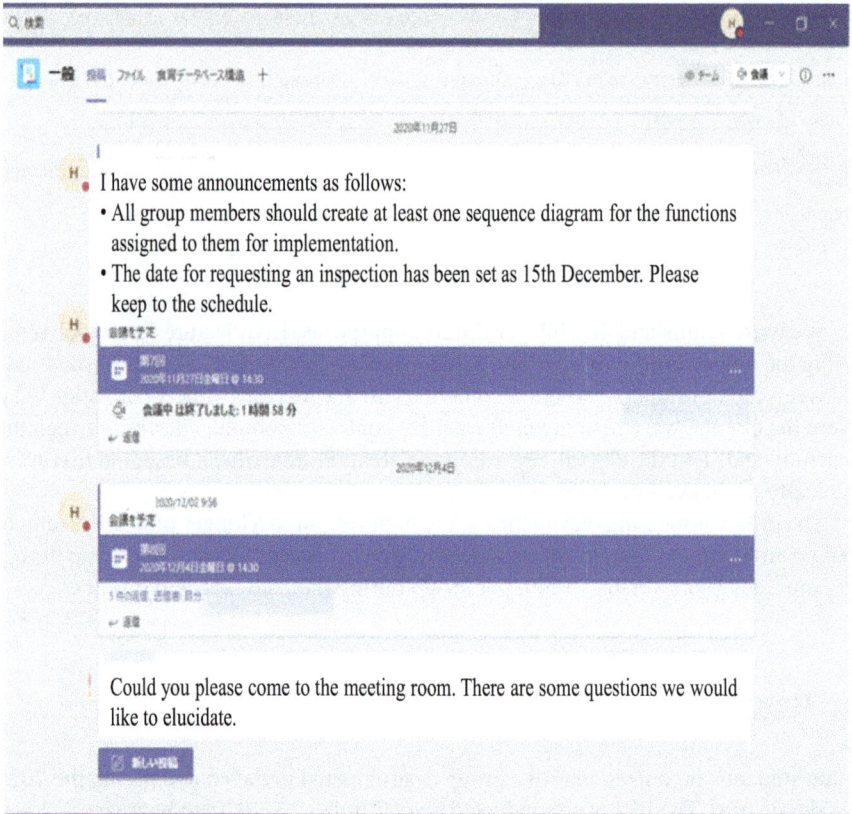

Fig. 1 Screenshot of teams

As a part of the final assignment, the instructor asked the students to describe the "difficulties they encountered during their remote learning activities and their solutions to these problems."

As the results indicate, none of the students encountered any major difficulties. Some of the students' comments are as follows.

The positive opinions regarding remote learning activities are as follows:

- I had no problem not meeting with other members face to face and saw the possibility of working remotely.
- I did not have any difficulty working remotely.
- It was easy to prepare for group meetings (no need to travel to class).
- No delay was caused by working remotely.
- I experienced no difficulties because we were able to communicate through social networking services.

The students also noted some of the difficulties they experienced:

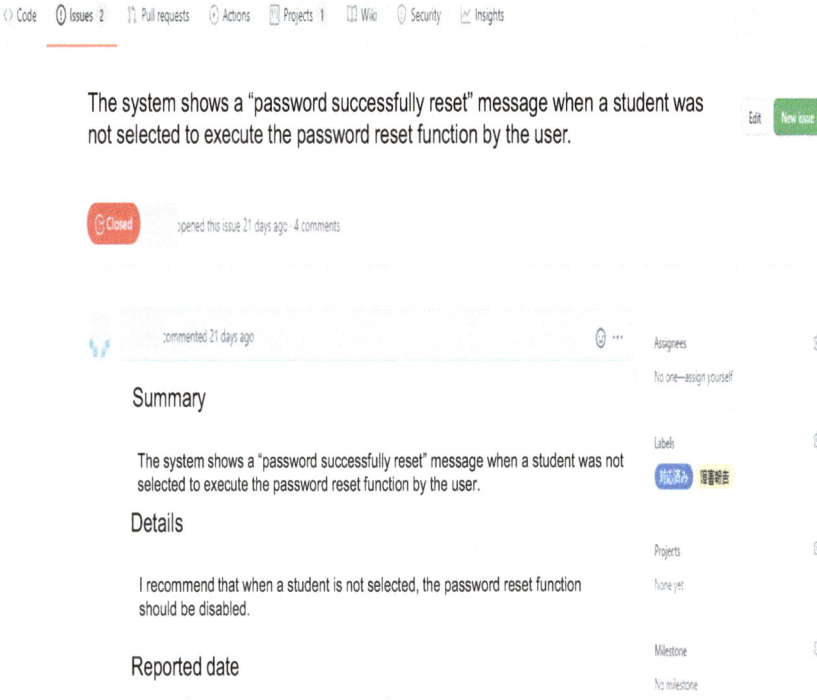

The system shows a "password successfully reset" message when a student was not selected to execute the password reset function by the user.

Edit   New issue

Closed   opened this issue 21 days ago · 4 comments

commented 21 days ago

Assignees
No one—assign yourself

Summary

The system shows a "password successfully reset" message when a student was not selected to execute the password reset function by the user.

Labels
対応済み 障害報告

Details

I recommend that when a student is not selected, the password reset function should be disabled.

Projects
None yet

Reported date

Milestone
No milestone

Fig. 2 Screenshot of the issues function provided by GitHub

- It was difficult to ascertain the progress of the other group members.
- I felt stress having text conversations because doing so takes hours and more time is required until a resolution is found.
- It was difficult for us to understand the requirements of the users. In addition, it was difficult to communicate my thoughts about the development to the other group members.
- It was difficult to fix my problems outside of class because the team members were not staying on campus.
- It was not easy to hold talks and/or chats (early chats may lead to group cohesion).
- I felt unmotivated because I was unable to view the environment surrounding the other members.

We asked the teaching assistants to answer the same question from their perspective and obtained the following opinions.

- I experienced no problems with remote PBL from either the student or the teaching staff side.
- I was able to ascertain the progress of the groups through the progress meeting held once a week during the lecture time.

- I could ascertain the artifacts in the upstream phase clearly because I participated in the inspection process as an inspector.
- Because the course does not introduce code inspection by the teaching staff, I experienced difficulty ascertaining the progress (this was also true during face-to-face PBL).

In general, the students who had less experience in software development conducted in a fully online remote PBL environment did not experience difficulties. Some students noted difficulties in ascertaining their progress, the extra effort needed to communicate, the difficulties that resulted from not being promptly supported when they needed help, and issues of motivation. These difficulties may arise even when this course is conducted in a collocated and not fully online remote environment because the group members do not meet frequently. Clearly, fully online remote environments make these difficulties tangible.

This evaluation shows that a project-based software development course can be conducted in an effective manner and without pauses in learning even when the students are unable to meet in a collocated manner. The evaluation results show that a learning environment in which GitHub is used in combination with a remote meeting system is effective in allowing groups to conduct software development activities.

7 Summary

This paper reported a practical implementation of PBL style software development education during the COVID-19 pandemic. Most students in our PBL course are novice software developers. As one of the characteristics of this learning style, the teaching staff provide feedback regarding the output of student groups, which is based on the results of the inspection process during the upstream phase, acceptance testing, and progress checking. Our PBL course was conducted before the COVID-19 pandemic, using GitHub for managing the artifacts and communication between the teaching staff and groups. Then, because of the COVID-19 pandemic, the weekly face-to-face lectures were also moved to a remote meeting system. No additional changes were made to the course during the pandemic. The PBL course of the 2020 academic year was successfully completed. The students experienced no serious difficulties regarding their fully online remote project-based software development course. In our opinion, the process for providing feedback, as outlined above, may be effective. In addition, we are also of the opinion that artifact management using GitHub and communication between the teaching staff and the groups using GitHub and Teams, which played an important role in this situation, may be effective in a fully online remote environment. COVID-19 has not yet been eradicated worldwide. Therefore, strong measures such as lockdowns may be required again in the future. Our approach may be effective in such situations.

In the development of modern software, such as open-source software, artifacts are created in an electronic format, communications are exchanged electronically, and the development is conducted in a distributed fashion. These characteristics are resilient to the conditions enforced by the COVID-19 pandemic. Our PBL has these characteristics. In addition, the proportion of activities that require face-to-face communication is low, and therefore, the transition to PBL can be achieved smoothly.

This paper described the interaction processes between the teaching staff and the student groups from the viewpoint of the teaching staff. We plan to further investigate the micro-processes within the student groups that occurred during the COVID-19 pandemic.

Acknowledgements This study is partially supported by the Grant-in Aid for No. (C) 20K12089 from the Ministry of Education, Science, Sports and Culture of Japan. The authors thank Professor Michiko Minami and Professor Sakura Kushiyama and all the students who participated in this course for their contributions in the 2020 academic year.

References

1. Bao W (2020) COVID-19 and online teaching in higher education: a case study of Peking University. Hum Behav Emerg Technol 2(2):113–115
2. Luburić N, Slivka J, Sladić G, Milosavljević G (2021) The challenges of migrating an active learning classroom online in a crisis. Comput Appl Eng Educ 1–25
3. Vogel-Heuser B, Bi FLK, Trunzer E (2020–2021) Transitions in teaching mechanical engineering during COVID-19 crisis. Interact Design Archit J 27–47
4. Serhan D (2020) Transitioning from face-to-face to remote learning: students' attitudes and perceptions of using zoom during COVID-19 pandemic. Int J Technol Educ Sci 4(4):335–342
5. Aristovnik A, Keržič D, Ravšelj D, Tomaževič N, Umek L (2020) Impacts of the COVID-19 pandemic on life of higher education students: a global perspective. Sustainability 12(20):34 pages
6. Ghazi-Saidi L, Criffield A, Kracl CL, McKelvey M, Obasi SN, Vu P (2020) Moving from face-to-face to remote instruction in a higher education institution during a pandemic: multiple case studies. Int J Technol Educ Sci 4(4):370–383
7. Rashid CA, Salih HA, Budur T (2020) The role of online teaching tools on the perception of the students during the lockdown of Covid-19. Int J Soc Sci Educ Stud 7(3):178–190
8. Mahmood S (2020) Instructional strategies for online teaching in COVID-19 pandemic. Hum Behav Emerg Technol 199–203
9. Hazeyama A (2000) An education class on design and implementation of an information system in a university and its evaluation. In: Proceedings of the 24th annual international computer software and applications conference (COMPSAC2000). IEEE, pp 21–27
10. Miyashita Y, Yamada Y, Hashiura H, Hazeyama A (2020) Design of the inspection process using the GitHub flow in project based learning for software engineering and its practice. arXiv: 2002.02056
11. Barr M, Nabir SW, Somerville D (2020) Online delivery of intensive software engineering education during the COVID-19 pandemic. In: Proceedings of the 2020 IEEE 32nd conference on software engineering education and training (CSEE&T2020). IEEE, pp 1–6
12. Schmiedmayer P, Reimer L, Jovanović M, Henze D, Jonas S (2020) Transitioning to a large-scale distributed programming course. In: Proceedings of the 2020 IEEE 32nd conference on software engineering education and training (CSEE&T2020). IEEE, pp 1–6

13. Kanij T, Grundy J (2020) Adapting teaching of a software engineering service course due to COVID-19. In: Proceedings of the 2020 IEEE 32nd conference on software engineering education and training (CSEE&T2020). IEEE, pp 1–6
14. Mues M, Howar F (2020) Teaching a project-based course at a safe distance: an experience report. In: Proceedings of the 2020 IEEE 32nd conference on software engineering education and training (CSEE&T2020). IEEE, pp 1–6
15. Plewnia C, Steffens A, Wild N, Lichter H (2021) A lightweight collaborative approach for teaching software project labs with industry partners, In: Joint proceedings of SEED & NLPaSE, CSUR-WS, vol 2799, pp 1–8. http://ceur-ws.org/Vol-2799/
16. Bringula R, Geronimo S, Aviles A (2021) Programming project in an undergraduate software engineering in the new normal: challenges and proposed solutions. In: Joint proceedings of SEED & NLPaSE, CSUR-WS, vol 2799, pp 29–37. http://ceur-ws.org/Vol-2799/
17. Motogna S, Marcus A, Molnar AJ (2020) Adapting to online teaching in software engineering courses. In: Proceedings of the 2nd ACM SIGSOFT international workshop on education through advanced software engineering and artificial intelligence. ACM, New York, pp 1–6
18. Yamada Y, Furukawa K, Hazeyama A (2021) Conducting a fully online education of a software engineering course with a web application development component due to the COVID-19 pandemic and its evaluation. In: Joint proceedings of SEED & NLPaSE, CSUR-WS, vol 2799, pp 20–28. http://ceur-ws.org/Vol-2799/

Chapter 17
A Tale of Two Zones: Pandemic ERT Evaluation

Enamul Haque, Tanvir Mahmud, Shahana Shultana, Iqbal H. Sarker, and Md Nour Hossain

Abstract During the COVID-19 pandemic, many educational institutes switched from in-person to all virtual classes. Few educational institutes were well prepared for emergency remote teaching (ERT), whereas many others faced considerable problems in terms of preparation and delivery. As we know that remote teaching is highly dependent on technology infrastructure and this is not evenly accessible in developed and developing countries, we, while working in different universities, noticed differences in both teachers' and students' attitude toward ERT and so decided to investigate ERT between these two zones. We first identified problematic factors and then evaluated and compared two zones in terms of their ERT programs from well-known CIPP (context, input, process, product) model perspective. Most of the analysis results are presented here following standard data science visualization techniques. We also conducted qualitative and quantitative research to find the gaps and improvement opportunities for both zones.

Keywords ERT · SEM · CIPP · COVID-19 · Education

E. Haque (✉)
University of Waterloo, Waterloo, Canada
e-mail: enamul.haque@uwaterloo.ca

T. Mahmud
Harrisburg University of Science and Technology, Harrisburg, USA
e-mail: tmahmud@my.harrisburgu.edu

S. Shultana
Prairie View A&M University, Prairie View, USA
e-mail: sshultana@pvamu.edu

I. H. Sarker
Chittagong University of Engineering and Technology, Chittagong, Bangladesh
e-mail: iqbal@cuet.ac.bd

M. N. Hossain
Indiana University Kokomo, Kokomo, USA
e-mail: mhossai@iu.edu

1 Introduction

The first epidemic of the twenty-first century was in the year 2002, and it was the severe acute respiratory syndrome (SARS). Like COVID-19, a kind of coronavirus was responsible for SARS, which was known as SARS-CoV in the literature. If we compare the year 2019 to 2002 from a technological advancement perspective, we will clearly see that the world was much better prepared for ERT. This can be shown through countless facilities that provided smooth online program options. Though, initially many educational institutes assumed the impact of COVID-19 might not reach elsewhere other than closer countries of China, when on March 11, 2020 WHO declared COVID-19 pandemic, many institutes found themselves not even prepared for them, and few of them were planning for alternative ways to run classes since the beginning of 2020. After the pandemic declaration, in the early spring of 2020, many educational institutes across the globe were shut down. If we consider only schools, around 1.2 billion children were out of the classroom in 186 countries around the world. The educational institutes that predicted and prepared accordingly managed to switch to emergency remote teaching (ERT) [30]. Those institutes which were not well prepared suffered the most. In this research, we focus particularly on ERT, the term which is coined to define temporary shift of instructional delivery under crisis circumstances which is obviously different from the regular online offering of courses.

The difference between developed and developing countries is measured basically from the Industrial Base and Human Development Index (HDI) relative to other countries. According to Investopedia [18], other factors like gross domestic product (GDP), gross national product (GNP), per capita income, level of industrialization and scale of technological infrastructure are also used to assess the classification of countries as developed or developing. The implementation of face masks, social distancing, and in-home quarantine have forced students and teachers alike to rely on technology at a much higher level than usual. The scale of deployment or coverage of such communication technologies are not the same everywhere, and teachers and students in different zones have different levels of affordability. In such a setting, it is worth exploring and evaluating user experience and preparation of the overall program from institutional perspective.

In this research, we have used survey techniques to gather data from school, teachers, and students. Our questionnaire was designed in such a way so that it could capture both quantitative and qualitative data. We asked specific questions so that we could measure their stress, performance, types, satisfaction, course load etc. These also helped us to identify the problematic factors in both zones which are hindering the overall success of the program. We planned for an ERT conceptual framework for students and overall ERT program evaluation, so the questions were carefully selected to serve that purpose.

The conceptual model helps us identify the relationships between unobserved constructs or latent variables and observable variables. We use data analysis techniques to measure the observed variables. Then, using the connections in conceptual model, we applied structural equation modeling (SEM) to measure the latent variables and thus validated our model [29]. Program evaluation is a systematic method for collecting, analyzing, and using information to answer questions about a program's strengths, weaknesses and effectiveness [31]. As suggested in [9], we considered tailored CIPP (context, input, process, product) evaluation model [34] to evaluate the ERT programs in developed and developing zones. Thus, we came up with the following research questions:

1. What are the identified factors in ERT and how do they affect learning in developed and developing countries?
2. How is the evaluation of ERT effort compared between the zones?
3. What are the gaps from a technical perspective and how can those be mitigated?

In the next few sections, we discuss relevant literature, then describe the methodology we followed to the answer these research questions. Then, we present the evaluation, results, and finally the conclusion and future research direction at the end.

2   Literature Review

COVID-19 situation is just a year old, yet the topic managed to be the subject of 211,756 scientific articles alone in the WHO database [39]. As we are focusing on the pandemic's impact on the teaching and learning side, we have identified some major publications related to education during pandemic.

There is a study of how COVID-19 has dramatically reshaped the way global education is delivered [6]. In this study, the researchers mentioned education as the most affected area in this pandemic. Another study [26] aims to explore the acceptance of remote study to continue knowledge spreading and gathering. The graduate students easily adopted the system, and the result was similar to other related studies. To be more specific, higher education students have low resource settings from the perspective of developing countries. The study [2] expands more on low resource student engagement and the effect of it. In [20], the authors discussed the best practices for implementing remote learning during a pandemic. This study [21] describes a local upper secondary school in Finland that applied the same technique to high school students during COVID-19. The results were surprisingly similar for different parts of the developed country. Despite a lot of negative results coming out so far, there was not an alternative method of engaging mobile and electronic devices. Educators were seeking various other ways to engage students.

The American Journal of Pharmaceutical Education [17] shows some analysis which can be used to get a clearer picture of the current situation. One of the findings is that different studies and the issues that hinder acceptance of criteria for ERT systems are not the same. Another problem is that the study and skill development are not proportional. Also, hands-on experience is missing in remote education. The article [3] shows drawbacks and possible ways of overcoming remote education obstacles. The study of Sub-Saharan Africa regarding device using efficiency [22] gives us the picture of how a lack of skill set in technology use can degrade effort down to zero. The socioeconomic situation of that area is not strong enough to buy mobile phones, desktop computers, or tablets for all students. The practical experiments are far from general acceptability. The article [10] explores the newly emerged interaction between education and health within the context of COVID-19. This provided ideas to develop plausible solutions for remote study. One goal was to find a proven method of study that can be applied as a model around the world. The theoretical only lecture is not relevant to most fields where an effective alternative for these studies is expected. A case study of Azerbaijan explains the attitude of students toward online education [5].

The report [1] by Amnesty International shows the impact of COVID-19 on a broken and unequal education system. This publication includes a scenario where teaching methods are failing due to non-accepted practices forced down from a governing system. This might be a normal case for people in developing areas because most of the people are not used to using mobile devices as a training platform. They expect mobile devices as communication and entertainment tools rather than to be learning from on a regular basis. This change of mentality requires a preparation of infrastructure that adopts the "New Norm" [4]. Another study in Indonesia presented similar results. In the republic of Indonesia (Kemendikbud), the authors investigated education acceptance criteria and concluded that it can be improved using different kinds of apps and collaborating software [12]. It is clear from their work that if the application has user-friendly features, then it can be a suitable and acceptable by the teachers and learners.

There are few examples in the works [33, 36] that illustrate how remote labs can reduce cost and increase the effectiveness of teaching methods. How modern software and Internet communication systems can help in remote learning is also explained in [19]. A full system transformation from offline to online lesson delivery system at the University of Sofia, and the introduction of a prototype in a laboratory of electrical engineering technology can be found in [23, 25]. A similar study in pharmaceutical education is also mentioned in [17].

Education has been highly affected by the COVID-19 pandemic. From kindergartens to universities, everyone involved in academia has struggled to keep going despite all the difficulties and restrictions [37]. Another research work [11] experimented for six months with communities and concluded that community study can be a very strong teaching method. Communicating in person is a much more interactive way of learning. This is due to most broadcasting and mobile medias being unidirectional. Moreover, in-person facial expression and nonverbal communications are absent there.

In [27], researchers are focused on comparison and development of innovative study methods. The Japanese education system introduced "Tencent classroom + vocational education cloud + QQ group" which is "suspended classes without suspension of learning" developed by Vocational colleges of Japan [16]. Many international organizations are working on this purpose. IEEE Region 9 (R9) Education Activities Committee (EAC) launched a new initiative to develop a generic platform to develop such systems [7]. The UAE Higher Education Institutions are also working with the same goal [24]. The research [38] contains some best practices which can be followed for remote education in a pandemic. Researchers in [2] depicted how engaging in emergency online learning in low resource settings can be effective in higher education. Abrupt change in system can also lead to unknown consequences [35]. Educators were also thinking about how to involve physical activities, i.e., "center of mass" (COM) concept and develop the method of evaluation (e.g., body mass index), to build a learning object that involves interactive modeling. The study and survey explored here is very helpful also in tracking progress [28].

The CIPP model is considered an acceptable method to gather this performance measurement [34]. Context, input, process, and product (CIPP) model is a standard model for program evaluation and which has been used successfully in the area of education [13]. In the work [13], researchers designed a survey that shows an evaluation of teaching from the student's perspective. For example, nursing programs must maintain a high-quality curriculum that graduates exemplary nurses. Systematic evaluation of key components of nursing education is discussed in [15]. Another example is in the study of entrepreneurship. As a brand-new educational concept and education mode, entrepreneurship education has been widely carried out in colleges and universities across the country [32]. Thus, the effect of emergency remote learning systems is not directly predictable, but rather determined by infrastructure, subject, technological readiness, and other socio-economical factors. Student feedback and their performance evaluation in the CIPP model can explain it better. Our primary hypothesis is among students who have participated in emergency remote teaching (ERT) during the COVID-19 pandemic, the outcome of education has a greater effect on countries with developed countries than developing countries.

3   Methodology

Measurement of quality can be different among various methods of evaluation. Conceptual perception can be vague and different in person to person. Usually, such perception can be measured by asking some questions to the attendee and also by assessing the overall system. In this research, we focused on the problems that students and teachers were facing during their ERT experience and overall evaluation of different phases of the program. If the number of problems could be minimized, we could assume that the ERT programs would be successful. In Fig. 1, we have outlined the basic steps we followed in conducting the research. This entire process contains five stages: (i) Data Collection and Preprocessing, (ii) Data Analysis, (iii) CIPP Evaluation Comparison, (iv) Analysis and SEM Modeling, and (v) Output Generation. We also depicted the interconnections among the steps and how that contributed in generating the analysis outputs.

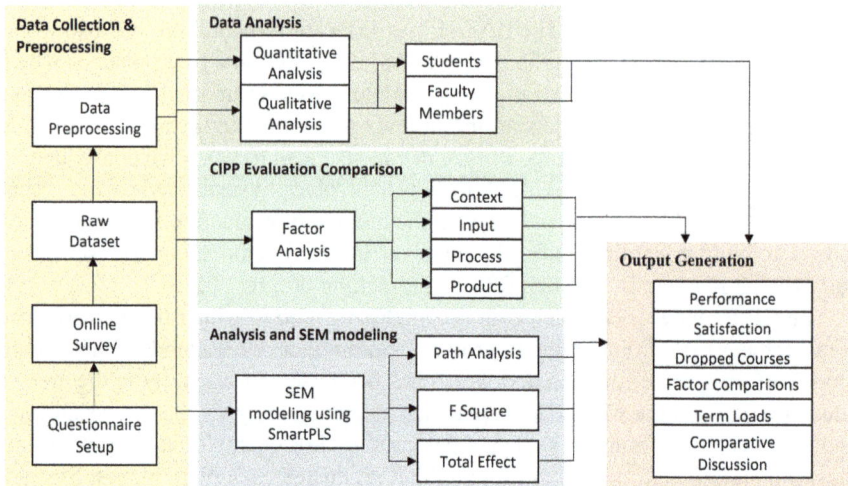

Fig. 1 Overview of the research steps

3.1 Data Collection and Preprocessing

We applied the mixed methodology approach where qualitative and quantitative methods are amalgamated in ways to reach the desired objective of the study, and here, data has been collected through survey questionnaire as well as from semi-structured analytical query. We collected data from developing (Bangladesh, Malaysia, and China) and developed (the United States of America, Canada, Germany, and Spain) countries. An online survey is performed based on the designed questionnaire. We used Google Forms to conduct this survey, and over 300 participants submitted their answers through the forms. The duration of collected data through this survey is more than a month, and we performed this survey in the middle of this pandemic era. We avoided collecting and revealing personal or sensitive information to assure that the participation is anonymous and spontaneous. Incomplete questionnaire data was removed where ratings and quantitative answers were replaced by regional averages if missing. We spent 40% of our time for data preparation and survey. We asked students various types of questions to identify their performance and satisfaction in this online learning mode during the pandemic situation. Students answered if they had any previous experience of online learning, were they missing their face-to-face learning, about the interactivity between online and in-person learning mode, about their course load and progress ratio, problems they faced in online learning mode, and which types of online learning mode they preferred. We were asking faculty members about their previous experience related to online learning mode, were they missing their face-to-face teaching, student–teacher interactivity between online and in-person learning mode, about their work load, student performance and course dropped ratio, and problems they faced in online teaching. In our designed survey, we collected data for these variables: country type, country name, joined online classes before, is synchronous learning, are they missing something in an online class, in-class interactiveness rating, online interactiveness rating, assignment performance, class load, class dropped, satisfaction, what are the improvement opportunities they feel about emergency remote teaching, even after going back to in-person classes which part they think can still be continued online, and ranking of the problems they faced. Each answer was an independent variable to construct our decision model after data collection. Then, we preprocessed the raw dataset that is used in three stages—Data Analysis, CIPP Evaluation Comparison and Analysis and SEM Modeling.

3.2   Data Analysis

Our survey was designed to identify the unique problems of two distinct zones; developing and developed countries. Therefore, the questions had a part to identify the problems and gaps and rank them. For quantitative analysis, we compared the satisfaction ratio, term loads, and performance of students and faculty members of developing and developed countries between online classes and in-person classes. For qualitative data analysis, we collected information from both teachers and students to determine improvement areas which should be addressed in such situations.

Other than that, students were asked to rank identified problems according to their priorities. We used these data in SEM modeling too. The problem list:

- Internet Connectivity
- Internet Speed
- Group works
- Financial Constraint
- Lack of proper devices
- Motivational Support
- Office Time productivity
- Time Management or Scheduling
- Special needs student access
- Home/Work–life balance Distractions
- Academic Integrity (Avoiding plagiarism and maintaining academic honesty).

3.3   CIPP Evaluation Comparison

CIPP is a well-known program evaluation method developed by Daniel Stufflebeam and his colleagues in the 1960s [34]. CIPP is an acronym for context, input, process, and product. "Context" deals with the goals of the program and seeks answers for the questions like: "What should be focused?." Then "Input" deals with the plans and answers about "How the necessary things are arranged?." "Process" is connected to the actions, and answers questions like "Are we doing as planned?" and finally "Product" is deals with the outcome and answers questions like "Did the program actually deliver?." After setting the questions and measuring the weighted answers, we were able to compare CIPP score of ERT for both the zones.

3.4 SEM Modeling and Analysis

First, we imported a preprocessed, error-free dataset into smart PLS. As per our model hypothesis, we took three latent variables: Student type, stress, and impact. Figure 3 shows the diagram of our proposed model. There are five observed variables in student type—zone, experience with online learning, like asynchronous mode, online interaction and in-person interaction. All these five observed variables are in formative manner. Now coming to the second latent variable which is stress. Observed variables in stress include, workload, missing face-to-face, and other problems mentioned and ranked in Fig. 2. All these observed variables of stress are also in formative manner too. Finally, coming to the last latent variable which is impact. From impact, three observed variables come out in a reflective manner including, DroppedCourses, Performance, and Satisfaction. There were total of three hypotheses in this entire process. One hypothesis was from student type to stress, another one was from student type to impact and lastly from stress to impact. In this overall

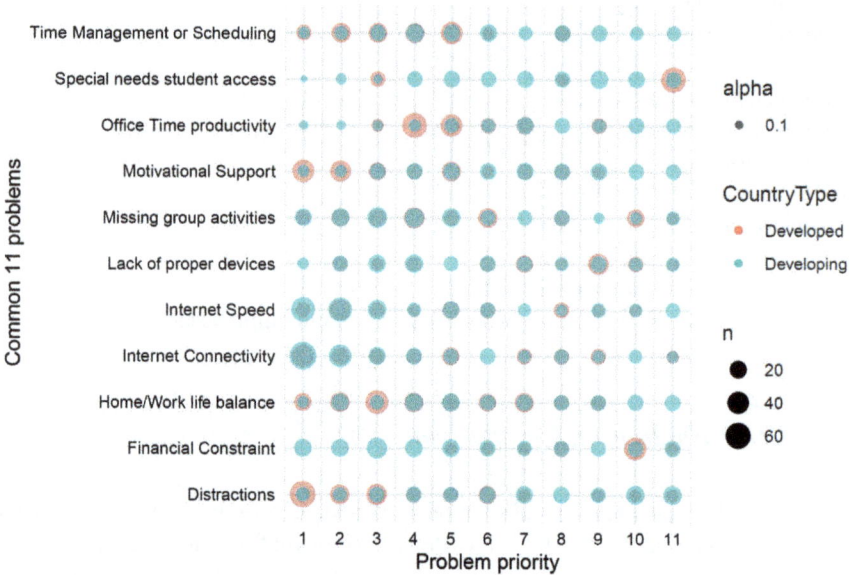

Fig. 2 Problem frequency

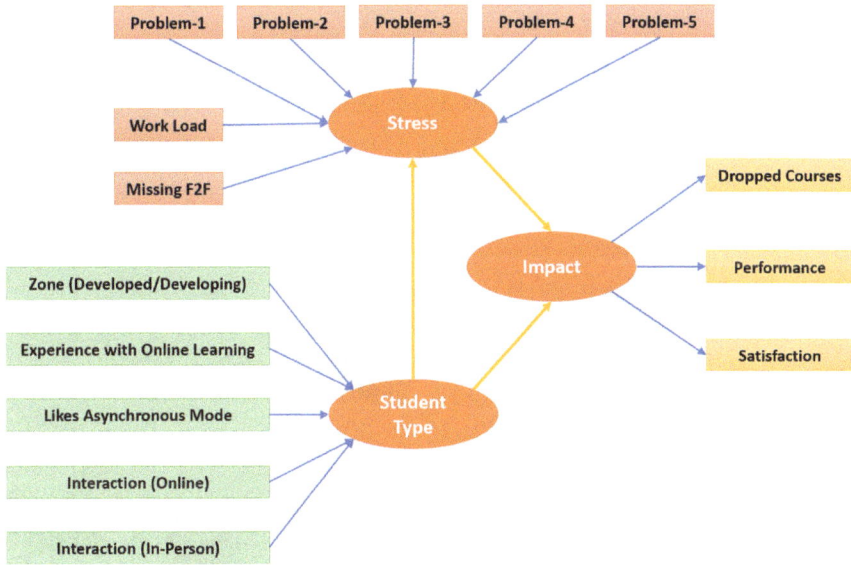

Fig. 3 ERT conceptual framework

process, from student type and stress we found our third latent variable which is impact.

We used the Consistent PLS algorithm in SmartPLS to find three impacts that were used. This was our choice because only two types of indicators were present: formative and reflective. After applying the Consistent PLS algorithm, we measured the value of path square, F Square, and total effect. Apart from this, we also found out the values of row-alpha for each of the three latent variables.

4 Results and Evaluation

We used different data mining techniques, structural equation modeling (SEM) and finally compared the CIPP evaluation for both zones to answer our research questions. We used python, R, SmartPLS-SEM, and Excel in our data analysis. Here, we present the outcomes of our described analysis procedures.

4.1 SEM of the Student

To test the aftermath of exogenous variables on the endogenous variable, the structural model is examined [8]. Our present study consists of three independent vari-

Table 1 Study of the structural model

Hypotheses	Path coefficient (β)	F-Square	Total effect
H1: Student type $\Rightarrow$ Impact	0.537	0.557	0.656
H2: Student type $\Rightarrow$ Stress	−0.338	0.129	−0.338
H3: Stress $\Rightarrow$ Impact	−0.354	0.242	−0.354

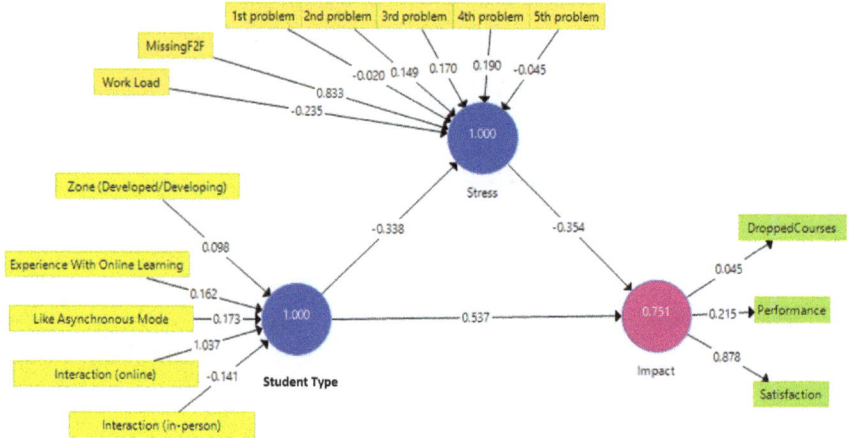

Fig. 4 Path analysis

ables (student type, impact, and stress). Independent variable student type has five dependent variables which are zone (developing/developed), experience with online learning, like asynchronous mode, interaction (online), and interaction (in-person). Another independent variable stress has three dependent variables which are missing face-to-face, work load, and problems (frist, second, third, fourth, fifth). The independent variable impact has three dependent variables (performance, satisfaction, and dropped courses). Table 1 shows the results of the SEM model of our current study that contains the path coefficient (β), F-Square and total effect.

The first hypothesis (H1) assessed the relationship between student type, and impact. The findings represented that student type has a remarkable positive influence on impact ($\beta = 0.537$) which represents a positive relationship between these two latent variables. Accordingly, the second hypothesis (H2) assessed the relationship between student type and stress. The findings represented that student type has a negative effect on stress ($\beta = -0.338$) which represents a negative relationship between these two latent variables. Finally, the third hypothesis (H3) assessed the relationship between stress and impact. The findings represented that stress has a negative effect on impact ($\beta = -0.354$) which also represents a negative relationship between these two latent variables. In Fig. 4, we can see the path analysis values of our model (Table 2).

Table 2 Rho alpha values

Latent variable	rho_A
Impact	0.751
Student type	1.000
Stress	1.000

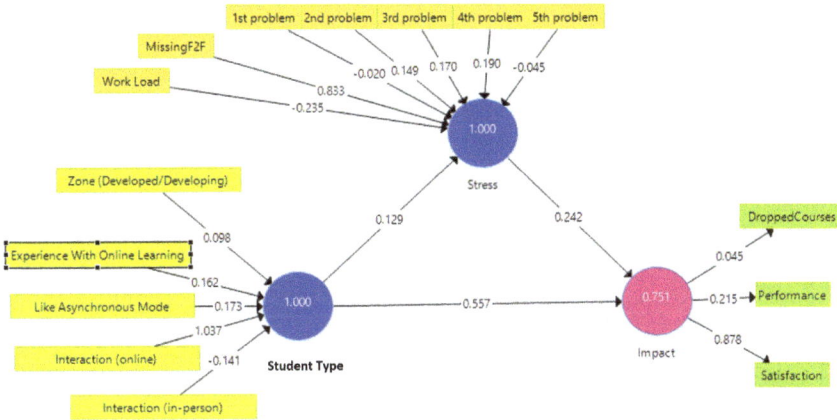

Fig. 5 F-Square

In linear regression, we used F-Square for calculating the effect size. Our current study consists of three hypotheses H1, H2, and H3. The F-Square values of these three hypotheses are represented in Fig. 5.

H1 has an F-Square value of 0.557 which indicates a large effect. H2 has an F-Square value of 0.129 which indicates a small effect. H3 has an F-Square value of 0.242 which indicates a medium effect.

Total effect indicates the summation of the direct and indirect effects of a relation. Figure 6 represents the total effect of our proposed model. In Table 1, the first hypothesis (H1) has a total effect of 0.656 and from the path analysis, we can find the direct effect of H1 is 0.537. So the indirect effect of H1 is 0.656-0.537=0.119. The second hypothesis (H2) has a total effect of -0.338 which is equal to its direct effect. So H2 has no indirect effect. The third hypothesis (H3) has a total effect of -0.354 which is also equal to its direct effect. So H3 also has no indirect effect.

Table 2 represented the rho alpha values of latent variables where rho alpha is a reliability coefficient which is used for examining composite reliability in structural equation modeling. For latent variable impact, we get rho_A value of 0.751. For the remaining two latent variables, i.e., student type, and stress, we get rho_A value of 1.000.

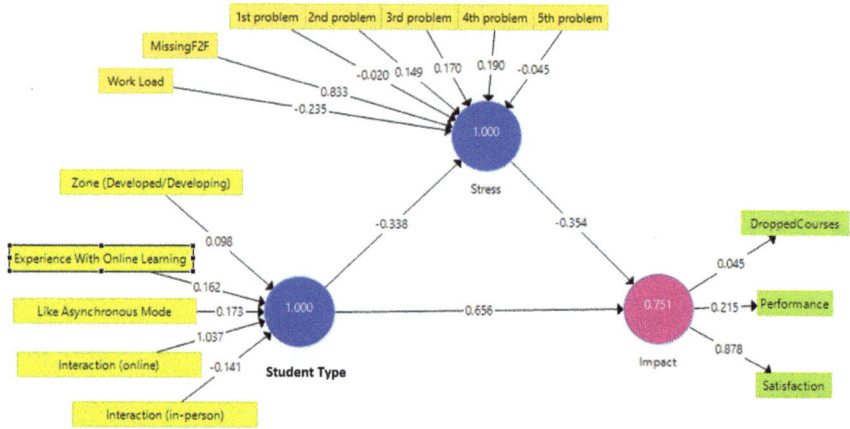

Fig. 6 Total effect

4.2 In-Class Versus Online Student's Satisfaction

We wanted to know and compare the satisfaction level of students while undergoing ERT. Figure 7 is providing us a histogram of satisfaction for both developed and developing countries. The red color represents the ratings of the online class participants, and the blue color represents in-class participants of course work. Length of bars is a representation of the higher satisfaction score of that number. Moving average density graphs is also showing a similar tendency. The red-colored or online class has more tendency to rate around five where blue-colored or in-class participants tend to rate their satisfaction near 8 where their density cloud is in the peak. We can conclude that those who have participated in-class courses have more satisfaction than online.

4.3 Online Versus In-Person Missing

The answer to the question "Is online education a replacement of regular in-class teaching or this is just the habit of participation among students?" we produced the next chart (Fig: 8) which is a mosaic chart where area of the rectangle distinguishes between smaller and bigger values for comparison. It shows the group of overlapped categories "JoinedOnlineClassesBefore" and "MissingSomething". By comparing the area of the chart, we can see that whoever had not attended an online class before was missing something that those who had attended one possessed. Therefore, we can say this is a habituation within the system.

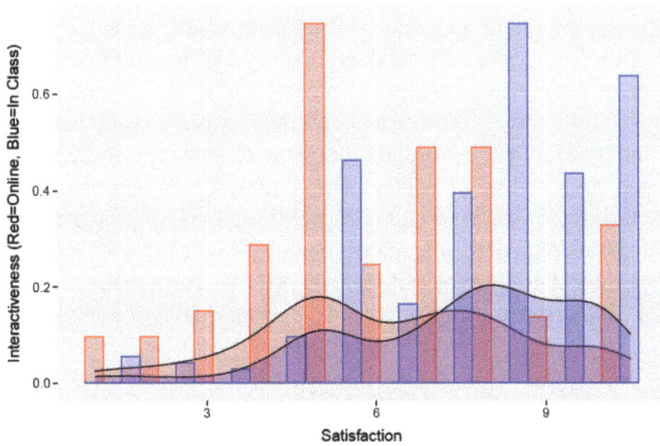

Fig. 7 Student's satisfaction score

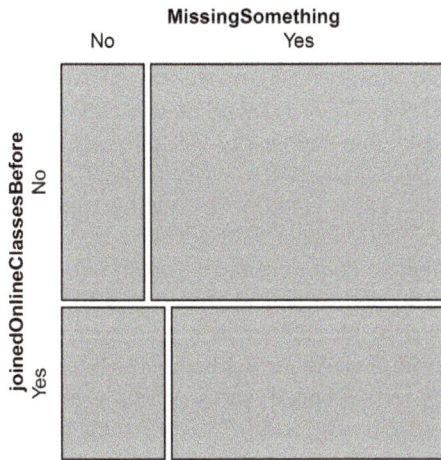

Fig. 8 Missing in-person classes

4.4 Problems Prioritization Frequency

Our initial investigation was in two major fields: effects of ERT and to find what are the hindrance behind execution of ERT. After surveying and asking students to rank the difficulties they were facing in ERT, we identified the problems. Figure 2 is showing problems and the count of their occurrence came up in relative frequency. The X-axis represents the priority, and the y-axis contains the list of 11 common problems in online classes. The developed country is represented in red color and developing country is in the blue color. The radius of intersection is the severity of that problem or the problem mentioned in the y-axis coming up in priority. Priority

1 means primary problem, and priority 11 means least appearance in consideration. Number one problem of a developed country was found to be distraction, and most frequent problem was Internet connectivity and speed. Several other problems like motivational support (in developed countries), missing group activities (in both developed and developing countries), office time productivity (mostly in developed countries), and home/work–life balance (also in developed countries) were also seen as significant. Surprisingly, financial constraints and lack of proper devices did not come up in the ranking. Also, time management was in medium-level priority, while special needs student access was in low-level priority. We also kept one open-ended comment for 12th problem which we considered for qualitative analysis.

4.5 Comparison of Factors

We were also interested to know different factors and the relation between them. The next plot (Fig. 9) is showing some fact comparison between teachers and students of developed and developing countries. Developing countries had less experience in online classes before (faculty 36% and students 38%) than developed countries (faculty 67% and students 43%). Asynchronous mode of education shows mixed results from a developed country (faculty 0% and students 27%) and developing country (faculty 27% and students 19%). So, in developed country, faculties did not like asynchronous at all, wherein in developing countries, faculties liked this mode of learning more than students. In pandemic situations, feedback for missing face-to-face is very high in both developed (faculty 67% and students 70%) and developing countries (faculty 100% and students 77%). This ratio is also mixed. Students are missing face-to-face classes more than teachers in developed countries, where it was the opposite in developing countries. All teachers were missing face-to-face presence. The course dropped ratio was higher always in teachers feedback than students in both developed (faculty 33% and students 30%) and (faculty 36% and students 7%). Course dropout was comparatively lower in developing countries than the developed ones.

4.6 Term-Wise Loads Perception

Online teaching systems require more concentration to achieve success because in-class teaching is more interactive, and the chance of distraction is less there. Term loads (Fig. 10) show the mixed results in two types of countries. In the current COVID-19 pandemic situation, in developed (faculty 53% and students 100%) and developing (faculty 49% and students 55%) countries, majority of people were thinking their term loads were more than usual. 100% of developed country faculties thought their term load had increased. In developed (faculty 0% and students 27%) country got less "no change" feedback than the developing countries (faculty 18%

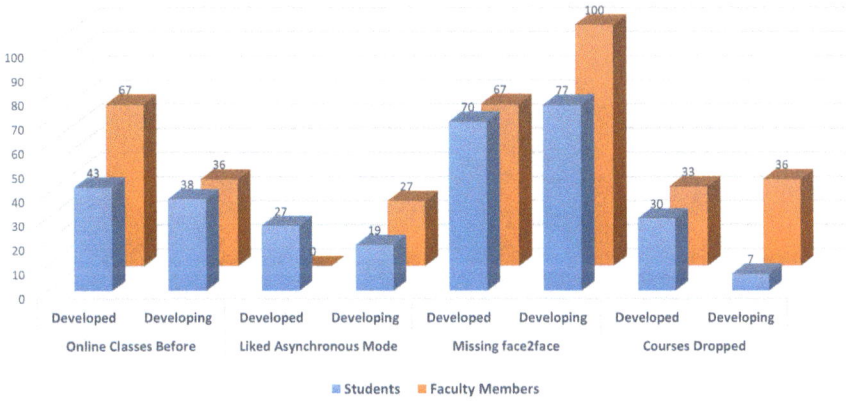

Fig. 9 Comparison of factors

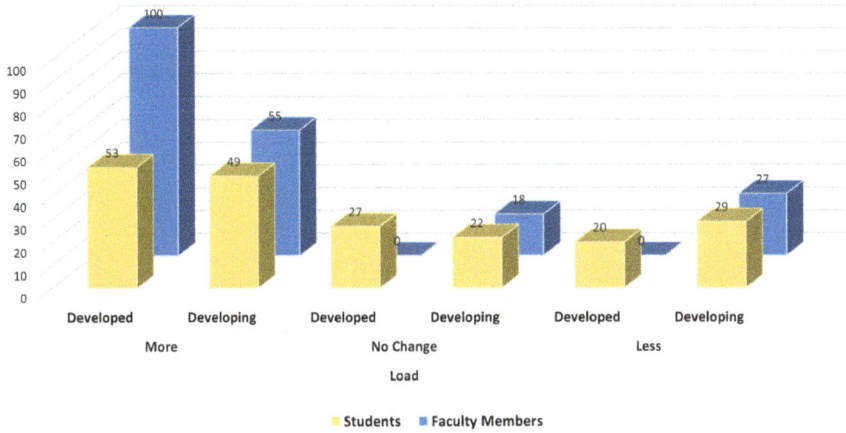

Fig. 10 Term-wise loads

and students 22%). A smaller number of faculty and students felt their course pressure was less in ERT system: developed (faculty 0% and students 20%) and developing (faculty 27% and students 29%).

4.7   *Comparison of Student's Performance*

Student performance feedback (Fig. 11) is mixed from two zones. The graph shows student's self-assessment of how they were doing and what faculties were thinking about their performance. In both developed (faculty 33% and students 3%) and developing (faculty 45.5% and students 26%) countries, students were thinking less about

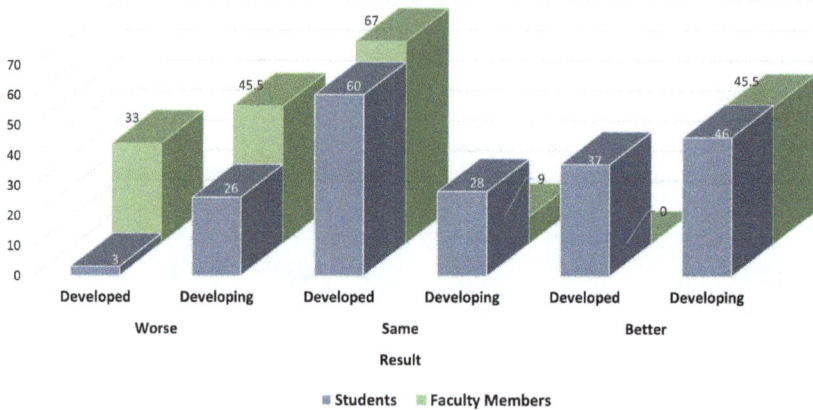

Fig. 11 Comparison of student's performance

worsening their performance where faculties were thinking their performance was degrading. We see the mixed results in "same performance feedback." In developed countries (faculty 67% and students 60%) faculties thought that student's performance was same but in developed countries, (faculty 9% and students 28%) students thought that their performance was same as before in ERT. From those who were thinking their performance was increasing in developed countries (faculty 0% and students 37%), no faculty thought their performance was better in the new system, but in developing countries, (faculty 45.5% and students 46%) both faculties and students in similar ratio thought that students' performance was better in remote learning.

4.8   CIPP Evaluation in both Zones

As we can see in the evaluation spider chart (Fig. 12), in terms of the context that is goal and objective, developed countries (blue) were way ahead than the developing countries (orange). They knew what was coming, whereas developing countries were waiting to see what could be done when it arrived. Process wise the approach was similar, as developed countries had clear cut plans, they could smoothly follow that, whereas developing countries suffered during execution due to lack of proper preparation. Input shows that when the lockdown was in effect, developing countries quickly planned and started executing faster than the developed countries. But at the end, product shows end performance of the program was similar though the scores suggests it could have been better for both.

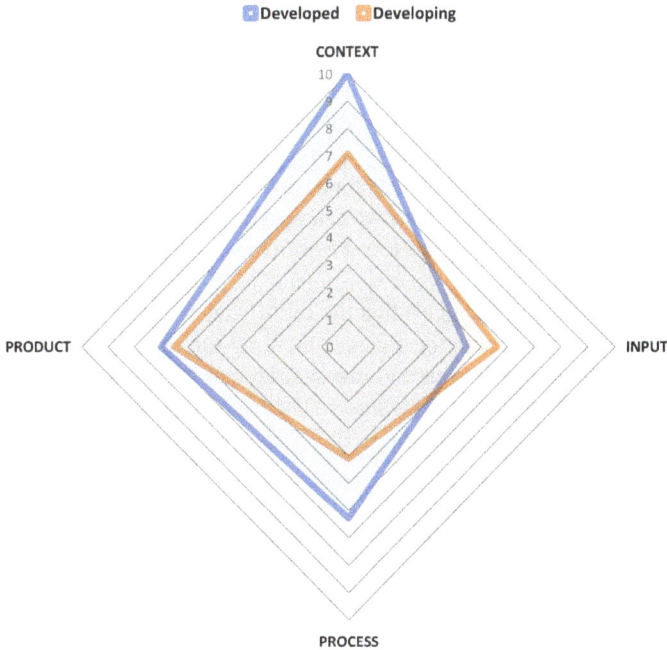

Fig. 12 Zone's CIPP evaluation

4.9 *Qualitative Analysis*

Almost 100% of educational institutes in developed countries use a student learning management system, e.g., blackboard, canvas, and most of them offer online degrees. A decent Internet connection and devices were available to the majority of the students but not all. On the downside, students missed better workload distribution and recording of lectures. They also noticed the absence of adequate innovative methods/tools for functional collaboration/group work and a successful run of a lab class.

To properly conduct a lab class, teachers found existing tools inadequate. Besides, teachers found recording only the key sections of a lecture very useful and convenient. Teachers also noticed the lack of advanced technology and training to monitor student's attentiveness, participation, and assessment/evaluation/exam. Faculties from developing countries also shared the same experience.

Besides, developing country teachers and students felt they needed better Internet infrastructure, access to student learning management systems, better communication technologies, and tools to make the class more engaging (student–teacher interaction). Students found the recording of the class lecture extremely useful, but class materials were not well prepared.

5   Conclusion

This COVID-19 pandemic has taught us an important lesson: "Good planning always pays off." The educational institutes that planned thoroughly about ERT faced comparatively less problems during execution than the others who did not plan ahead. Technological advancement favors developed countries most of the time, but from the teaching and learning perspective, as we can see from the outcomes of this research, developing countries can catch up too. Priorities and problems can also be different between these two groups, yet requirements to tackle them and their demands are similar in nature which suggest that the useful and effective uses of the sustainable technology will help both the zones in the long run. Institutes no matter which zone they are in should plan well and invest sufficient budget to ensure better ERT experiences for the instructors and learners. The analysis outcomes presented in this research can easily be incorporated into the strategies developed by the educational institutes.

We are also well aware that there are number of machine learning applications available in online teaching and exams systems [14] that may have significant potential in such settings. In the future, we are planning to study effectiveness and adoption of AI in ERT and suggest what measures should be taken to make ERT a better experience in different zones.

References

1. Amnesty International: South Africa—failing to learn the lessons? the impact of covid-19 on a broken and unequal education system (2021). https://www.amnesty.org/en/documents/afr53/3344/2021/en/
2. Abou-Khalil V, Helou S, Khalifé E, Chen MA, Majumdar R, Ogata H (2021) Emergency online learning in low-resource settings: effective student engagement strategies. Educ Sci 11(1):24
3. Aseey AA, Andollo AA (2019) Electronic mobile devices, transformative pedagogy and learning: higher education and changing times in Kenya. J Educ Soc Res 9(3):54
4. Attallah B (2020) Post covid-19 higher education empowered by virtual worlds and applications. In: 2020 seventh international conference on information technology trends (ITT). IEEE, pp 161–164
5. Chang CT, Hajiyev J, Su CR (2017) Examining the students' behavioral intention to use e-learning in Azerbaijan? The general extended technology acceptance model for e-learning approach. Comput Educ 111:128–143
6. El Said GR (2021) How did the covid-19 pandemic affect higher education learning experience? an empirical investigation of learners' academic performance at a university in a developing country. Adv Human-Comput Interact 2021
7. Guerra MA, Gopaul C (2021) IEEE region 9 initiatives: supporting engineering education during covid-19 times. IEEE Potentials 40(2):19–24
8. Hair JF, Hult GTM, Ringle C, Sarstedt M (2016) A primer on partial least squares structural equation modeling (PLS-SEM). Sage publications
9. Hodges C, Moore S, Lockee B, Trust T, Bond A et al (2020) The difference between emergency remote teaching and online learning. Educause Rev 27:1–12
10. Hu R (2021) Covid-19 challenges: health and education in Chinese society. Int J Soc Sci Educ Res 4(3):358–366

11. Indrajit RE, Wibawa B (2020) Portrait of higher education in the covid-19 period in a digital literacy perspective: a reflection on the online lecture process experience. In: 2020 Fifth international conference on informatics and computing (ICIC). IEEE, pp 1–5
12. Kamil M, Rahardja U, Sunarya PA, Aini Q, Santoso NPL (2020) Socio-economic perspective: mitigate covid-19 impact on education. In: 2020 Fifth international conference on informatics and computing (ICIC). IEEE, pp 1–7
13. Kim S, Choi J (2020) Development of evaluation criteria for forest education using the CIPP model. J Forest Environ Sci 36(2):163–172
14. Kučak D, Juričić V, Dambić G (2018) Machine learning in education-a survey of current research trends. Ann DAAAM Proc 29
15. Lippe M, Carter P (2018) Using the CIPP model to assess nursing education program quality and merit. Teach Learn Nurs 13(1):9–13
16. Liu L, Liu K, Zhao J (2020) Development of online flipped blended teaching mode in higher vocational education during covid-19 outbreak: a case study. In: 2020 Ninth international conference of educational innovation through technology (EITT). IEEE, pp 193–198
17. Maine LL (2020) American journal of pharmaceutical education 2020: It is all about people. Am J Pharm Educ 84(3)
18. Majaski C (2021) Investopedia article—developed economy. https://www.investopedia.com/terms/d/developed-economy.asp#axzz1legO8olO
19. Mohammed AK, El Zoghby HM, Elmesalawy MM (2020) Remote controlled laboratory experiments for engineering education in the post-covid-19 era: Concept and example. In: 2020 2nd novel intelligent and leading emerging sciences conference (NILES). IEEE, pp 629–634
20. Morgan H (2020) Best practices for implementing remote learning during a pandemic. The Clearing House: A J Educ Strat, Issues Ideas 93(3):135–141
21. Niemi HM, Kousa P et al (2020) A case study of students' and teachers' perceptions in a finnish high school during the covid pandemic. Int J Technol Educ Sci
22. Pete J, Soko J (2020) Preparedness for online learning in the context of covid-19 in selected sub-Saharan African countries. Asian J Dist Educ 15(2):37–47
23. Poliakov M, Rida I (2020) Remote laboratories for engineering education: status and prospects. In: 2020 advances in science and engineering technology international conferences (ASET). IEEE, pp 1–6
24. Potluri RM, Anjam M (2021) Knowledge transfer in UAE higher education institutions during covid-19 pandemic: Learners' cannot learn surgery by watching. In: 2021 2nd international conference on computation, automation and knowledge management (ICCAKM). IEEE, pp 25–30
25. Radonov R, Angelov G, Rusev R (2020) Remote education applications in the technical university of sofia. In: 2020 XXIX international scientific conference electronics (ET). IEEE, pp 1–6
26. Rafidiyah D, Nadia H et al (2020) The emotional experiences of Indonesian phd students studying in Australia during the covid-19 pandemic. J Int Students 10(S3):108–125
27. Rai J, Tripathi R, Gulati N (2020) A comparative study of implementing innovation in education sector due to covid-19. In: 2020 9th international conference system modeling and advancement in research tTrends (SMART). IEEE, pp 94–97
28. Santana OA, das Braga G, de Sa Braga JO, Carvalho H (2020) Interactive model tool about center of mass during covid-19 pandemic: a new learning path in stem for k-12 education. In: 2020 IEEE international conference on teaching, assessment, and learning for engineering (TALE). IEEE, pp 503–508
29. Sarstedt M, Ringle CM (2020) Structural equation models: from paths to networks (westland 2019)
30. Schlesselman LS (2020) Perspective from a teaching and learning center during emergency remote teaching. Am J Pharm Educ 84(8)
31. Shackman G (2008) What is program evaluation? A beginners guide. The Global Social Change Research Project. http://www.ideas-int.org

32. Shi X (2018) Research on performance evaluation system of college entrepreneurship education level based on CIPP model. Educ Sci: Theory Pract 18(5)
33. Sibirskaya E, Popkova E, Oveshnikova L, Tarasova I (2019) Remote education versus traditional education based on effectiveness at the micro level and its connection to the level of development of macro-economic systems. Int J Educ Manage
34. Stufflebeam DL (1971) The relevance of the CIPP evaluation model for educational accountability
35. Tang SK, Lei P, Tse R, Lam CT, Cheong CWL (2020) Overcoming the sudden conversion to online education during the covid-19 pandemic: a case study in computing education. In: 2020 IEEE international conference on teaching, assessment, and learning for engineering (TALE). IEEE, pp 17–22
36. Vitliemov P, Bratanov D, Marinov M (2020) An approach to use virtual and remote labs in mechatronics education based on cloud services. In: 2020 7th international conference on energy efficiency and agricultural engineering (EE&AE). IEEE, pp 1–4
37. Vladoiu M, Constantinescu Z (2020) Learning during covid-19 pandemic: online education community, based on discord. In: 2020 19th RoEduNet conference: networking in education and research (RoEduNet). IEEE, pp 1–6
38. Wenham C, Smith J, Morgan R (2020) Covid-19: the gendered impacts of the outbreak. The Lancet 395(10227):846–848
39. WHO (2021) Covid-19 global literature on coronavirus disease. https://search.bvsalud.org/global-literature-on-novel-coronavirus-2019-ncov/

Part IV
Adapting for Improved Resilience

Chapter 18
Anticipating and Preparing for Future Change and Uncertainty: Building Adaptive Pathways

Jeremy Gibberd

Abstract The COVID-19 pandemic has highlighted how ill-prepared the building sector has been in anticipating and responding to change. Further major disruptions related to climate change, emerging technologies such as artificial intelligence and new business models, are anticipated. It is necessary, therefore, to prepare for this change in the way we plan and manage built environments. This paper investigates the nature of anticipated future change and its implications for buildings. It proposes a structured approach to prepare for, and respond to, change in a proactive, structured way. This methodology is called building adaptive pathways and is illustrated and tested through application to a case study. Findings indicate that methodology provides useful insights into how change and uncertainty can be addressed in built environments and recommends that further work on the approach be undertaken.

Keywords Uncertainty · Change · Building adaptive pathways

1 Introduction

In February 2021, it was estimated that there were 112 million coronavirus cases globally and almost 2.5 million deaths [1]. In the worst-hit areas, there were 493 deaths per 100,000 people [2]. In badly affected countries, death rates were 188 per 100,000 people in Belgium, 187 in Slovenia and 176 in the UK [3]. In many areas of the world, hospitals struggled to cope with the surges of coronavirus patients. In February 2021, US hospitals in states like California and Georgia were under 'extreme stress' and had more than 30% of their ICU beds are filled by COVID-19 patients [4]. Shifting population demographics and the extreme stress imposed on buildings are examples of changes that built environments need to cope with because of the coronavirus epidemic.

Climate change is bringing other changes, such as higher temperatures, increasingly erratic weather conditions and storms and droughts [5]. Existing infrastructure

J. Gibberd (✉)
Council for Scientific and Industrial Research, Pretoria, South Africa
e-mail: jgibberd@csir.co.za

© The Author(s), under exclusive license to Springer Nature Singapore Pte Ltd. 2022   255
R. J. Howlett et al. (eds.), *Smart and Sustainable Technology for Resilient Cities and Communities*, Advances in Sustainability Science and Technology,
https://doi.org/10.1007/978-981-16-9101-0_18

and systems in many cities and urban areas have not been adapted for this change, and services such as water and electricity supplies are likely to become increasingly unreliable [6].

At the same time, new technologies are being applied to buildings. Globally, there has been a rapid increase in onsite energy generation using photovoltaic systems. Local energy production has enabled the development of markets in which producers and consumers of electricity trade this through peer-to-peer local energy transactions and microgrids [7]. New business models are finding ways of using latent building capacity. An example is Airbnb which rents out unused spaces and rooms and provide owners with revenue for this [8].

Rapidly changing environments and increased stresses being placed on buildings mean that there is an urgent need for these to be managed differently. The nature of this change means that it is difficult to predict what will happen. However, as the impact of these changes is very significant, change must be planned for and addressed effectively.

Building management and planning must therefore anticipate change and develop strategies that address this. This responsive approach is the focus of the Building Adaptive pathway methodology. This paper presents the methodology and critically evaluates whether this provides an effective planning tool for addressing issues such as climate change, the COVID-19 pandemic and new technologies and business models that have become part of a rapidly changing built environment.

2 Adaptive Pathways

Adaptive pathways refer to a methodology that assesses the adaptability potential of management strategies into the future [9]. The approach was conceptualized in 2010 and developed initially for climate change adaptation [10, 11]. It has been applied in three main ways. First, it has been used to understand climate change adaptation and develop plans to address this. Second, it has been used to promote collaborative learning and adaptive planning processes. Third, it has been used as a structured way of managing complexity and long-term change [11].

A key tool within the methodology is the visual representation of potential actions and sequencing (paths) that can be implemented. Key thresholds or tipping points are identified which will potentially change the conditions of stable systems into another state [12]. By predicting these points, the methodology can steer around these by developing plans which chart, evaluate and implement different courses of action before disruptive events occur [13].

The approach considers system vulnerabilities to ensure that these are integrated and addressed in a proposed adaptive pathway plan [14]. Developing adaptive pathway plans requires insight into the type and size of potential future challenges. It also requires an understanding of how these challenges can be avoided or overcome through physical measures or policy instruments. In addition, there must be a good understanding of the likely impacts of physical measures and policy instruments to

Action A
Action B
Current policy
Action C
Action D

0   10   70   80   90   100
                                              years

○ Transfer station to new action
| Adaptation Tipping Point of an action (Terminal)
━ Action effective in all scenarios
▪ ▪ Action not effective in scenario X

Path actions		Relative Costs	Target effects	Side effects
1	○	+++	+	0
2	○ ○	+++++	0	0
3	○ ○	+++	0	0
4	○ ○	+++	0	0
5	○	0	0	-
6	○ ○	++++	0	-
7	○ ○	+++	0	-
8	○ ○	+	+	---
9	○	++	+	---

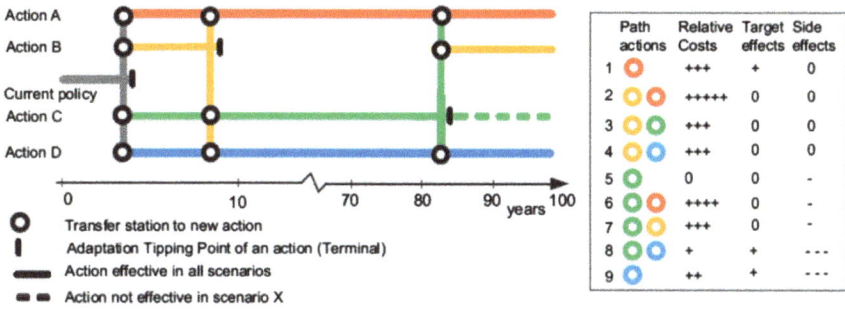

Fig. 1 Adaptive pathway map and pathway scorecard [16]

ensure that these are appropriate and sufficient for future challenges and do not have unintended consequences [15].

Figure 1 provides an example of an adaptive pathways map [16]. This shows the nine different possible pathways that can be pursued, starting from the current situation. It shows that within 4 years the current situation will reach a tipping point and therefore changes need to be made to avoid this. Actions A and D enable tipping points to be avoided for the next 100 years. If Action B is chosen, this will lead to a tipping point being achieved in about 9 years. This will require a change to one of the other pathways (follow the orange lines). If action C is chosen, this will require a shift in about 85 years to Action A, C or D. The scorecard pathways provide an evaluation of the different actions in terms of cost, target effects and side effects.

The adaptive pathways methodology has been applied to flood risk planning [15, 17], developing resilient waterfronts [13], sustainable development planning [9], the development of small-scale PV systems [18] and water supply planning [16, 19]. This study explores how the adaptive pathway approaches can be applied to the planning and management of buildings to respond to economic uncertainty, climate change and impacts related to the COVID-19 pandemic. The methodology for the study is outlined below.

3 Methodology

The methodology follows an exploratory research approach. This is suitable when research is at a preliminary stage and aspects of the topic are not known. It is appropriate for carrying out an initial analysis of a new topic and generating new ideas. It is used to answer questions like what, why and how and does not aim to achieve conclusive results. The methodology aims to provide a basis for future research by exploring a new approach to addressing uncertainty and change within built environments. It investigates how the adaptive pathway approach can be developed and applied to built environments as a way of planning for future change and uncertainty.

Applying the adaptive pathway approach follows an eight-step process that leads to the development of a building adaptive pathway plan. The eight steps of the methodology are briefly introduced below. First, the system is described. This includes understanding the system's characteristics, the objectives of the system, the constraints in the current situation and potential constraints in future situations. Second, alternative future situations, opportunities and vulnerabilities are identified. Third, possible actions that can be taken to address future situations are identified. Fourth, these actions are evaluated in terms of how they effectively address future situations and create opportunities. Fifth, information from the earlier steps is used to create an adaptation map of the different actions. Sixth, the adaptation map is evaluated to develop preferred pathways. Seventh, a contingency plan with corrective actions in case of unexpected events is developed. Eighth, the earlier stages are used to develop and implement an adaptive pathway plan.

These steps are applied to a case study building and the results are presented below. The building was selected because it is typical of many office buildings globally. These buildings were developed 20–40 years ago in suburban campus settings. Their locations mean that they are poorly served by public transport and are typically accessed by car. HVAC and water systems within the buildings in many cases have not been upgraded and are now considered highly inefficient. Change and increasingly stringent demands on buildings have meant that the inefficiency and poor sustainability performance of this building type are being questioned. It therefore is a highly appropriate building type for the application of the methodology, and findings from the study will be relevant to similar building types globally.

Data required for the study were obtained from the organization's facilities management unit and publicly available information. This included data from websites of the organization, the local municipal, water and power utilities and Google maps (for the site).

4   Results

The results of the study are presented below and follow the eight steps of the methodology.

4.1   Describe the Study Area

The case study building is an 8409 m^2 building located on a research campus in a suburb in the east of Pretoria as shown in Fig. 2. The site is surrounded by residential buildings and has limited access to public transport.

Fig. 2 Aerial photograph of the building and site

The building is owned and occupied by a research organization and is used for commercial research activities, laboratory work and some training. The accommodation consists of single offices between 15 and 40 m^2, shared kitchenettes, boardrooms and toilets as well as a large double story laboratory as shown in Fig. 3.

As a result of the COVID-19 pandemic, the organization has requested that most employees work from home. The pandemic has also affected public transport, making access to the building and site more difficult. This has particularly affected junior research and administrative staff who commute up to 50 km from areas with affordable accommodation. Utilization rates are estimated to have been around 10% for the period April 2020 to April 2021. At the same time, the organization is restructuring to work more closely with industry. The new strategic plan includes objectives that focus on increased collaboration with industry partners and the redevelopment of the campus and buildings to support this. As with many other sectors, the organization has been badly affected by the economic downturn resulting from the pandemic and needs to achieve significant reductions in operational costs as income has been reduced.

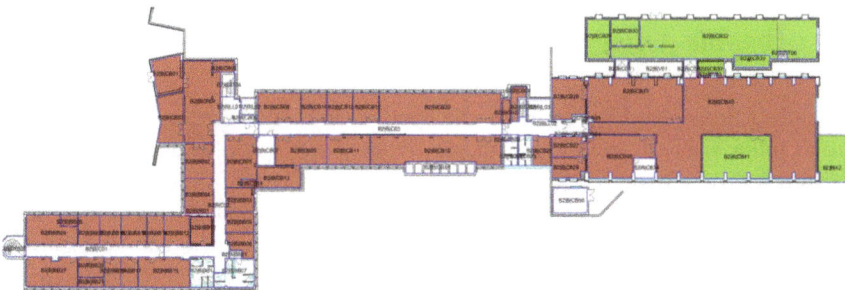

Fig. 3 Plans of the case study building

Significant constraints face the organization and the management of the building. The organization has identified the need for improved capacity to embark on its new strategy. However, it has found it difficult to attract new staff as there is significant competition for suitable staff from local and international corporations, research organizations and universities.

South Africa has also been affected by energy and water shortages and projections from the energy utility indicate that there may be outages because of constrained capacity for at least the next 5 years [20]. Climate change projections indicate that temperatures will increase heightening water scarcity. Both issues have led to rapid increases in energy and water tariffs, significantly increasing operational costs. Equipment installed in the building is dated and inefficient leading to high energy and water consumption. Building and space utilization rates are low.

4.2 Problem Analysis

The following possible future scenarios, opportunities and vulnerabilities are identified. Possible future scenarios are as follows. Firstly, the organization may fail as income declines and operational costs increase. Secondly, the organization may decline because of the difficulty in attracting and retaining staff. Thirdly, the organization may suffer from disruption and reduced productivity because of power and water supply interruptions.

Opportunities identified for the project are as follows. Converting buildings to residential and co-living areas could be carried out. Spaces can be refurbished and rented to partners and organizations in linked industries to create synergies and improved cooperation. Energy-efficient and renewable energy technologies could be installed. Water-efficient, greywater and rainwater harvesting technologies could be installed.

The current vulnerabilities identified are as follows. There may be an unwillingness by management to address issues and implement change. Capital costs of changes may be deemed too high.

4.3 Possible Actions

Possible actions identified based on the opportunities and vulnerabilities identified are as follows. Firstly, residential and co-living spaces could be integrated into the existing building to accommodate research and support staff. Resulting affordable accommodation in an attractive campus can be used to attract and retain staff. Building operating costs can also be shared. Secondly, partners can be identified and encouraged to take up space in the building. Having partners close by can be used to support improved collaboration and share operating costs. Thirdly, an energy system upgrade can be developed with backup power for 1–2 h. Fourthly, a water

Table 1 Pathway scorecard for the project (author)

Actions	Vulnerability	Opportunity	Date required
Co-living	+	+	1–2 years
Partner tenants	++	+++	1–2 years
Energy efficiency	++	+	1–2 years
Water efficiency	++	+	1–5 years
Off-grid energy system	+++	++	1–5 years
Off-grid water systems	+++	++	5–10 years
Key			
Address vulnerabilities fully	+++	Creates new opportunities	+++
Partially addresses vulnerability	++	Partially creates new opportunities	++
Does not address vulnerability	+	Does not create new opportunities	+

efficiency system upgrade can be installed to improve water efficiency and provide backup for 1–2 days. Fifthly and sixthly, full off-grid energy and water systems could be installed.

4.4   Evaluate Actions

The actions identified in 'Actions' are evaluated in this stage. This assesses the action in terms of vulnerability, opportunity and date required. The vulnerability assessment reflects the extent to which the proposed action resolves the vulnerabilities identified. The opportunity assessment evaluates whether new additional opportunities are created because of the action. The 'date required' assessment indicates the date by which the action should be taken to address the vulnerability. This evaluation is shown in Table 1.

4.5   Assembly of Pathways

The information from previous steps is used to assemble pathways and create an adaptation map with a portfolio of actions that address vulnerabilities and create opportunities for the building. This is shown in Fig. 4.

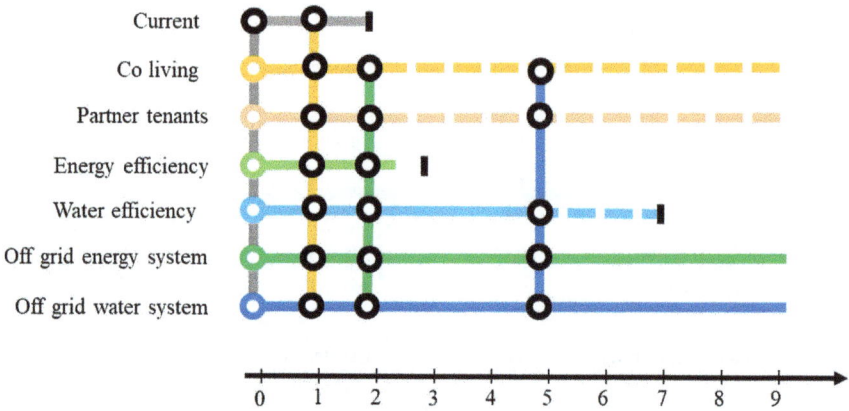

Fig. 4 Building adaptive pathways for the project (author)

4.6 Preferred Pathways

A review of Fig. 4 is used to identify a preferred pathway. This is as follows. Firstly, partner tenants and co-living actions would be taken and be completed within the first 1–2 years. This ensures that the operating costs of the building are shared, reducing costs for the organization. It also promotes collaboration, increasing the competitiveness of the organization and improved access to markets. The co-living action can be used to attract the suitable capacity required to implement the new strategy and ensure that existing junior researchers and support staff are retained through attractive living environments. Secondly, improvements in energy efficiency would be carried out and an off-grid system would be installed in the first 2 years. This would avoid business disruption as an electricity supply would be maintained. Operational costs would also be controlled and could be shared between tenants in the building. Thirdly, water efficiency could be improved, and an off-grid water system installed in the years 3–5. This would enable the organization to avoid disruptions associated with water shortages and outages. Water costs would also be controlled and shared between users of the building.

4.7 Contingency Planning

A contingency plan would be developed to enable corrective actions to be undertaken to stay on track in case of more rapid change or unforeseen events. This plan would include triggers and contingency actions that responded to these events. Key actions and associated triggers and responses are outlined in Table 2.

This shows that if existing spaces within the building continued to have utilization of below 30%, the partial conversion of office spaces to residential units and

Table 2 Contingency plan for the project (author)

Actions	Trigger	Response
Co-living	0–30% utilization rates for 12 months	Fastrack conversion
Partner tenants	20–50% utilization rates for 12 months	Fastrack identification of partners
Energy efficiency	1–2-h interruptions/month	Fastrack energy system upgrades
Water efficiency	1–2-day interruptions/month	Fastrack water system upgrades
Off-grid energy system	3–4-h interruptions/month	Full off-grid energy system installation
Off-grid water systems	2+ day interruptions/month	Full off-grid water system installation

tenants should be fast-tracked. This avoids further wastage of space and ensures that the organization can share building operating costs with others. Energy and water outages of longer than 1–2 h would trigger the fast-tracking of energy and water installations. Where outages were longer and the building experienced 3–4-h electricity interruptions per month and/or over 2-day water interruptions, full off-grid system installations would be fast-tracked.

4.8   Dynamic Adaptive Plan

The results from the early steps can then be developed into a building adaptive pathway plan. This plan would include the objectives, the actions and phasing of the plan. Costs and implications of not implementing the plan would also be provided.

5   Discussion

A review of the methodology indicates that it generates interesting and challenging options. This includes actions such as the incorporation of residential accommodation and full off-grid energy and water systems that are significantly departured from the norm. While these actions may seem radical, they may be necessary to sustain the organization. The methodology, therefore, provides a useful way of ensuring that organizations and decision-makers make difficult decisions that are necessary for business survival and continuity. This is valuable as it could reduce business failures and the wastage and lost opportunity associated with unproductive assets.

The methodology could also help the organization and decision-makers understand future change and uncertainty in a way that can be used to inform strategic plans and decisions. By identifying vulnerabilities and opportunities, the methodology encourages the development of more lateral thinking which enable better and different models and solutions to be developed. The contingency plan development included in the plan enables dynamic responses to changed circumstances [21]. This supports rapid decision-making and avoid delays which could have significant negative implications for businesses. The building adaptive pathway plan developed through this process, therefore, enables organizations to face uncertainty and future challenges and plan for this in a structured proactive way. Developed plans offer the potential for major disruption and failure to be avoided, enabling organizations and buildings to be more resilient and able to deal with future change when this happens.

However, the quality of the plan requires a rigorous approach to understanding the current situation, future change and uncertainty and being able to identify vulnerabilities and generate opportunities. As potential actions may require significant deviations from the norm, independent thinking and decisive action are needed for implementation.

The methodology could be criticized for being insufficiently scientific. This is a valid criticism where all the influencing factors are known or can be established. However, increasingly, under conditions related to climate change, or a global pandemic, the exact nature of influencing factors is not known, and while specialist studies can be used to obtain more detail, a changing context may result in findings rapidly becoming redundant. Given this context, the building adaptive pathway methodology provides a valuable tool for thinking about how uncertainty is managed in buildings. The approach provides a structured way of addressing vulnerabilities and drawing on opportunities to develop flexible plans that can be used to deal with future change.

6 Conclusion and Recommendations

The building adaptive pathway methodology was developed and applied to investigate how future change and uncertainty could be planned for in the built environment. The methodology was applied to a case study building to show how climate change, economic uncertainty and failing municipal services could be addressed and planned for. The study provides valuable insight into how plans and actions can be structured to respond to uncertainty and future change. It indicates that the adaptive pathway approach appears to offer significant potential as means for planning for future change and uncertainty in built environments and recommends that this be developed further.

References

1. Worldometers, 2021. Coronavirus Up-date (Live): 112,263,225 Cases and 2,485,386 Deaths from COVID-19 Virus Pandemic—Worldometer. https://www.worldometers.info/cor

onavirus/?utm_campaign=homeAdvegas1? Last accessed 23/2/2021
2. South African Medical Research Council, 2021. Report on weekly deaths in South Africa|South African Medical Research Council. https://www.samrc.ac.za/reports/report-weekly-deaths-south-africa?bc=254. Last accessed 23/2/2021
3. John Hopkins University (2021) Mortality Analyses—Johns Hopkins Coronavirus Resource Center. https://coronavirus.jhu.edu/data/mortality. Last accessed 23/2/2021
4. National Public Radio (2021) Is Your Hospital Overwhelmed With COVID-19 Patients? Look It Up Here: Shots—Health News: NPR. https://www.npr.org/sections/health-shots/2020/12/09/944379919/new-data-reveal-which-hospitals-are-dangerously-full-is-yours. Last accessed 23/2/2021
5. Engelbrecht F (2016) Detailed projections of future climate change over South Africa, CSIR Technical Report
6. Brikké F, Vairavamoorthy K (2016) Managing change to implement integrated urban water management in African cities. Aquat Procedia 6:3–14
7. Mengelkamp E, Notheisen B, Beer C, Dauer D, Weinhardt C (2018) A blockchain-based smart grid: towards sustainable local energy markets. Comput Sci Res Develop 33(1–2):207–214
8. Airbnb (2019) Vacation rentals, homes, experiences & places. Available at: https://www.airbnb.co.za/. Last accessed 16/11/2019
9. Sadr SM, Casal-Campos A, Fu G, Farmani R, Ward S, Butler D (2020) Strategic planning of the integrated urban wastewater system using adaptation pathways. Water Res 182:116013
10. Werners SE, Wise RM, Butler JR, Totin E, Vincent K (2021) Adaptation pathways: a review of approaches and a learning framework. Environ Sci Policy 116:266–275
11. Lin BB, Capon T, Langston A, Taylor B, Wise R, Williams R, Lazarow N (2017) Adaptation pathways in coastal case studies: lessons learned and future directions. Coast Manag 45(5):384–405
12. Butler JRA, Bohensky EL, Suadnya W, Yanuartati Y, Handayani T, Habibi P, Puspadi K, Skewes TD, Wise RM, Suharto I, Park SE (2016) Scenario planning to leap-frog the sustainable development goals: an adaptation pathways approach. Clim Risk Manag 12:83–99
13. Kingsborough A, Borgomeo E, Hall JW (2016) Adaptation pathways in practice: mapping options and trade-offs for London's water resources. Sustain Cities Soc 27:386e397. https://doi.org/10.1016/j.scs.2016.08.013
14. Van Veelen PC, Stone K, Jeuken A (2015) Planning resilient urban waterfronts using adaptive pathways. Proc Inst Civ Eng Water Manag 168:49e56. https://doi.org/10.1680/wama.14.00062
15. Jeuken A, Haasnoot M, Reeder T, Ward P (2015) Lessons learnt from adaptation planning in four deltas and coastal cities. J Water Clim Change 6:711e728. https://doi.org/10.2166/wcc.2014.141
16. Klijn F, Kreibich H, De Moel H, Penning-Rowsell E (2015) Adaptive flood risk management planning based on a comprehensive flood risk conceptualisation. Mitig Adapt Strat Glob Change 20(6):845–864
17. Lawrence J, Haasnoot M (2017) What it took to catalyse uptake of dynamic adaptive pathways planning to address climate change uncertainty. Environ Sci Policy 68:47–57
18. Michas S, Stavrakas V, Papadelis S, Flamos A (2020) A transdisciplinary modelling framework for the participatory design of dynamic adaptive policy pathways. Energy Policy 139:111350
19. Cradock-Henry NA, Blackett P, Hall M, Johnstone P, Teixeira E, Wreford A (2020) Climate adaptation pathways for agriculture: insights from a participatory process. Environ Sci Pol 107:66e79. https://doi.org/10.1016/j.envsci.2020.02.020
20. Tshuma N (2021) Eskom thanks SA for load shedding patience but says power cuts still 'highly possible'. https://www.iol.co.za/capeargus/news/eskom-thanks-sa-for-load-shedding-patience-but-says-power-cuts-still-highly-possible-680c0e28-c82a-46f1-83f5-25fc79da4ede. Last accessed 2/10/2021
21. Haasnoot M, Middelkoop H, Offermans A, Beek E, van Deursen WPA (2012) Exploring pathways for sustainable water management in river deltas in a changing environment. Climatic Change 115:795e819. https://doi.org/10.1007/s10584-012-0444-2

Chapter 19
A Health-Energy Nexus Perspective for Virtual Power Plants: Power Systems Resiliency and Pandemic Uncertainty Challenges

Sambeet Mishra and Chiara Bordin

Abstract This chapter introduces and discusses a novel "health-energy nexus under pandemic uncertainty" concept that arises as a consequence of the current pandemic that we are experiencing worldwide. In light of the pandemic implications on the power and energy systems, we discuss how the global health conditions are tightly connected with the energy consumption needs and how the two areas closely interact with each other. A real-world dataset from the Estonian energy consumption over three years (2018, 2019, 2020), together with information gathered in the recent literature, will be illustrated to motivate the foundations behind the health-energy nexus concept. Opportunities and challenges that lie behind the interaction between health and energy will be outlined, and ways to address the changes in the power systems resiliency due to pandemic conditions will be discussed. Virtual power plants will be presented, as a way to address the pandemic challenges and improve the systems' resiliency and reliability. A novel concept of Cyber-Physical Health-Energy Systems will be discussed. Moreover, the value of interdisciplinary education and research, together with the novel interdisciplinary domain of energy informatics, will be proposed as key pathways to overcoming the challenges posed by the novel health-energy nexus under pandemic uncertainty.

Keywords Virtual power plants · COVID-19 · Energy informatics · Optimization

1 Introduction

The recent COVID-19 pandemic outbreak has significantly changed how society functions at various levels. For power and energy systems, both the commercial and the residential load demands are experiencing a dramatic change in the pattern of

S. Mishra (✉)
TalTech, Tallinn University of Technology, Tallinn, Estonia
e-mail: sambeet.mishra@ttu.ee

C. Bordin
UiT, The Arctic University of Norway, Tromsø, Norway

© The Author(s), under exclusive license to Springer Nature Singapore Pte Ltd. 2022   267
R. J. Howlett et al. (eds.), *Smart and Sustainable Technology for Resilient Cities and Communities*, Advances in Sustainability Science and Technology,
https://doi.org/10.1007/978-981-16-9101-0_19

consumption. The load demand, both heating and electric, used to be distributed with peak demand at the office buildings during the day, and residential demand peaks during the early morning and late afternoon/evening. Due to the "work-from-home" requirements during the pandemic, the electric demand has been flattened. Most people live indoors, even for working hours; therefore, the overall consumption stays flat. The office buildings normally have a higher volume of energy consumption due to their bigger size. These types of commercial demands are no longer relevant as they were before. The other sector which experienced a high degree of disruption is the transportation sector. The energy consumption in the transportation sector came to a stand-still. The energy production sector has also experienced a significant change in the utilization of the total capacity. The energy production from behind the meter generation units, such as small PV or battery banks, is mostly consumed locally instead of through grid injection. The large generation units are not fully utilized, since the total volume of consumption is reduced, due to a smaller consumption from office buildings and transportation.

The authors in [1] present the impact of COVID-19 within the power and energy sector. The climatic conditions have improved dramatically with the reduction in the NO_2 and CO_2 levels during the lockdown periods. The paper also outlines how the total volume of electric demand has decreased in this pandemic period, compared to previous years. Moreover, a report developed by the International Energy Agency outlines that the renewable energy generation in Europe was higher than that of fossil-fuel-based generation during the lockdown [2]. However, in the USA, the balance is gradually shifting toward higher fossil-fuel utilization. In Europe, the same trend was observed—as the total electricity demand falls, the share of renewable energy in the total generation mix increases, while the share from non-renewable resources falls. Since the overall electric demand has fallen in volume, a higher portion is met by renewable-based resources in comparison with fossil fuel-based. Indeed, new record generation from solar and PV resources is registered during the lockdown in Europe. At the same time, natural gas generation has increased due to low prices and high carbon prices.

1.1 Objectives and Key Contributions

The objective of this work is to introduce and discuss a novel "health-energy nexus under pandemic uncertainty" concept that arises as a consequence of the current pandemic that we are experiencing worldwide. Traditionally, the concept of "health-energy nexus" in literature has been utilized merely to investigate the energy consumption within the healthcare system [3]. However, other studies proposed a health-energy nexus concept to investigate the effect of climate change on the overall state of health of the population and accounting for the health impacts from electricity generation to justify the decarbonization needs [4]. The key contribution of this work is to propose a novel health-energy nexus perspective that widens the scope: In light of the pandemic implications on the power and energy systems, we discuss how the global health conditions are tightly connected with the energy consumption needs and how the two areas closely interact with each other. Real-world dataset [5–8] from the Estonian energy consumption over three years (2018, 2019, 2020), together with information gathered in the recent literature, will be illustrated to motivate the foundations behind the health-energy nexus concept. Opportunities and challenges that lie behind the interaction between health and energy will be outlined, and ways to address the changes in the power systems resiliency due to pandemic conditions will be discussed. Virtual power plants will be presented, as a way to address the pandemic challenges and improve the systems' resiliency. A novel concept of Cyber-Physical Health-Energy Systems will be discussed. Moreover, the value of interdisciplinary education and research, together with the novel interdisciplinary domain of energy informatics, will be proposed as key pathways to overcome the challenges posed by the novel health-energy nexus under pandemic uncertainty. Finally, the role of energy policies within the nexus will be outlined.

2 The Novel "Health-Energy Nexus, Under Pandemic Uncertainty"

As outlined in the previous section, the recent pandemic outbreak of COVID-19 is adding new challenges and it is already affecting the energy and power sector as a whole. Therefore, new solutions should be able to address the effects of extraordinary scenarios, like the one we are currently experiencing with the COVID-19 pandemic. Utilities are looking at the impact of lessening the demand for power from commercial and industrial enterprises, and the possible rise in consumption from the residential sector, with schools and businesses closed and people ordered to work from home [9]. The pandemic outbreak opens the doors to "a new health-energy nexus," showing that health and energy are linked to one another, and events in one particular area can have effects on the other area. Due to the shifting of energy demand from commercial areas toward residential areas, pandemics like the current COVID-19 can lead to an outage, and thereby, they represent a reliability and resiliency concern for power systems. Indeed, during the outbreak, the power network resiliency is threatened, and power companies should know which areas of the network need more attention, and which ones least.

Figure 1 summarizes the key concepts of the proposed novel health-energy nexus.

The health and energy sectors are tightly interconnected, as is shown by the blue circle of health that intersects the green circle of energy. The pandemic uncertainty is the issue that lies at the intersection between health and energy. Indeed, global energy consumption has radically changed due to the pandemic situation that introduced new working and living habits for people. Thus, a health-energy nexus arises that poses both challenges and opportunities to be addressed. The main challenge is that the power and energy systems must adapt to the new energy needs that arise as a

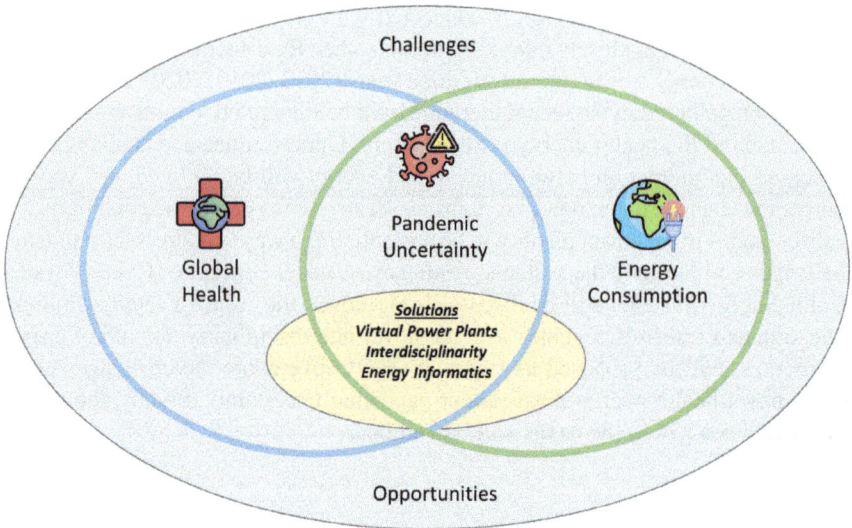

Fig. 1 Key concepts of the novel health-energy nexus

consequence of the pandemic, with particular regard to new demand curves, new demand peaks, and new demand concentrations.

The figure shows that it is possible to focus on three main concepts to address the health-energy nexus challenges posed by the pandemic situation: VPP, interdisciplinarity, and the novel domain of energy informatics.

These concepts represent the main opportunities that we discuss in this work. Indeed, they are the three paths that jointly can lead to solutions to the challenges of the health-energy nexus.

3   Energy Demand Patterns and Other Pandemic Implications

Figure 2 shows a schematic representation of the cascaded implications of a pandemic, from societal work/life habits' change toward new energy consumption trends that generate new emerging technical issues for power systems, leading to the need for new approaches and solutions. This is the key path that lies behind the health-energy nexus proposed and discussed in this chapter.

As shown in the figure, the overall pandemic situation has three main implications, which are all tightly linked to the energy and power systems field: new working habits due to home office, higher use of online meetings due to social distancing requirements, and different traveling habits due to the overall safety restrictions.

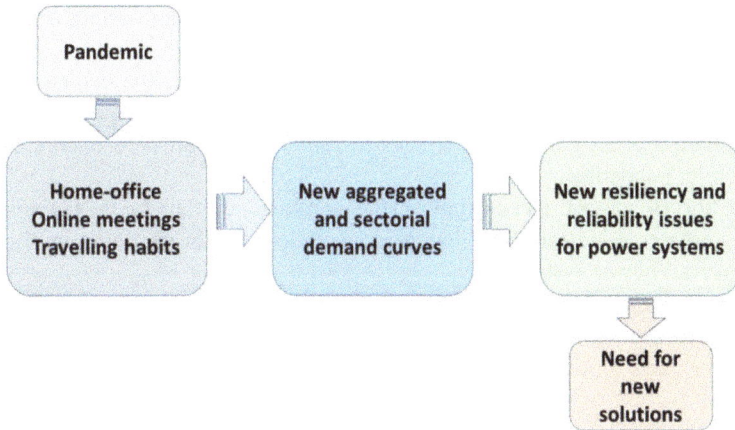

Fig. 2 Cascaded implications of a pandemic: from health to energy

The new advent of home office (namely, smart-working or home-working) generates new demand peaks and new demand concentrations, due to the shift of workers from industrial and commercial areas to residential areas. This means that it is not only the aggregated energy demand of a country that is affected by a pandemic, but it is also the sectorial energy demand that is subject to variations [10].

Another reason why the energy demand patterns change under pandemic conditions is due to the energy demand being more scattered during the day. Indeed, while before the demand was concentrated in industrial/commercial areas where offices and working places were accessible, now it is scattered in wider residential areas due to home-working requirements [11, 12]. In sum, commercial and industrial demands decline, but residential demand increases [1].

Few works in the literature have recently investigated the impacts of stay-home living patterns on the energy consumption of residential buildings. The study in [13] shows that in the USA, the overall electricity demand is lower because the lockdown impacts negatively the activities within commercial buildings and manufacturing sectors. However, the energy consumption for the housing sector increased by 30% during the full 2020 lockdown period. This is because of the higher occupancy patterns during daytime hours, which led to higher use of energy-intensive systems such as heating, air conditioning, lighting, and appliances.

Limited analyses have so far been conducted to evaluate the impact of lockdown and stay-at-home orders on energy use for various sectors. The study in [13] discusses sectorial demand variations pre- and post-COVID-19. The data gathered among Argentina, Australia, United Kingdom, Ireland, and Texas suggest that the overall trend is an increase in the residential energy demand and a decrease in the commercial and industrial energy demand. The available data are not enough to make a thorough comparison between the demand patterns in the commercial and industrial sectors.

Authors in [14] discuss the different energy consumption of industrial and commercial areas with commuting, compared to a home office. They show that consumption is interrelated, where the decrease in energy consumption in offices will lead to an increase in energy consumption at home. However, an important observation of this study is also that the degree of increment and decrement, which contribute to the net consumption, is not equivalently shifting between the options. So far, there are not yet enough studies to fully understand such energy demand trends under pandemics.

In addition to the few data available in the literature, new data for the Estonian case will be presented and discussed below, to better understand the pandemic implications on the overall power system.

In Fig. 3, the total electricity demand in Estonia during the years 2018–2020 is presented. In Fig. 4, annual electricity consumption and the cumulative number of COVID-19 cases officially registered are presented.

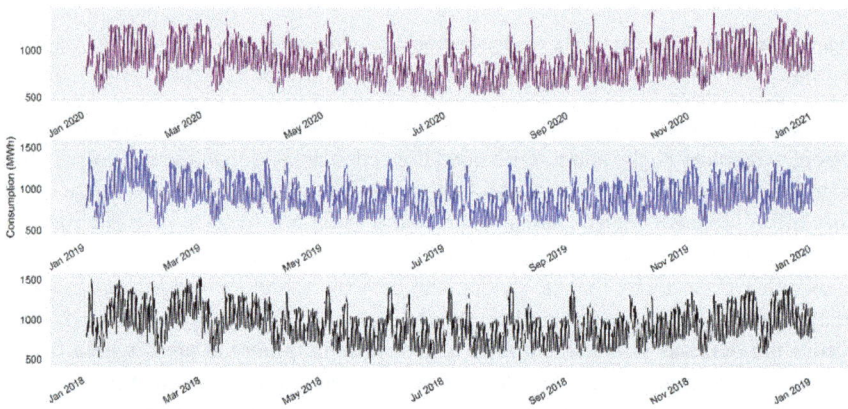

Fig. 3 Total electricity consumption with hourly resolution from 2018 to 2020 in Estonia

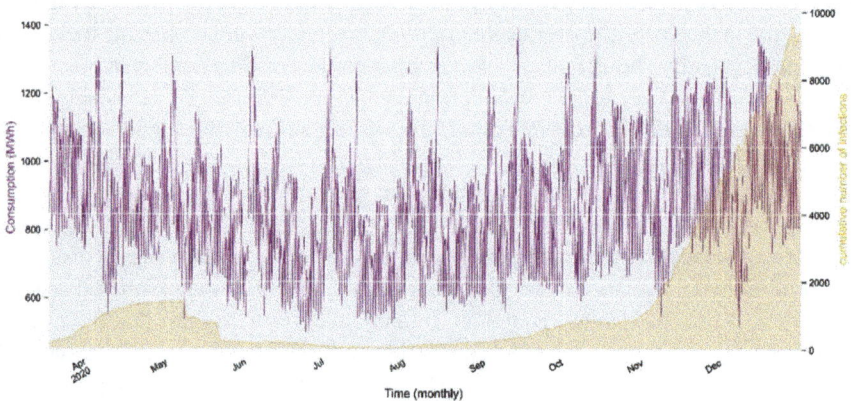

Fig. 4 Number of covid cases and energy consumption in 2020 in Estonia

The time-series data are obtained from the Estonian transmission system operator Elering [5]. Time-series data describe the pattern and trends in a dataset. However, electricity demand is highly correlated with the weather but also dependent on various other variables such as holidays, appliances, etc.

Three consecutive years are compared to validate if a trend is temporary or persistent. From the observations, the total volume of consumption has dropped below 1400 MWh in 2020. In the past two years, the consumption volume peaked at 1500 MWh. This validates the fact that total volume has dropped. The factor that could describe this change is the pandemic outbreak.

On March 13, 2020, the Estonian government has declared a state of emergency until 1 May which was later extended until 17 May. Looking more closely at the daily consumption patterns on 1 May over the three years, in Fig. 5, it is evident that the trend has changed. Apart from the volumetric change, the trend has smoothened during the pandemic. The peak demand hours from 15:00 to 20:00 are highlighted where the trend can be observed.

Moving to monthly demand patterns during May over the three years presented in Fig. 6, there is a consistent change in patterns. For example, during 12, 17, and 25 May, the patterns are similar during the years 2018 and 2019, while in 2020 it is different. The patterns were also observed during the first week of May as in Fig. 7. During days 4 and 5.

Histograms of the consumption are presented in Fig. 8 which demonstrates the overall pattern of the consumption by how frequently certain volumes of consumption were attained. In 2020, the peak demand was 800 MWh which occurred a little over 500 times. In 2019 and 2018, the peak demand was 1000 MWh which occurred 480 times. Through this investigation, it is established that the total volume of consumption was reduced, and the pattern of consumption was changed.

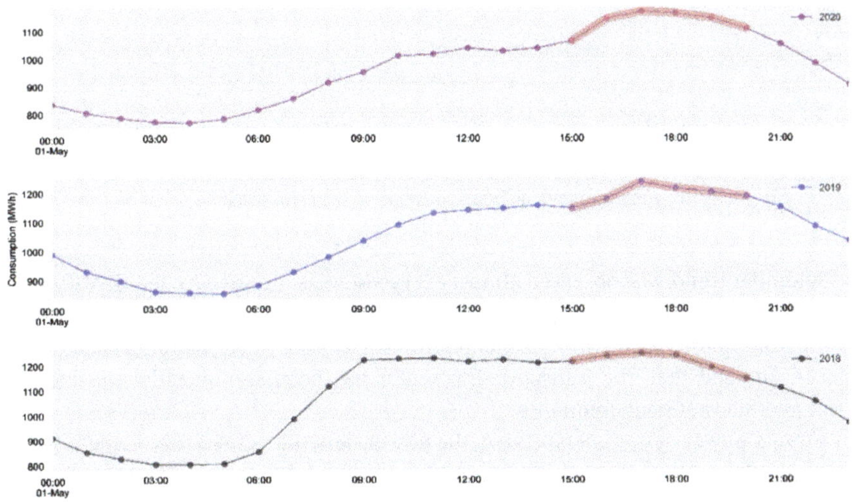

Fig. 5 Electricity consumption with an hourly resolution during a day

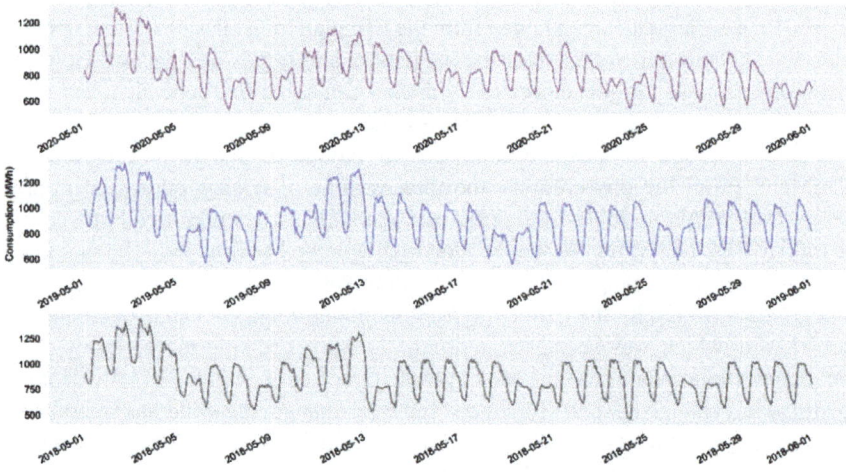

Fig. 6 Electricity consumption with an hourly resolution during a month

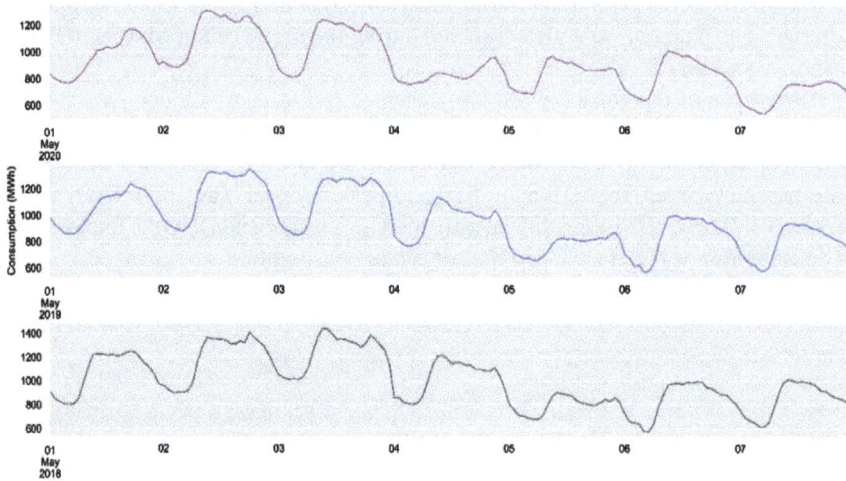

Fig. 7 Electricity consumption with an hourly resolution during a week

A similar trend was observed almost everywhere as reported by the IEA [2]. The plausible explanation is the pandemic that has impacted the way energy is consumed. Work from home has led to less energy consumption in office buildings and public places. Beyond that, the transportation sector has been significantly impacted by local and international commutes.

From a power system perspective, the power network is unevenly loaded. While normally the network has a high capacity made available to the office buildings or public places, the concentration of demand has now shifted to purely residential areas. Certain transformers are now always loaded due to the consistent demand.

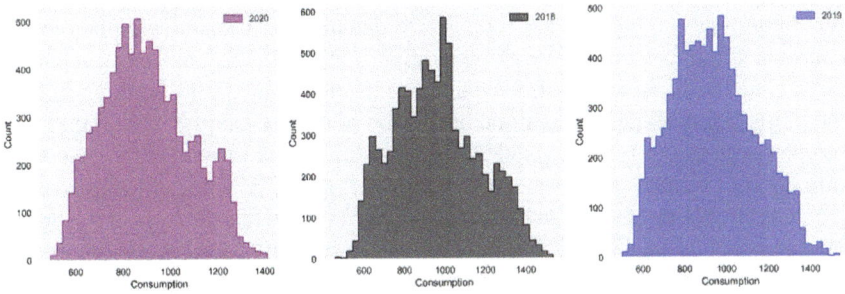

Fig. 8 Distribution of electric consumption

This has adverse effects on the life cycle of the power apparatus and might lead to voltage imbalances, brownouts, or even blackouts. The demand-side flexibility has become even more important given that network replacement is a very expensive choice for system operators specifically at a distribution level.

As outlined at the beginning of this section in Fig. 2, another implication of a pandemic is the higher use of data, due to heavy use of online meetings and online resources to support the home-working and social distancing requirements. The need for information and communications technologies to support digitalization has a direct impact on the energy consumption of data centers that increases heavily [14].

Finally, there are significant consequences observed in the transportation sector. From the road to rail, air and maritime transportations have experienced a fall in volume. Electric transportation through cars, buses, or rail not only consumes but also stores energy to be discharged at a later stage. The consequence is that traditional fossil energy demand declines, but renewable energy demand increases [2].

The new demand patterns discussed above create new challenges for the energy and power systems, regarding resiliency and reliability. Even though the total consumption decreases, the geographical consumption is still a problem because the distribution lines for residential consumption are overloaded, while the industrial and commercial buildings with a higher capacity of connections are underused. The pandemic implications outlined in the previous paragraphs generate a novel health-energy nexus that requires the development of new solutions, able to tackle the resulting challenges on the energy and power systems. The following sections will address three pathways to tackle the health-energy nexus, VPP, the need for interdisciplinary research, and energy informatics as a key domain.

4   Value of Virtual Power Plants Within the Novel Health-Energy Nexus

The pandemic has shifted the focus to localized energy consumption and regeneration. Consequently, the demand-side participation to provide flexibility represents a great opportunity to balance the supply with demand. A virtual power plant (VPP) is formed through a collection of generation units from various sizes which are geographically dispersed. It also includes demand-side flexibility in the portfolio. One of the key challenges is optimal utilization of the non-dispatchable and renewable power generation units. A VPP can facilitate access to renewable generation units often in remote locations. This in turn will result in better utilization of the resources. Furthermore, the grid resiliency can be fostered through many small-scale units acting interim a VPP. Now that the residential consumption is peaking during the pandemic, a VPP can provide active participation for demand-side flexibility. In [15], the authors report that there is a sharp decrease in the electricity from renewable resources during the pandemic. Furthermore, the investment projects for new generation units have been suspended due to a sharp fall in the electricity price. Alongside, the challenges for the distribution system operator have increased due to voltage imbalance, accurate demand projections, and flexibility reserve.

As outlined in the previous section, energy consumption has changed in both volume and pattern due to the pandemic. While health concerns have triggered government policies which resulted in the change, a VPP can aid in maintaining the balance while creating economic opportunities. Demand response could provide an immediate solution to the network capacity problem. Price signals are considered as among the clear motivators for demand shifting. Coordinating each price signal to reach a certain volume of consumption is rather complex. Then, predicting the consumption for the next day or week might be a too short time to draw attention. A VPP formed locally with a district or region could commit to a sizable volume of energy consumption while relaxing the coordination challenge. Then again, prosumers or small-scale producers participating within a VPP might provide flexibility in terms of both production and consumption sides.

Consequently, VPP can also reinforce the network resiliency and security by balancing the consumption and production locally to the requirements of the overall power grid. For example, the voltage imbalance due to excessive PV injection can be avoided by negative price signals [16]. In addition, the reactive power consumption or production can be shifted with a better price offer. A local energy market through a VPP can aid in maintaining overall grid resiliency and thereby enhance reliability.

Islands, suburban, rural, and other remote corners of the grid often act either as energy injection through PV or wind resources or demand. Through VPP, these points can take an active role in the overall grid resiliency. During pandemic or other disaster management scenarios, they can take a more active role in balancing the overall grid. For example, if a certain plant experiences an operational failure in one of the networks, a VPP which has geographically distributed resources can provide alternative solutions how to balance the grid. Peer-to-peer interactive energy transactions [17] are gaining pace recently with energy tokenization and consumer participation. A VPP can act as a local trading platform aggregating a small region that might be geographically distributed. Through many VPPs within and across distribution system operators, the market competition would derive better value for the end-user. Unlocking economic potential and enabling participation are among the key outcomes of transactive energy which could be realized through a VPP.

Aggregating the individual consumption and production potential to a certain volume would enable trading and ease of control. This phenomenon was also observed in the transportation sector through companies like Uber, Bolt, and Ola. However, any free market would require regulation and policies to avoid coalitions and motivate competition. Gradually, the share of renewable resources is growing in both volume and type making the generation distributed across the geographic region. System operators would continue to own networks when the VPP can provide a local trading platform for peer-to-peer energy transactions while reinforcing the network operations.

The uncertainty arising from renewable resources and catastrophic situations such as the pandemic can also be addressed at different levels—neighborhood, city, and regional levels. A VPP can facilitate uncertainty handling through various scenarios within and beyond.

Finally, a VPP platform can enable geographic expansion and flexibility pool to the power system. Scheduling the energy discharge to peak hours matching the variable energy production from renewable resources can be facilitated through a VPP. The same balancing effect can also be coordinated in air travels to different locations through scheduling flights through a VPP. Consequently, the dependence on fossil fuels could be further reduced while increasing the share of renewables in the overall energy system.

5 Energy Informatics and Interdisciplinarity to Tackle the Novel Health-Energy Nexus

The novel health-energy nexus is an interdisciplinary concept in itself since it touches upon the two main disciplines of health and energy. However, such a concept requires an even wider interdisciplinary approach since the topics of health and energy are nowadays tightly connected to many areas of computer science. Energy informatics, in particular, is a novel domain that lies at the intersection of energy systems, power systems, economics, computer engineering, and computer science. As such, energy informatics represents a valuable subject to study and address the resiliency and reliability challenges that arise within the power systems. An energy informatics perspective is therefore needed to tackle the novel health-energy nexus under pandemic uncertainty as well as implement VPP solutions for it. Interdisciplinarity, together with energy informatics, is therefore proposed in this section as the key instrument to address the challenges of the health-energy nexus in general and enhance the role of VPP in particular.

The key subjects of energy informatics have been identified in [18]. The paper outlines how mathematical optimization, in general, and smart energy and power systems modeling, in particular, lie at the heart of the energy informatics domain. The paper also identifies cyber-physical energy systems (CPES) and the Internet of Things (IoT), together with mathematical models, as the three main dimensions of energy informatics. The same three dimensions are key within the novel health-energy nexus as well. Health and energy represent physical spaces tightly interconnected with each other, where a wide variety of issues arise as a consequence of their interaction. The physical space comprising the sectors of health and energy can be investigated, understood, and controlled through modern mathematical and computer science techniques, which altogether are part of a so-called cyberspace. A cyberspace can successfully function through four main tasks that are strongly interconnected:

- Learn: understand the data
- Predict: forecast and generate new data
- Model: build technological mathematical optimization models
- Optimize: utilize the data and the models to make optimal decisions that can positively influence the physical space.

Figure 9 represents this concept. By combining the physical space and the cyberspace outlined above, a novel Cyber-Physical Health-Energy System (CPHES) arises as a direct consequence of the pandemic.

Once the learning and prediction tasks are over, the key is how to utilize this new knowledge to build mathematical models that represent the physical system. The knowledge, the data, and the models can be utilized within optimization tools, to make optimal decisions that can positively impact the physical space where the

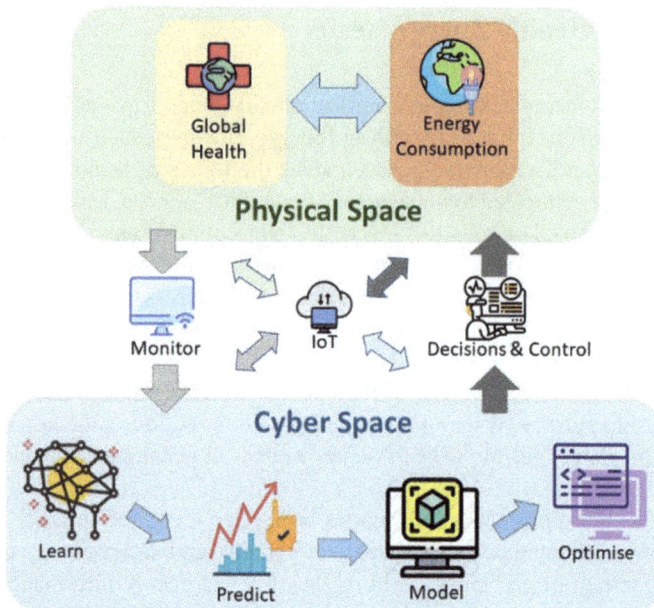

Fig. 9 A novel cyber-physical health-energy system (CPHES) arising from the health-energy nexus as a consequence of the pandemic

health and energy sectors are located. The concept of developing tools for predictions (i.e., machine learning) and utilizing mathematical optimization to identify optimal decisions (both short term and long term) over such predictions marks the transition from a limited purely predictive analytics approach, toward a more advanced and complete prescriptive analytics approach [19].

The physical space and the cyberspace introduced above are linked through two main tasks. "Monitoring" allows transferring data from the physical space into cyberspace. While "Decision and Control" take the decisions developed in cyberspace, and it implements them back into the physical space for the optimized management of the health-energy nexus. On top of the CPHES defined above, it is possible to add the Internet of Things (IoT). Through IoT, the CPHES can be connected to the Internet, and decisions can be automatized and enhanced.

It is clear that mathematical modeling, with its intrinsic capability to contribute to decisions support systems tools in general, is a key subject within the health-energy nexus. Mathematical optimization has already been successfully applied to solve both power and energy systems-related problems [20]. In addition, mathematical optimization can be successfully utilized for optimal investment decision-making and operational management of VPP [21, 22] within the nexus. The uncertainty behind the pandemic forthcoming developments can be tackled as well by optimization models. Resiliency and reliability issues that arise from different future energy demand projections can be addressed in particular with stochastic, multi-horizon optimization [23, 24] and long-term scenario development at a qualitative level.

However, the complex and interdisciplinary nature of the nexus requires the interconnection of different computer science subjects to tackle all the related challenges. Interdisciplinary approaches are therefore a must to address the main cascaded implications of the pandemic as discussed in the previous sections. An example of an interdisciplinary approach is presented in [25] where mathematical optimization is linked to other computer science subjects to address the consequences of data centers' energy consumption within the power systems. This can have strong implications on the health-energy nexus as well, since an increased energy consumption within data centers is one of the cascaded implications of the pandemic, as further discussed previously.

6 Toward an Extended Education-Health-Energy Nexus

Figure 10 shows an extended nexus concept that arises in form of education-health-energy nexus. Indeed, education and consequently the knowledge that grows and spreads through education, play an important role to understand, investigate, and address the current and future challenges posed by worldwide disasters like a pandemic.

Increased people's knowledge and awareness lead to better actions both for personal health and for sustainable energy consumption. Better healthy habits

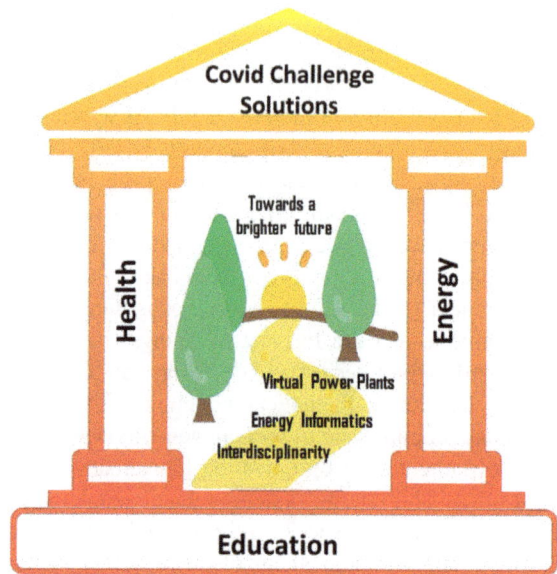

(namely, physical activity and nutrition) lead to a stronger immune system and therefore mitigate the pandemic spreading [26] and the consequent effects on human energy behavior. Better human energy behavior leads to better use of resources, reduced energy waste, and increased sustainability. It, therefore, mitigates the challenges on the power grids. Formal education, non-formal education, and informal education [27] are all key to lead the population toward healthier habits as well as more sustainable choices.

Raising awareness among the population is not enough, since the pandemic challenges on the health sector and the power system sector need to be addressed also at more specialized technical levels, by researching and investigating new solutions and pathways. Higher education plays an important role from this point of view, providing skills and a proper mindset to understand advanced topics relevant for the health-energy nexus.

As shown in Fig. 10, three main pathways branch off from the education foundation: interdisciplinary teaching and research in general, the novel interdisciplinary domain of energy informatics in particular, as well as the key concept of VPP. The first two are tightly interconnected. The latter cannot be addressed alone but requires the first two as preparation paths to be fully understood and implemented.

Education within the energy informatics domain is still in its infancy, but recent works in literature have highlighted and discussed the importance of educating the future generation of energy informatics specialists, to address the future challenges of energy and power systems as well as the increasingly interdisciplinary needs of both research and industry [28].

7   The Role of Energy Policies Within the Education-Health-Energy Nexus

Energy policy is the set of measures through which the government addresses issues related to energy growth and usage. The latter includes energy production, distribution, and consumption. While it has been discussed that the health implications of COVID-19 affected the overall energy use and availability, it must be highlighted that a good portion of these changes was caused by the governments' responses and the policies that were pursued during the pandemic outbreak. Energy security of supply, as well as the quality and efficiency of energy services, is among key aspects that should be addressed by energy policies, especially under the threats of pandemic outbreaks like the one experienced with COVID-19.

Governments should identify appropriate strategies when responding to a pandemic outbreak so that short-term policy goals aimed at tackling an emergency will not negatively impact medium- to long-term policy goals aimed at ensuring the security of supply and high-quality energy services.

Governments should also increase their awareness of the implications of their measures not only at a national level but also at an international and global level. Indeed, during the pandemic outbreak, the governments' responses and policies while being fairly ubiquitous were not the same everywhere. Therefore, some countries have been hit much harder than others. However, it must be highlighted that the world nowadays is tightly interconnected, and no country is an island. Therefore, the different policies do not only directly affect the single country where they are developed, but they also indirectly affect the surrounding countries that are connected to it. This is particularly true when it comes to mobility constraints, which have been playing an important role in the energy consumption changes, as well as job allocation, during and after the pandemic. From the perspective of an interconnected world, even though policies might have been slightly different in each country, still their holistic effect impacted the overall energy system as a whole.

8   Conclusions

This work is about a topical issue, regarding pandemic implications on the energy and power systems, and the solutions to such issues, identified in VPP, energy informatics, and interdisciplinary education and research. The first part of the chapter aimed at reflecting on the implications of the COVID-19 pandemic on the energy sector. By understanding and verifying that indeed, certain health-related events affect energy consumption, the ambition was to introduce a new nexus concept to discuss how the two sectors can be interconnected. The implications of health problems linked to the pandemic uncertainty on the power and energy systems have been identified as a novel health-energy nexus. Such nexus has been illustrated and discussed. In addition, the new concept of Cyber-Physical Health-Energy Systems (CPHES) has been discussed and a further extension of the nexus toward education-health-energy nexus has been introduced.

COVID-19 is one specific health-related event that has been used to demonstrate that health can affect energy. However, the ambition of the chapter is to widen this aspect by reflecting on the broad implications that health and energy have on each other and expand the discussion toward key pathways that can help to overcome the challenges that may arise when health and energy meet each other. While the COVID-19 pandemic is one ongoing and topical instance that creates awareness of the health-energy nexus, many other different and unexpected events in the health sector may arise in the future causing similar implications on the energy sector. The ambition is therefore to learn from the pandemic, to identify broad pathways suitable to tackle future events within the health-energy nexus that may have similar implications.

It is clear that the proposed health-energy nexus has important implications for the energy and power systems as a whole. Therefore, the health-energy nexus as identified in this paper should be given more attention in the near future, and it should be considered as a priority at policy and political levels. This can be done by devolving funds to projects and research centers that tackle both sectors, health and energy, and that propose interdisciplinary solutions, with computer science as the main instrument to address the energy challenges. Suitable funding programs within the health-energy nexus should be developed and made accessible both for research purposes in academia and for applied purposes in the industrial world (namely, supporting start-ups aimed at developing novel products able to address the challenges of the health-energy nexus).

As mentioned in the previous sections, research and industry are tightly connected to education. Therefore, an education-health-energy nexus arises, where interdisciplinary education is the foundation to understand the subjects of health and energy and identify pathways to address the pandemic challenges. From this perspective, new interdisciplinary study programs should be developed to educate the new generation of energy informatics specialists, with adequate skills and mindset to take on the current and future challenges of energy and power systems under health-related pandemic uncertainty.

As discussed, energy policies played an important role during and after the pandemic outbreak. Since the world is highly interconnected, energy policies have not only direct national implications but also indirect international implications. From the perspective of an interconnected world, even though policies might have been slightly different in each country, still their holistic effect impacted the overall energy system as a whole. Therefore, the conclusions in terms of the relevance of a novel education-health-energy nexus as well as the value of pathways like VPP, energy informatics, and interdisciplinary teaching and research are universally valid.

Acknowledgements This work was supported by the Estonian Research Council Grant PUTJD915. Icons have been procured from flaticon.com

References

1. Madurai Elavarasan R, Shafiullah GM, Raju K, Mudgal V, Arif MT, Jamal T, Subramanian S, Sriraja Balaguru VS, Reddy KS, Subramaniam U (2020) COVID-19: impact analysis and recommendations for power sector operation. Appl Energy 279:115739. https://doi.org/10.1016/j.apenergy.2020.115739

2. I.E.A. (IEA) (2020) Covid-19 impact on electricity—analysis—IEA, IEA 1–10. https://www.iea.org/reports/covid-19-impact-on-electricity

3. Sudarshan H, Trust K (2018) Decentralized renewable energy in healthcare: the energy-health nexus Karuna Trust in partnership with SELCO Foundation

4. Howard DB, Soria R, Thé J, Schaeffer R, Saphores JD (2020) The energy-climate-health nexus in energy planning: a case study in Brazil. Renew Sustain Energy Rev 132:110016. https://doi.org/10.1016/j.rser.2020.110016

5. Elering (2021) Production and consumption database of transmission system operator in Estonia (Elering), Elering Database

6. Guidotti E, Ardia D (2020) COVID-19 data hub. J Open Source Softw 5:2376. https://doi.org/10.21105/joss.02376

7. Hasell J, Mathieu E, Beltekian D, Macdonald B, Giattino C, Ortiz-Ospina E, Roser M, Ritchie H (2020) A cross-country database of COVID-19 testing. Sci Data 7:1–7. https://doi.org/10.1038/s41597-020-00688-8

8. Takaya H (2021) CovsirPhy Kaggle notebook. COVID-19 data with SIR model

9. Asma Aziz AT (2020) COVID-19 implications on electric grid operation. IEEE Smart Grid, IEEE Power Energy Mag

10. Wang B, Yang Z, Xuan J, Jiao K (2020) Crises and opportunities in terms of energy and AI technologies during the COVID-19 pandemic. Energy AI 1:100013. https://doi.org/10.1016/j.egyai.2020.100013

11. Cheshmehzangi A (2020) COVID-19 and household energy implications: what are the main impacts on energy use? Heliyon 6:e05202. https://doi.org/10.1016/j.heliyon.2020.e05202

12. fei Chen C, Zarazua de Rubens G, Xu X, Li J (2020) Coronavirus comes home? Energy use, home energy management, and the social-psychological factors of COVID-19. Energy Res Soc Sci 68. https://doi.org/10.1016/j.erss.2020.101688

13. Krarti M, Aldubyan M (2021) Review analysis of COVID-19 impact on electricity demand for residential buildings. Renew Sustain Energy Rev 143:110888. https://doi.org/10.1016/j.rser.2021.110888

14. Jiang P, Van Fan Y, Klemeš JJ (2021) Impacts of COVID-19 on energy demand and consumption: challenges, lessons and emerging opportunities. Appl Energy 285:116441. https://doi.org/10.1016/j.apenergy.2021.116441

15. Zhong H, Tan Z, He Y, Xie L, Kang C (2020) Implications of COVID-19 for the electricity industry: a comprehensive review. CSEE J Power Energy Syst 6:489–495. https://doi.org/10.17775/CSEEJPES.2020.02500

16. Piao L, de Vries L, de Weerdt M, Yorke-Smith N (2021) Electricity markets for DC distribution systems: locational pricing trumps wholesale pricing. Energy 214:118876. https://doi.org/10.1016/j.energy.2020.118876

17. Mishra S, Bordin C, Tomasgard A, Palu I (2019) A multi-agent system approach for optimal microgrid expansion planning under uncertainty. Int J Electr Power Energy Syst 109:696–709. https://doi.org/10.1016/j.ijepes.2019.01.044

18. Bordin C, Håkansson A, Mishra S (2020) Smart energy and power systems modelling: an IoT and cyber-physical systems perspective, in the context of energy informatics. In: Procedia Comput. Sci., Elsevier B.V., pp 2254–2263. https://doi.org/10.1016/j.procs.2020.09.275

19. Bertsimas D, Kallus N (2020) From predictive to prescriptive analytics. Manage Sci 66:1025–1044. https://doi.org/10.1287/mnsc.2018.3253

20. Bordin C (2015) Mathematical optimization applied to thermal and electrical energy systems, PhD thesis, University of Bologna

21. De Filippo A, Lombardi M, Milano M, Borghetti A (2017) Robust optimization for virtual power plants. Lecture notes in computer science (Including Subseries Lecture notes in artificial intelligence. Lecture notes in bioinformatics). 10640 LNAI (2017) 17–30. https://doi.org/10.1007/978-3-319-70169-1_2

22. Zheming L, Guo Y (2016) Robust optimization based bidding strategy for virtual power plants in electricity markets. In: IEEE power energy society general meeting. IEEE computer society. https://doi.org/10.1109/PESGM.2016.7742043

23. Bordin C, Mishra S, Palu I (2021) A multihorizon approach for the reliability oriented network restructuring problem, considering learning effects, construction time, and cables maintenance costs. Renew Energy 168:878–895. https://doi.org/10.1016/j.renene.2020.12.105

24. Bordin C, Tomasgard A (2019) SMACS MODEL, a stochastic multihorizon approach for charging sites management, operations, design, and expansion under limited capacity conditions. J Energy Storage 26:100824. https://doi.org/10.1016/j.est.2019.100824

25. Ziagham Ahwazi A, Bordin C, Mishra S, Hoai Ha P, Horsch A, st Ziagham Ahwazi A, th Hoai Ha P, th Horsch A (2021) EasyChair preprint VEDA-moVE DAta to balance the grid: research directions and recommendations for exploiting data centres flexibility within the power system, EasyChair

26. Lange KW, Nakamura Y (2020) Movement and nutrition in COVID-19. Mov Nutr Heal Dis 4:89–94. https://doi.org/10.5283/mnhd.33

27. Dib CZ (2008) Formal, non-formal and informal education: concepts/applicability. In: AIP conference proceedings. AIP Publishing, pp. 300–315. https://doi.org/10.1063/1.37526

28. Bordin C, Mishra S, Safari A (2021) Educating the energy informatics specialist: opportunities and challenges in light of research and industrial trends. Springer Nat Appl Sci

Chapter 20
A Sustainable Nutritional Behavior in the Era of Climate Changes

Gavrilaş Simona

Abstract First of all, this chapter briefly summarizes general aspects regarding climate changes and their causes. An enlarged section dedicated to the influence of the natural background modifications on the different ecosystems follows the introductory passages. The three principal directions considered were human health, the food industry, and production durability. Several closing remarks, suggestions, and conclusions end the topic approach. The nineteenth century might be the starting point in scientific analysis of climate change monitoring. During that time appear many theories of the negative influence of *greenhouse gas emissions* and *human activity* on various environmental aspects. Mathematical modeling can help connect probable causes and the effects of multiple factors on environmental degradation. Generally, many elements affect climate, including geographic location, airflow, characteristic topography, and the greenhouse gases: CO_2, CH_4, N_2O, *fluorinated gases*. These are due to *intensive* activities, such as *burning fossil gas* and *fuel, deforestation, animal husbandry*, nitrogen fertilizers, and *fluorine-based gases*. Based on their direct and immediate impact, effects on climate change influence human *health, behavior,* and the *environment*. The results of climate change are visible on a global scale. Regardless of the cause, any new situation will influence people's eating behavior and the environment. The current pandemic highlights the necessity and importance of a short, secure food supply chain. The present sanitary crisis raises questions about the possibility of people providing food. Therefore, each state tries to restrict food exports, trying to meet the food needs of its population as much as possible from its production. Such an approach can represent internal and external challenges: internal to stimulate domestic production and outward to persuade partner countries to maintain the level of food and/or related exports to a certain extent.

Keywords Environmental changes · Human health · Durable actions

G. Simona (✉)
Faculty of Food Engineering, Tourism and Environmental Protection, "Aurel Vlaicu" University of Arad, 2-4 E. Drăgoi Str, Arad, Romania
e-mail: simona.gavrilas@uav.ro

1 Climate Changes

1.1 General Aspects

It is essential to understand the differences between the climate and the weather. In contrast to the environment, the weather describes the atmospheric characteristics for a limited period. The specialists have delimitated six zones characterized by specific features, considering the solar radiation incidence angle: *tropical*, *dry*, *subtropical*, *continental*, *polar*, and *highlands*.

Climate changes modify the weather paradigm observed for longer in a considered geographic area. The situation came to the researchers' attention in the early 1980s with the rise in temperatures. Since then, scientists have made other connections between atmosphere-specific characteristics alteration and different social and economic transformations. Most of these cause-effect associations are difficult to prove scientifically. Generally, they are results of statistic modulation. Based on this, climate change can have *direct* or *indirect* consequences.

Further, we emphasized several direct repercussions: *increase of atmosphere and ocean temperature, sea level, and severe precipitations enhance, glaciers reduction,* and *permafrost de-icing.*

The *indirect* outcomes are related mainly to the impact on the people and the soundings. The two aspects directly influence all country's economy. Till now it was concluded that the developing countries are more negatively affected than the developed ones. All climatic modifications might be the starting point of the *water* and *food crisis.* These, along with high temperature and heat waves, contribute to human health degeneration.

Primarily, the agricultural sector ensures the population's nutritional factors. Atmospheric inconstancy, which contributes to pests and pathogen's proliferation, might seriously affect it. The authorities must grant special attention to implement optimum regimentations to limit such effects.

The impact on the biosphere might be considerable. The time required for all species to adapt to new conditions is well known. Limited biodiversity negatively affects all trophic chains. For example, the increase in atmospheric carbon dioxide concentration does not affect only the terrestrial activity but also the marine one by pH decrease due to an acid–base imbalance determined by the presence of bicarbonate or carbonate ions. The two result from carbonic acid dissociation. The acid is the product that results from the reaction between the atmospheric CO_2 present above the water surface.

1.2 Initiators of Climate Modification

Scientists consider greenhouse gases to be an essential factor in climate change. Their constant increase level determined by the industrial development might determine the meteorological modifications [1]. Under this title are the gases that can assimilate and reemit the infrared radiation released from the Earth. The greenhouse effect results from this mechanism. Examples of such compounds are CO_2, CH_4, *vaporized water*, O_3, N_2O, and *F-gases*. As mentioned, the connection between climate modification and possible causes is based mainly on statistical correlations among the greenhouse gas concentrations and the cold or the warm intervals.

The greenhouse gases have different natural and anthropic sources. They could be a *tectonic activity, ample water or land stretches, fossil fuel burn, nitrogen fertilizers use, deforestation, animal husbandry*, etc. In some cases, these could be limited.

Water vaporization has an essential role in greenhouse effect formation. Its action mechanism is different from the other listed gases, acting as a system response to climate. Human activity has a lower influence on atmospheric water quantity.

There is a direct dependence between the water volatilization yield and the Earth's surface temperature. In the inferior atmospheric layer, a high water vapor concentration determines a high degree of infrared radiation retransmitted.

The most harmful greenhouse gas is *carbon dioxide*. It has three primary sources: degradation or combustion of organic compounds, vulcanos outgassing, and aerobic respiration. An important mechanism in carbon dioxide reduction is photosynthesis. A clear benefit in CO_2 amount decrease has reforestation and increased green areas in this situation.

The carbonate ions react with different metals resulting in the marine animals' shells or sedimentary minerals, which in time are released into the atmosphere as CO_2 as a result of volcano outgassing. During fermentation, carbon dioxide, and methane form. They can be atmosphere released or rock incorporated, determining fossil fuel formation. The result of their burning is CO_2 and H_2O atmosphere discharge.

Even if the biological, anthropogenic, and geochemical paths regarding the carbon cycle are similar, their yield presents essential differences, the first two having a significant implication in CO_2 formation.

Another critical greenhouse gas is *methane*. Its destructive potential is higher than the carbon dioxide due to its radioactive force released and the specific wavelengths absorbed. Its atmosphere concentration and resistance are lower compared to the CO_2. Higher methane quantities are in the tropical and northern wet Earth zones. Other sources are methane-oxidizing microorganisms, volcanos, or continental compounds. In the troposphere, it reacts with hydroxyl, forming water and carbon dioxide.

It is necessary to differentiate the *ozone* formed as a consequence of air pollution from the stratospheric one. The ozone generated in the troposphere due to carbon monoxide, nitrogen oxides, or volatile organic compounds photochemical reaction produces the greenhouse effects.

There are production mechanism similarities independent of the initial reactant. In all cases, after oxidation, CO_2 and H_2O result. For example, carbon monoxide primarily reacts with a hydroxyl radical (1) [2]. The intermediate radical formed is unstable and is immediately oxidized. The product is a peroxy radical (2), which contributes to nitrogen dioxide formation in the presence of nitrogen monoxide (3). Further, the NO_2 is photolyzed by radiation with a wavelength lower them 400 nm with the appearance of atomic oxygen (4). According to relation (5), ozone will result in the presence of molecular oxygen,

$$CO + \dot{O}H \rightarrow \dot{H}OCO \tag{1}$$

$$\dot{H}OCO + O_2 \rightarrow H\dot{O_2} + CO_2 \tag{2}$$

$$NO + H\dot{O_2} \rightarrow NO_2 + \dot{O}H \tag{3}$$

$$NO_2 \xrightarrow{h\nu} NO + O(^3P) \tag{4}$$

$$O(^3P) + O_2 \rightarrow O_3 \tag{5}$$

As a result of industrialization, *nitrous oxide* and *fluorinated gases* are present in the atmosphere at increasing levels. If the first one can result from natural water and soil reactions, the second one has only industrial origins.

2 Climate Change Influence on Environmental Modifications

2.1 Human Health Impact

Due to the extreme temperatures registered for long intervals, the number of deaths determined by this situation increases every year. Another aspect to consider is the different disease vectors that may chaotically multiply in cases of trophic chain disruption.

The weather fluctuation inevitably affects all biological systems which need to find the resources to adjust to the ecosystems' perturbations. The effort implied will generate a supplementary need for good quality assets to ensure good health. All climate modifications raise the pressure on every individual. Factors such as pollution or temperature modification increase the tension causing alterations of the individual reaction.

Modifications of the environmental pattern affect all economic sectors, from agriculture to zootechnics, pisciculture, and altogether human health. The local specificity, geo position, informational resources, educational status, or well-being can predict the impact [3].

Climate modifications determined the increase of people's vulnerability to weather changes. The environmental changes could also influence the different health disorders' progress, the organism reaction being unpredictable in such cases. Social programs could limit such effects, especially on the defenseless [4], to increase awareness of the importance of health protection. For example, it is essential to ensure the optimum house temperature, the consumption of an adequate amount of daily liquids, or the use of a suitable wardrobe [5], independent of the temperature level increase or decrease compared to the known dynamics.

Another essential aspect is extreme meteorological phenomena, which lead to floods and/or uncontrolled wastewater discharges. The release of vast amounts of water from rivers or torrential rains usually affects essential agricultural areas and directly impacts a vital raw material source for the food industry.

Another aspect of industrialization refers to increasing wastewater quantity and specific parameter modification. Unfortunately, the collecting and purge systems are usually not well dimensioned, thus contributing to the emergence of ecological disasters. Lately, high amounts of heavy metals and different pathogens are present in the wastewaters. According to recent publications, climate modification can determine resistant microorganism's appearance in the effluents [6]. Considering that water from rivers is used as a source for drinking or irrigation, special attention needs to be paid to its quality to limit the possibility of using it as a disease vector. The recent pandemic underlined the necessity of increasing public awareness regarding environmental protection's importance [7].

Another factor that influences human health is air quality. More significant problems are present in urban agglomerations where pollution usually exceeds the allowed limits. Significant influence also has the air movement, which could contain infectious vectors. Implementing well-directed environmental policies and reducing the minimum admission level of the suspended particles could help human health improvement independent of the social level apartness [8, 9].

An essential result of climate change is the increase of different metabolic illnesses and malnutrition. The general knowledge level and nutritional education also influence both aspects. Inhabitants' awareness about the importance of an equilibrated diet and the possibility to ensure an ecological and sustainable eating attitude minimize the health risks. Different severe weather conditions such as water or dust storms, desertification, or air pollution contribute to food nutritional value decreases and/or its quantity.

An aspect that should concern each decision-maker is related to the country's development level. A direct dependency between life length, youngs death rate, and weather impact is present, analyzing the human health dynamic for half. People from poor regions are more susceptible to health problems [10]. Access to information improves human health, acting in multiple ways: sanitation enhances, nutrition reinforcement, and ecosystem protection.

Clinicians play an essential role in raising awareness of the health risks posed by climate change. Studies revealed that the professional's knowledge regarding the links between health and climate changes influences their disponibility to be involved in such educational projects. Only a third of those interviewed in an analysis consider that incidence of illness caused by food or water inappropriate quality is related to climatic modifications [11]. More than half of the persons still think that is a direct dependency only between the illnesses and poor air quality [12].

Statistical evaluations revealed the possibility of health decline due to climate change in the next period, even in the most developed countries. Any pandemic situation tends to decrease the population's well-being level. One of the synergic factors is the aging population, which cannot adapt to the new climatic and socioeconomic conditions. Many extreme phenomena have affected the economy, properties, and infrastructure, weakening the person's immunity systems. The psychological stress added to the thermal one negatively influences the organism's defense capacity to fight against the new pathogens [13].

2.2 Food Industry Influence

Climate changes have multiple implications in different industrial and life sectors. It is a permanent struggle to ensure safe and nutritional food products, for the food industry is no exception.

Various studies highlighted the possible negative influence of temperature inconsistancy and extreme weather phenomena on food quality. It starts from the raw material's characteristics and continues with all food chains that influence different percentages. In this regard, the nourishment is more susceptible to microbial proliferation and degradation. In this context, we must again underline the impact of wastewater's specific characteristics. Their uncontrolled discharge in different effluents affects all marine species with the potential of metal, toxin, and pathogen contamination.

The tendency to use dehydrated food products or fresh ones is well known, whiteout any thermal preparation or refrigeration. Such products could represent a risk to human health from many points of view. For example, temperature shock and its variations in fresh products will contribute to mycotoxins formation, microbial degradation, or zoonose transmission.

All climate modifications observed, global warming, drought prolonged periods, torrential rains and hail, oceans and sea-level increase, influence the raw sectors for food products: agriculture, animal husbandry, or fishing. Further, food processing and delivery are directly or indirectly affected to environmental changes [14]. Insurance of safe food for populations must be a priority for each state. Implementing sustainable policies in all food chains can lead to achieving this purpose. Encouraging the local production and applying the principles of the short food chain determine the easy overcome of many natural or difficult medical obstacles.

Another sector of human alimentation regards the catering services. Research in the domain underlined the necessity of public awareness of its implications as a source of greenhouse gases. It is expected as an informed consumer or employee to take proper actions and decisions to limit or eliminate the negative impact of its activities on the environment. An important aspect is related to the possibility and opportunity to implement efficient policies for waste management.

It is a known fact that this field can be a significant consumer of fresh raw food products. Specific trends that recently are gaining ground are related to local and traditional product consumption, green production, and minimizing red meat consumption [15]. All this supports the idea of a sustainable industry and limits the factors considered to promote climate change.

Specific technologies have been present in the last years in this field. All contribute to energy use reduction and ensure the dishes' high nutritional value. The climate changes indirectly influence the hospitality sector through the possible interruptions in the supply raw food chain. These are reflected in poor quality, limited production, or even its lack. Such situations demand imports increase and inevitably higher costs. In the case of the foods exclusively delivered by other countries, situations such as the present sanitary crisis inevitably attract supply distributions. Such a situation admits as feasible the trend of using local products. Such an approach stimulates the community to deliver fresh supplies, increasing their income but decreasing the consumer purchase price.

The development of local small green farms may be a feasible solution to limit the negative impact of the crisis on the ecosystem and avoid insufficient food supply [16].

When we think about the food behavior of tourists/customers, we have in mind the attitude they have when they serve food in a public catering unit and what actions they take to limit the amount of waste that will reach the environment. Field research shows a lack of research on the impact of climate change through healthy behavior. Awareness of the client's eating attitude regarding climate change determines their food waste behavior [17].

One new trend in human nutrition is the vegan one. Even if there are many pros and cons of this, from the environmental point of view, this attitude has a positive influence, limiting the formation of greenhouse gases, which, as we said, are among the causes of climate change. Going further, also the nourishment is green obtained without synthetic fertilizers or pesticides. That is a second way in which the negative effect on the natural ecosystems reduces. Another possibility to decrease the human impact on the environment may consist of by-product utilization as raw materials for different goods: dietary supplements, processed foods, or cosmetics. A study from Great Britain underlines the subjects' positive attitude regarding the possibility of using seafood by-products to limit food waste and improve human health and the environment. As in other research studies, also this investigation revealed the people's necessity to be informed in the domain and the implications that the daily activities and attitudes may have on the surroundings and the future [18].

As mentioned above, one of the observed effects of climate change is the increase in extreme weather patterns. Situations such as droughts or prolonged floods are factors that cause food insecurity. Fruitful cooperation among all local and national decision-makers can limit catastrophic situations. They can establish good practice models, provide agricultural workers with urgent information about meteorological phenomena on time, and adopt the best technical solutions to limit losses under

extreme conditions [19]. Such an approach can limit or eliminate momentary decisions in crises, which can decrease productive efficiency and thus increase food
insecurity.

There is a growing trend on the North American continent to use vegetable proteins
instead of animal ones, even if we think of classic dishes such as burgers. There is an
example of dietary changes under the influence of climate change on human nutrition.

Regardless of the cause, any new situation will influence people's eating behavior
and the environment. The pandemic underlined the necessity and importance of a
short, secure food supply chain [20].

As mentioned above, the climatic modifications significantly impact the fisheries
sector is. Fishes and seafood are known as vectors to biotoxins formations and heavy
metals chelation. It is a significant provider for both the hospitality and processing
industries. So it is necessary to underline the importance of water quality in the
fishing zone and the treatments applied after catching.

The tendency to reduce red meat consumptions and the world population increase
put pressure on the aquaculture sector, the importance of a quality protein in the diet
being undisputed. The bays and the estuaries ensure proper conditions to promote
such activity [21]. Implementation of a similar project requires determining all the
factors influencing its performance. Knowing all synergetic aspects improves the
ecosystem capacity to ensure the proper production conditions. The climatic changes
may negatively affect such an approach through the water level and/or temperature
fluctuations or extreme phenomena caused.

Another vital aspect determined by the climate changes that could go unnoticed
is the possibility of salty seawater infiltrating the groundwater [22]. The first sector
affected will be agriculture, and secondly the food, and the economy in such a
situation. Also must be mentioned the implication at the level of national security in
respect of ensuring the drinking water supply for the population. As a consequence,
rationalization could be necessary. In consideration of agriculture, a possible solution
is the use of species adapted to slated soils. A complex problem raises in respect of
the livestock sector.

The water, soil, and food system will always be interconnected and influenced
by the clime conditions. An integrating map helps to determine all the synergetic,
additive, and antagonistic relations between them. The system's complexity and
unpredictability determine such an approach. The studies performed in this direction
aimed to present a sustainable alternative for all resource management [23].

by rain instability are other connections between environmental changes and
human nutrition is the long dry periods followed by rain instability influence the
environment dynamic and thus the human nutrition. Such situations could endanger
the agriculture sector. Low production and increased demand could lead to food
insecurity [24]. A well-planned national and international strategy is necessary to
decrease the possibility of such occurrence. The concept of sustainable resource
management has to be its central point.

2.3   Production Sustenance

As mentioned above, a growing number of individuals are heading toward ecological eating concepts. Such a concept also means fewer intensive farms and livestock and production conditions ensured as natural as possible. Regarding husbandry units, this refers to the possibility of using the pasture in ecosystems less or minimally affected by human intervention. An aspect that can raise problems is the natural degradations determined by the climate perturbations and/or intensive exploitation. Significant effects of the meadow deterioration consist of productivity decrease, erosion increase, and carbon sequestration reduction [25]. All this is indirectly influencing human well-being status. The critical aspects concern the uncontrolled use of nitrogen fertilizers, pesticides, or herbicides for the agriculture units.

A vital aspect determined relates to the influence of the climatic changes regarding soil pH modification and its nitrous balance modification. Implementing a durable policy in respect of meadow exploitation can improve their productivity, the carbon, and nitrogen balance and limit the production of greenhouse gases [25].

Different states have already adopted intelligent approaches to limit the negative effect on output and indirectly in the economy, based on the interference of the environmental modifications on production. The procedures implemented have as second impact the reduction of greenhouse gases emission [26].

Adopting the proper actions to limit the effects of climate change will also reduce the possibilities of increasing food insecurity. Long-term policy implementation in respect of waste and resource management ensures the achievement of these objectives. However, studies in the domain showed the connections between the two issues.

3   Conclusions

The effects of climate changes are visible in many domains. Regardless of whether these are direct or indirect, we can limit or preferable stop their negative impact by adopting the proper and immediate measures. The authorities have an essential role in establishing the legislative framework.

It is necessary to know as much as possible about all the dependencies between the different ecosystems. They all work as a unitary whole, based on synergetic, additive, and antagonistic relations. Specialists from different domains have to work closely to retain the negative industrial revolution mark on the environment. An integrated approach ensures the success of such an objective.

Research and innovation in every sector have to begin from an environmentally friendly premise. The limitation of greenhouse gasses production and emission is necessary. It is essential to take into consideration also the equilibrium principle for all mechanisms projected in this sense. A gradual inclusion of all changes will ensure a better adjustment and/or readjustment of all constituents.

Depending on the specificity, the return to the initial conditions may not be entirely possible, if not even impossible. Some examples are natural landslides, coastal erosion, soil increase salinity, desertification, or deforestation. Such situations may represent premises for the projection of sustainable anthropic intervention in natural sites. During times, uncontrolled mining excavations determined soil collapse, thus causing changes in terrestrial ecosystems.

Also, if we think to adopt principles of healthier eating habits, an initial step consists of using organic products, meaning cleaner agriculture procedures. Such an attitude, without a doubt, will induce the consumer's responsibility toward a green environment.

References

1. Kaddo JR (2016) Climate change: causes, effects, and solutions. A with honors projects 164. http://spark.parkland.edu/ah/164
2. Reeves CE, Penkett SA, Bauguitte S, Law KS, Evans MJ, Bandy BJ, Monks PS, Edwards GD, Phillips G, Barijat H, Kent J, Dewey K, Schmitegen S, Kley D (2002) Potential for photochemical ozone formation in the troposphere over the North Atlantic as derived from aircraft observations during ACSOE. J Geophys Res Atmos 107(D23):4707, ACH 14 https://doi.org/10.1029/2002jd002415. ISSN: 0148-0227
3. McMichael JA, Haines A (1997) Global climate change: the potential effects on health. BMJ 315:805–809
4. Meeting of the Advisory Committee (ACM) to review technical matters to be discussed at the Sixty-first Session of the Regional Committee WHO/SEARO, New Delhi, 30 June-3 July 2008
5. Haines A, Patz JA (2004) Health effects on climate change. JAMA 291(1):99–103
6. Sterk A, de Man H, Schijven JF, de Nijs T, de Roda Husman AM (2016) Climate change impact on infection risks during bathing downstream of sewage emissions from CSOs or WWTPs. Water Res 105:11–21. https://doi.org/10.1016/j.watres.2016.08.053
7. Christos Tsagkaris C, Moysidis DV, Papazoglou SA, Louka AM, Kalaitzidis K, Ahmad S, Essar MY (2021) Detection of SARS-CoV-2 in wastewater raises public awareness of the effects of climate change on human health: the experience from Thessaloniki, Greece. J Clim Change Health 2:100018
8. Garg A (2011) Pro-equity effects of ancillary benefits of climate change policies: a case study of human health impacts of outdoor air pollution in New Delhi. World Dev 39(6):1002–1025. https://doi.org/10.1016/j.worlddev.2010.01.003
9. Wright CY, Kapwata T, du Preez DJ, Wernecke B, Garland RM, Nkosi V, Landman WA, Dyson L, Norval M (2021) Major climate change-induced risks to human health in South Africa. Environ Res 196:110973
10. Meierrieks D (2021) Weather shocks, climate change and human health. World Develop 138:105228
11. Lee H, Pagano I, Borth A, Campbell E, Hubbert B, John Kotcher J, Maibach E (2021) Health professional's willingness to advocate for strengthening global commitments to the Paris climate agreement: findings from a multination survey. J Clim Change Health 2:100016
12. Kotcher J, Maibach EW, Miller J, Campbell E, Alqodmani L, Maiero M, Wyns A (2021) Views of health professionals on climate change and health: a multinational survey study. Lancet Planet Health. https://doi.org/10.1016/S2542-5196(21)00053-X
13. Harper SL, Cunsolo A, Babujee A, Coggins S, De Jongh E, Rusnak T, Wright CJ, Domínguez Aguilar M (2021) Trends and gaps in climate change and health research in North America. Environ Res 199:111205

14. Misiou O, Konstantinos Koutsoumanis K (2021) Climate change and its implications for food safety and spoilage. Trends Food Sci Technol. https://doi.org/10.1016/j.tifs.2021.03.031
15. Dagmar Lund-Durlacher D, Gossling S (2020) An analysis of Austria's food service sector in the context of climate change J Outdoor Recreation Tourism 100342. https://doi.org/10.1016/j.jort.2020.100342
16. Tirado MC, Cohen MJ, Aberman N, Meerman J, Thompson B (2010) Addressing the challenges of climate change and biofuel production for food and nutrition security. Food Res Int 43(7):1729–1744
17. Kim MJ, Hall CM (2019) Can climate change awareness predict pro-environmental practices in restaurants? Comparing High Low Dining Expenditure Sustain 11:6777. https://doi.org/10.3390/su11236777
18. Altintzoglou T, Pirjo Honkanen P, Whitaker RD (2021) Influence of the involvement in food waste reduction on attitudes towards sustainable products containing seafood by-products. J Cleaner Prod 285:125487
19. Müller A, Bouroncle C, Gaytán A, Girón E, Granados A, Mora V, Portillo F, Etten F (2020) Good data are not enough: understanding limited information use for climate risk and food security management in Guatemala. Clim Risk Manag 30:100248
20. Barman A, Das R, Kanti De P (2021) Impact of COVID-19 in food supply chain: disruptions and recovery strategy. Curr Res Behav Sci 2:100017
21. Chapman EJ, Byron CJ, Lasley-Rasher R, Lipsky C, Stevens JR, Peters R (2020) Effects of climate change on coastal ecosystem food webs: implications for aquaculture. Mar Environ Res 162:105103
22. Omar MM, Moussa AMA, Hinkelmann R (2021) Impacts of climate change on water quantity, water salinity, food security, and socioeconomy in Egypt. Water Sci Eng 14(1):17–27
23. Yue Q, Zhang F, Wang Y, Zhang X, Guo P (2021) Fuzzy multi-objective modelling for managing water-food-energy-climate change-land nexus towards sustainability. J Hydrol 596:125704
24. Chen L, Chang J, Wang Y, Guo A, Liu Y, Wang Q, Zhu Y, Zhang Y, Xie Z (2021) Disclosing the future food security risk of China based on crop production and water scarcity under diverse socioeconomic and climate scenarios. Sci Total Environ 148110. https://doi.org/10.1016/j.scitotenv.2021.148110
25. Dong S, Shang Z, Gao J, Boone RB (2020) Enhancing sustainability of grassland ecosystems through ecological restoration and grazing management in an era of climate change on Qinghai-Tibetan Plateau. Agric Ecosyst Environ 287:106684
26. Teklewold H, Gebrchiwot T, Bezabih M (2019) Climate smart agricultural practices and gender differentiated nutrition outcome: an empirical evidence from Ethiopia. World Dev 122:38–53

Chapter 21
The Development of a Smart Tunable Full-Spectrum LED Lighting Technology Which May Prevent and Treat COVID-19 Infections, for Society's Resilience and Quality of Life

U. Thurairajah ⓘ**, John R. Littlewood** ⓘ**, and G. Karani** ⓘ

Abstract This paper discusses developing a smart and sustainable full-spectrum LED lighting technology that could prevent and may treat COVID-19 infections and enhance the global society's resilience and quality of life (QoL). This research aims to develop a full-spectrum lighting system using light-emitting diode (LED) technology to provide the daylight effect for people in buildings, to improve Vitamin D during the day which should prevent and may treat COVID-19 infections, for society's resilience and quality of life. The first author is undertaking the proposed work as part of his Ph.D. at Cardiff Metropolitan University, UK, based in Canada. There is currently (2021) no such lighting method available to treat or prevent COVID-19 infections. This novel method is not replicated but is original and could be considered an innovative method which may prevent and may treat COVID-19 infections. This paper will be helpful for academics, researchers, scientists, medical doctors, engineers, consultants, architects, lighting designers, contractors, developers, financial institutions, and government agencies funding to upgrades the lighting system.

Keywords Built environment · Canada · Resilience · LED full-spectrum light source · Technology · Environment · COVID-19 prevention and treatment · Vitamin D · Health · Well-being · Quality of life

U. Thurairajah (✉) · J. R. Littlewood
Cardiff School of Art and Design, Sustainable and Resilient Built Environment Group, Cardiff Metropolitan University, Cardiff C5 2YB, UK

G. Karani
Cardiff School of Health Sciences, Environmental Public Health Group, Cardiff Metropolitan University, Cardiff C5 2YB, UK

1   Introduction

A new COVID-19 infection pandemic began in Wuhan, China, in late 2019, formerly called 2019-nCoV [1] and renamed COVID-19 by the World Health Organization (WHO) in February 2020. The primary reason for death is severe atypical pneumonia [2]. Vitamin D is involved in calcium absorption, immune function, and protecting bone, muscle, and heart health [3, 4]; it occurs naturally in some food and can also be produced by the human body when the skin is exposed to sunlight. So, it could be considered that the perfect daily balance for human QoL and resilience should include food rich in Vitamin D as the appetizer and sunlight as the main course.

Exposure to full-spectrum light (sunlight), heat, and humidity seems to weaken the coronavirus [5, 6]. The powerful effect solar light appears to have on killing the virus is both on surfaces and in the air. Figure 1 provides the COVID-19 cases in Ontario, Canada, from April 1, 2020, to January 13, 2021 [7]. The data clearly shows that the coronavirus spreads lower during the summer months from June 20, 2020, to September 22, 2020.

Vitamin D can be suitable because it has shown antiviral properties [8] and is freely available through sunlight exposure or the use of LED full-spectrum light sources in the indoor built environment. One disadvantage of sunlight is that it is not intense enough to provide adequate Vitamin D levels throughout the year in all countries due to different annual climatic cycles. For example, in the United Kingdom (UK), Vitamin D that people need for a healthy lifestyle is not adequate between October and the middle of March/April, so from 5.5 to six months of the year, so other Vitamin D sources are needed [9]. Thus, LED full-spectrum light sources in the indoor built environment could be one solution to prevent or protect people from the COVID-19 virus. The development of a smart tunable full-spectrum LED lighting

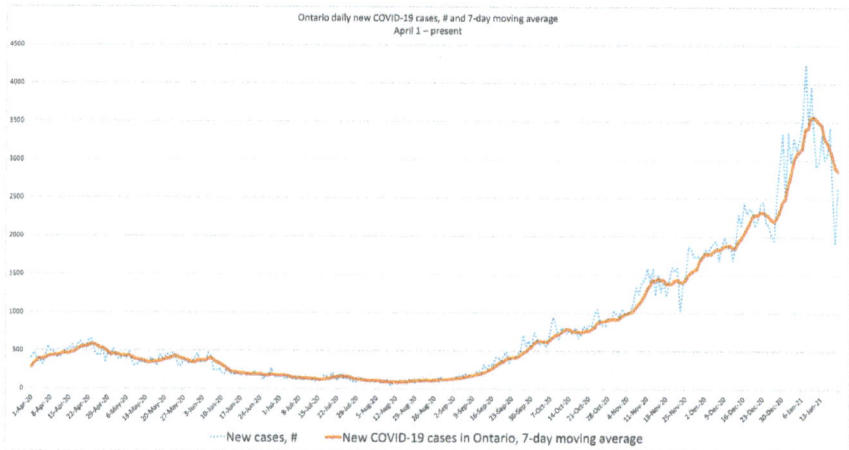

Fig. 1 Data from the province of Ontario—visualizations by John McGrath [7]

technology used to increase the Vitamin D level in the human body may prevent and treat COVID-19 infections.

2   Background and Literature Review

Cannell et al. [10] hypothesized that the winter peak was due in part to the conjunction with the season when solar ultraviolet B (UVB) doses and the 25-hydroxy vitamin D [25(OH)D] concentrations are lowest in most mid- and high latitude countries like Canada, Greenland (Denmark), Norway, and Alaska (United States of America (USA). As a result, mean serum 25(OH)D concentrations in the north and central regions of the United States of America (USA) are near 21 nanograms per milliliter (ng/ml) in winter and 28 ng/ml in summer. In contrast, they are nearer 24 ng/ml in the south region in winter and 28 ng/ml in summer [11]. The winter peak of influenza also coincides with low temperature and relative humidity conditions, which allows the influenza virus to survive longer outside the human body than under warmer conditions [12]. Seasonal influenza infections generally peak in winter [13].

The active form of vitamin D (1,25-dihydroxy vitamin D or 1,25 (OH)2D) could play a central role in protecting against respiratory virus infections by modulating the antiviral immune response via Vitamin D receptors. In winter, Vitamin D deficiency has been found to contribute to acute respiratory distress syndrome, COVID-19, and fatality rates increase with age, both of which are associated with lower 25(OH)D concentration [14]. A comprehensive review of the role of Vitamin D and influenza was published in 2018 [15]. On the positive side, Vitamin D-related innate and adaptive immune responses to viral infections exist. A Vitamin D randomized controlled trial (RCT) conducted in 2010 on school children in Japan reported significantly reduced incidence of influenza type A, but not influenza type B for children in the treatment who were taking 1200 international units per day (IU/d) of Vitamin D [16]. Influenza type A and type B are similar, but type A is more prevalent, sometimes more severe, and can cause epidemics and pandemics [17]. A 13-year study conducted between 2006 and 2018 in Milan, Italy, reported that summertime means of 25(OH)D concentrations reached about 33–35 ng/ml for both males and females, ~20 ng/ml for males, and 23 ng/ml for females in winter [18]. An analysis of standardized 25(OH)D concentration data from 14 European population studies indicated that 13.0% of the 55,844 European individuals had serum 25(OH)D concentrations <12 ng/ml on average in the year, with 17.7% and 8.3% in those sampled during the extended winter (October–March) and summer (April–November) periods, respectively [19]. A study in Italy published in 2014 that male chronic obstructive pulmonary disease cases had mean 25(OH)D densities of 16 ng/ml, and female patients had 25(OH)D concentrations of 13 ng/ml [20]. A significant body of evidence shows that higher 25(OH)D concentrations and vitamin D supplements reduced the risk of many chronic diseases, including cancers, cardiovascular disease (CVD), and diabetes mellitus [21].

When people are exposed to full-spectrum LED light during the day, in most cases when there are no underlying health conditions preventing sleep, they have an adequate sleep at night, same as sunlight exposure during the day [22]. Sufficient sleep increases melatonin production, enhancing the immune system to fight COVID-19 in anti-inflammation, anti-oxidation, and immune response regulation [23]. The authors [23] consider that melatonin has a high safety profile and is one way to boost the immune system and protect the body during this COVID-19 pandemic. Therefore, a night of sound sleep by COVID-19 patients would be highly beneficial in their fight and resilience against the virus and detrimental impact upon their QoL.

The proposed LED full-spectrum light source can be used every day. A blood test will reveal the individual's Vitamin D needs. The duration of exposure can be determined based on individual needs as per physician recommendation. For example, 60 min of full-spectrum light exposure during the day will not be harmful but unnecessary. Standard LED light has a higher blue spectrum and potentially causes sleep disruption if anyone is exposed at night. But the full-spectrum LED luminaires that emit white light are often referred to as "broad spectrum" lights similar to the sunlight. However, it is essential to have low level and warm color temperature light at night to create a cozy and relaxing environment, produce more melatonin, improve sleep quality, and prevent or treat COVID-19 infections. Figure 2 provides data from Johns Hopkins University—COVID-19 Map and Case Count [24].

The powerful effect that of solar light that appears to have on killing the COVID-19 virus is both on surfaces and in the air [5, 6]. Figure 2 indicates that the case count provides the COVID-19 cases from February 2020 to March 2021. The data clearly shows that the coronavirus spreads lower during the summer months (between June and September) in all four countries (i.e., UK, Canada, Italy, Germany).

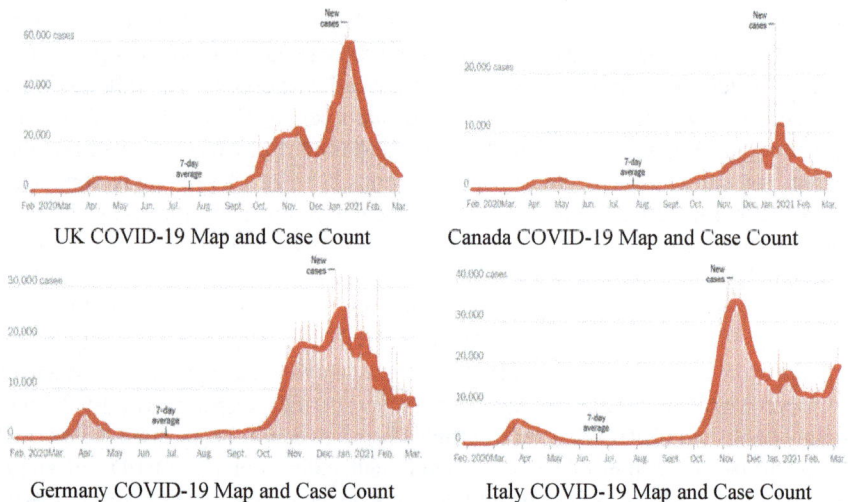

UK COVID-19 Map and Case Count   Canada COVID-19 Map and Case Count

Germany COVID-19 Map and Case Count   Italy COVID-19 Map and Case Count

Fig. 2 Data from Johns Hopkins University—COVID-19 map and case count [24]

3   Methodology

Over millions of years, humans and their ancestors evolved under natural daylight [25]. Anyone can witness the warm (2700 K) sunrise and 6500 K (cool-white) as the day progresses toward noon, where correlated color temperature (CCT) peaks [26]. As time goes toward the afternoon and evening, the color temperature decreases to around 2700 K again. Sunlight has well-balanced colors, a full spectrum, and has all the frequencies [26]. If anyone exposes to the sunlight correctly, it can give enough Vitamin D for our needs [9].

Various color temperatures provide different spectral power distributions (SPDs). Therefore, anyone can tune the LED light source from less than 3000 K to more than 6500 K to imitate the sunlight. This way, an individual can receive the required lighting for an individual needs. The single full-spectrum light source can be used for an individual need for a specific duration of exposure, or a smart tunable full-spectrum light source can provide adequate lighting for individual needs.

Figure 3 provides tunable LED gradual changes of color temperature from 2700–6500 K [26], and Fig. 4 includes two optional lighting systems (Option 1 and Option 2). If the LED is a tunable full-spectrum source, it can give different SPD based on the requirement. The tunable LED luminaire CCT increases gradually from left to right, and the blue increases. Perfect light at about 6500 K is almost an exact spectral distribution match of sunlight at noon. It is a pretty good representation of light from all the visible spectrums.

From Fig. 4, Option 1 is a 6500 K full-spectrum lighting used only for a fixed-time exposure to get enough Vitamin D. It can be used for individual needs. Option 2 is a tunable LED from 2700 to 6500 K full-spectrum lighting. It can be used for office and residential applications. This light treatment can be called the complementary and alternative medical electric light optional therapy (CAMELOT) method to prevent and treat COVID-19.

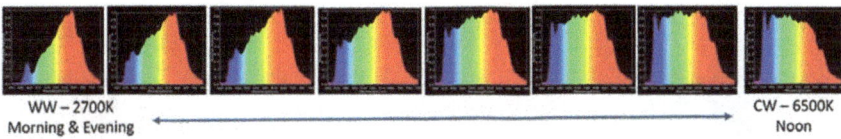

WW – 2700K   CW – 6500K
Morning & Evening   Noon

Fig. 3   Tunable LED gradual changes of color temperature from 2700 to 6500 K [26]

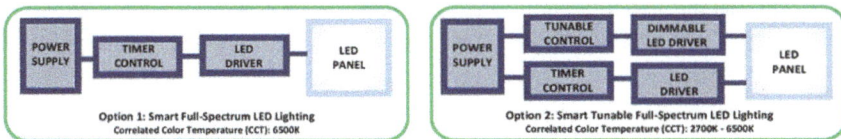

Fig. 4   Option 1 and option 2 lighting system

In this paper, the effect of the latest available LED luminaire has been used to suppress melatonin and increase Vitamin D and cortisol production during the day. The study was based on various SPDs of a test tunable LED luminaire from 2700 to 6500 K. The results may show a strong correlation between CCT and Vitamin D production. However, Vitamin D production may vary for different LEDs with the same CCT since the SPD varies based on the type of LED used. The SPDs for the test LED luminaire for various CCTs were presented in Figs. 5, 6, and 7, and their corresponding colorimetric parameters are described in Table 1.

LED technology and phosphor compositions bring a wide range of opportunities to create any required SPD of white light. For that reason, there is a theoretically unlimited number of ways to generate white light. The SPDs of LEDs with additional blue light radiation within the maximum melanopsin sensitivity range energize and produce cortisol and Vitamin D. Any other researchers for validation can repeat this method and confirm the usefulness of this CAMELOT treatment method.

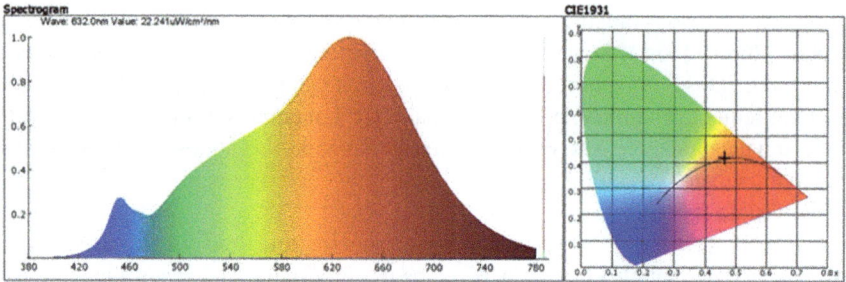

Fig. 5 SPD and chromaticity diagram for test LED luminaire for 2712 K

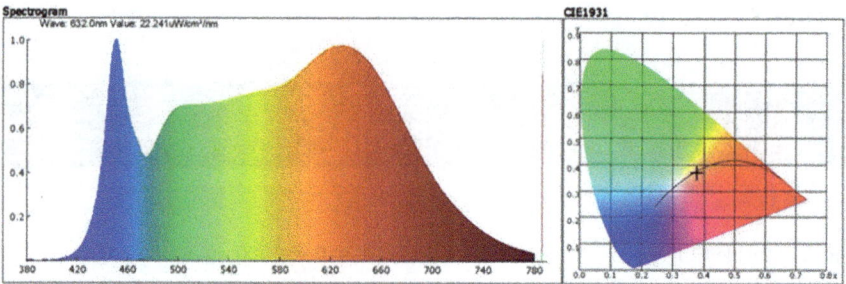

Fig. 6 SPD and chromaticity diagram for test LED luminaire for 4040 K

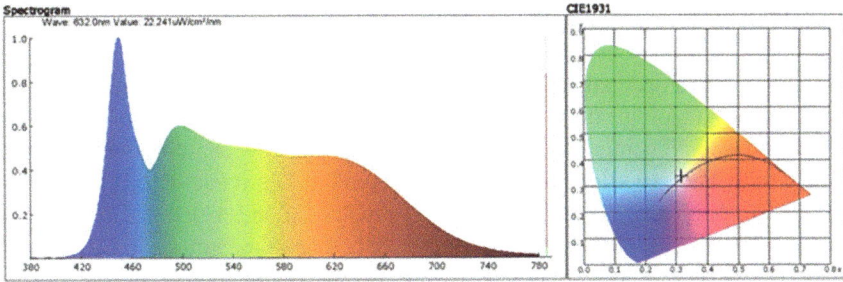

Fig. 7 SPD and chromaticity diagram for test LED luminaire for 6228 K

Table 1 Colorimetric parameters of the test LED luminaire

SPD	Test LED luminaire		
Source CCT (K)	2700	4000	6500
Measured CCT (K)	2712	4040	6228
CRI	97.9	96	97.5
X	0.4624	0.3767	0.3171
Y	0.4167	0.3680	0.3366
S/P	1.321	1.929	2.446
Peak (nm)	632.8	450.7	449.5
Red (%)	26.9	21.1	15.8
Green (%)	71.6	75.85	79.7
Blue (%)	1.5	3.1	4.5
Vitamin D[a]	Very low	Low	High

[a]There is no clinical trial completed to verify the amount of Vitamin D production at this time

4 Results

The scope of this paper is to develop and test an LED light model with various SPDs to imitate the sunlight CCTs. Therefore, the author carefully designed and tested the LED light model. The tests were conducted in May 2021 using the OHSP350 spectrometer, and the results of the tunable LED luminaire are shown in Table 1.

There is ambient light from the laboratory environmental and data taking distance influence while taking the above-noted data. Therefore, the corresponding source color temperatures 2700 K, 4000 K, and 6500 K were displayed as 2712 K, 4040 K, and 6228 K in Table 1. There is a clear relationship between the CCT of the test LED luminaire and its Vitamin D and cortisol production and alertness. Therefore, the colder light, like 6500 K CCT, might bring significantly higher Vitamin D production. On the other hand, researchers indicated that the dim lighting with the low CCT are most helpful light for relaxation and might not suppress much melatonin and gradually make a person fall asleep [27].

5 Discussion

The method discussed in this paper makes artificial light the most like the sunlight to keep the natural biological rhythm for humans. In contrast, such dynamic light might support boosting Vitamin D production, alertness, and concentration when needed or help relax and fall asleep to support fast recovery. Ultraviolet B (UVB) peek at 300 nm is more responsible for Vitamin D production and melatonin suppression peeks at 480 nm, accountable for alertness [28]. Therefore, stimulation by light is crucial in this area. For that reason, special LEDs with increased radiation of around 300 nm to 480 nm are needed for this application.

However, only 5% of the UVB wave reaches the earth from sunlight [29]. Therefore, too much UVB is not necessary for Vitamin D production. The clinical trials for Vitamin D production and validation will be covered under a separate paper. Additional light radiated around the maximum melanopsin sensitivity will influence melatonin suppression during the day. Consequently, such light will increase cortisol production, improve alertness, and energize during the day. But it is essential to mention that irresponsible use of this light during the night might disturb sleep and negatively affect the human body.

The melanopsin is a short-wavelength-sensitive pigment with a peak spectral sensitivity near 480 nm, rendering some intrinsically photosensitive retinal ganglion cells (ipRGCs) [30]. These ipRGCs mediate most effects of light on the circadian clock. Therefore, overexposure to the high color temperature light is not recommended during the night. The best way to control blue light from artificial light sources is to tune the LED luminaires at the application level below 3000 K. Therefore, special attention should be paid to the exposure time and duration while turning on and turning the light source.

6 Conclusion

There are two different approaches to light treatment: An intelligent full-spectrum LED lighting for individuals is Option 1, and an intelligent tunable full-spectrum LED lighting for office, or residential lighting applications is Option 2. Option 1 is used for schedules or fixed-time light exposure for individual needs. Option 2 provides the opportunity to raise or lower the color temperature and intensity based on the requirement. It has the flexibility to operate as a standard lighting system. It is not harmful if it runs over the time limit since it is not a UVB light but a full-spectrum lighting.

The ability to tune the lights to different light levels and color temperature significantly benefits residential and office applications. Anyone can use the required lighting levels and color temperature and use this lighting as therapy where required. The proper controls system will enhance lighting service while reducing operational costs and energy. This method contains a detailed analysis of all aspects of the design process. Therefore, this novel method can be regarded as an innovative method that may prevent and treat COVID-19. Therefore, this light treatment can be called the complementary and alternative medical electric light optional therapy (CAMELOT) method to prevent and treat COVID-19.

There are adequate research and clinical data to endorse Vitamin D used to prevent or treat COVID-19. However, ongoing, more observational studies evaluate the role of vitamin D in preventing and treating COVID-19. In order to validate the CAMELOT method, further clinical trials should be conducted to verify the Vitamin D production against the corresponding LED luminaires CCTs and SPDs.

Acknowledgements The first author wishes to acknowledge his employer's WSP Canada for the support of conducting Ph.D. research related to his professional practice as Professional Lighting and Electrical Engineer and Researcher.

References

1. Zhu N et al (2020) A novel coronavirus from patients with pneumonia in China, 2019. N Engl J Med. https://doi.org/10.1056/NEJMoa2001017. Cited at: https://www.nejm.org/doi/full/10.1056/nejmoa2001017. Accessed 25 Feb 2021 (available)
2. Yin Y et al (2018) MERS, SARS and other coronaviruses as causes of pneumonia. Respirology 23:130–137. Cited at: https://onlinelibrary.wiley.com/doi/full/10.1111/resp.13196. Accessed 25 Feb 2021 (available)
3. Khazai N et al (2008) Calcium and vitamin D: skeletal and extra skeletal health. Cited at: https://www.ncbi.nlm.nih.gov/pmc/articles/PMC2669834/. Accessed 25 Feb 2021 (available)
4. Norman PE et al (2014) Vitamin D and cardiovascular disease. Cited at: https://www.ahajournals.org/doi/full/10.1161/CIRCRESAHA.113.301241. Accessed 25 Feb 2021 (available)
5. Bryan W (2020) Sunlight, heat and humidity weaken coronavirus, US official says. Cited at: https://www.todayonline.com/world/sunlight-heat-and-humidity-weaken-coronavirus-us-official-says-1. Accessed 25 Feb 2021 (available)
6. Enwemeka CS et al (2020) Light as a potential treatment for pandemic coronavirus infections: a perspective. Cited at: https://www.ncbi.nlm.nih.gov/pmc/articles/PMC7194064/. Accessed 25 Feb 2021 (available)
7. TVO.Org. (2021) COVID-19: what you need to know for January 20, The latest coronavirus updates from across the province. Cited at: https://www.tvo.org/article/covid-19-what-you-need-to-know-for-january-20. Accessed 25 Feb 2021 (available)
8. Beard JA et al (2011) Vitamin D and the anti-viral state. Cited at: https://www.ncbi.nlm.nih.gov/pmc/articles/PMC3308600/. Accessed 25 Feb 2021 (available)
9. National health Service. How to get vitamin D from sunlight—healthy body. (2021). Cited at: https://www.nhs.uk/live-well/healthy-body/how-to-get-vitamin-d-from-sunlight/. Accessed 25 Feb 2021 (available)
10. Cannell JJ et al (2006) Epidemic influenza and vitamin D. Epidemiol Infect 134:1129–1140. Cited at: https://pubmed.ncbi.nlm.nih.gov/16959053/. Accessed 25 Feb 2021 (available)
11. Kroll MH et al (2015) Temporal relationship between vitamin D status and parathyroid hormone in the United States. PLoS One 10:e0118108. Cited at: https://pubmed.ncbi.nlm.nih.gov/25738588/. Accessed 25 Feb 2021 (available)
12. Shaman J et al (2010) Absolute humidity and the seasonal onset of influenza in the continental United States. PLoS Biol 8:e1000316. Cited at: https://pubmed.ncbi.nlm.nih.gov/20186267/. Accessed 25 Feb 2021 (available)

13. Hope-Simpson RE (2021) The role of season in the epidemiology of influenza. J Hyg (Lond)1981 86:35–47. Cited at: https://pubmed.ncbi.nlm.nih.gov/7462597/. Accessed 25 Feb 2021 (available)
14. Grant W et al (2020) Evidence dat vitamin D supplementation could reduce risk of influenza and COVID-19 infections and deaths. Cited at: https://pubmed.ncbi.nlm.nih.gov/32252338/. Accessed 25 Feb 2021 (available)
15. Gruber-Bzura BM (2018) Vitamin D and influenza-prevention or therapy? Int J Mol Sci 19. https://doi.org/10.3390/ijms19082419. Cited at: https://pubmed.ncbi.nlm.nih.gov/30115864/. Accessed 25 Feb 2021 (available)
16. Urashima M et al (2010) Randomized trial of vitamin D supplementation to prevent seasonal influenza A in schoolchildren. Am J Clin Nutr 91:1255–1260. Cited at: https://pubmed.ncbi.nlm.nih.gov/20219962/. Accessed 25 Feb 2021 (available)
17. Johnson S et al (2020) Influenza A vs. B: what to know. Cited at: https://www.medicalnewstoday.com/articles/327397#types. Accessed 25 Feb 2021 (available)
18. Ferrari D et al (2019) Association between solar ultraviolet doses and vitamin D clinical routine data in European mid-latitude population between 2006 and 2018. Photochem Photobiol Sci 18:2696–2706. Cited at: https://pubs.rsc.org/en/content/articlelanding/2019/pp/c9pp00372j#!divAbstract. Accessed 25 Feb 2021 (available)
19. Cashman KD et al (2016) Vitamin D deficiency in Europe: pandemic? Am J Clin Nutr 103:1033–1044. Cited at: https://academic.oup.com/ajcn/article/103/4/1033/4662891. Accessed 25 Feb 2021 (available)
20. Malinovschi A et al (2014) Severe vitamin D deficiency is associated with frequent exacerbations and hospitalization in COPD patients. Respir Res 15:131. https://doi.org/10.1186/s12931-014-0131-0. Cited at: https://pubmed.ncbi.nlm.nih.gov/25496239/. Accessed 25 Feb 2021 (available)
21. Charoenngam N et al (2019) Vitamin D for skeletal and non-skeletal health: What we should know. J Clin Orthop Trauma 10:1082–1093. Cited at: https://pubmed.ncbi.nlm.nih.gov/31708633/. Accessed 25 Feb 2021 (available)
22. Choi JH et al (2020) Relationship between sleep duration, sun exposure, and serum 25-hydroxyvitamin D status: a cross-sectional study. Cited at: https://www.ncbi.nlm.nih.gov/pmc/articles/PMC7060268/. Accessed 25 Feb 2021 (available)
23. Juybari K et al (2020) Melatonin potentials against viral infections including COVID-19: current evidence and new findings. Cited at: https://www.ncbi.nlm.nih.gov/pmc/articles/PMC7405774/. Accessed 25 Feb 2021 (available)
24. The New York Times (2021) Coronavirus world map: tracking the global outbreak. Cited at: https://www.nytimes.com/interactive/2020/world/coronavirus-maps.html. Accessed 25 Feb 2021 (available)
25. Münch M et al (2020) The role of daylight for humans: gaps in current knowledge. Cited at: file:///C:/Users/uthay/Downloads/clockssleep-02-00008-v2.pdf. Accessed 25 Feb 2021 (available)
26. LED Dynamics (2021) LED Dynamics 'perfeklight' technology tunes & corrects white light. Cited at: https://leddynamics.com/leddynamics-perfektlight-tunes-corrects-white-light. Accessed 25 Feb 2021 (available)
27. Lin J et al (2019) Several biological benefits of the low color temperature light-emitting diodes based normal indoor lighting source. Cited at: https://www.nature.com/articles/s41598-019-43864-6. Accessed 25 Feb 2021 (available)

28. Kalajian TA et al (2017) Ultraviolet B light emitting diodes (LEDs) are more efficient and effective in producing vitamin D3 in human skin compared to natural sunlight. Cited at: https://pubmed.ncbi.nlm.nih.gov/28904394/. Accessed 25 Feb 2021 (available)
29. Alexander, H (2019) What's the difference between UVA and UVB rays? Cited at: https://www.mdanderson.org/publications/focused-on-health/what-s-the-difference-between-uva-and-uvb-rays-.h15-1592991.html. Accessed 25 Feb 2021 (available)
30. Blume C et al (2019) Effects of light on human circadian rhythms, sleep and mood. Cited at: https://link.springer.com/article/10.1007/s11818-019-00215-x. Accessed 25 Feb 2021 (available)

Chapter 22
Energy-Efficient Technologies for Ultra-Low Temperature Refrigeration

Cosmin Mihai Udroiu, Adrián Mota-Babiloni, Carla Espinós-Estévez, and Joaquín Navarro-Esbrí

Abstract New vaccines have been developed in response to the current COVID-19 pandemic, and some of these require ultra-low temperature refrigeration (at $-80\,°C$). After their appearance, the number of ultra-low temperature freezers of different capacities has been increased worldwide. Sustainable transition is ongoing in many refrigeration and heat pump applications following what is established in national and international regulations. However, many have not controlled ultra-low temperature refrigeration because of the challenges associated with these systems' operation. The energy performance is low for this range of temperatures because of the distance between the heat sink and source temperatures. Moreover, a limited number of refrigerants are available because of restrictions in their normal boiling point and other challenges related to the lubricating oil. This chapter presents the main characteristics of several technologies (vapor compression cycle with element variations, sublimation, or absorption cycle) that can be applied for this range of temperatures, focusing on the constructing elements, advantages, and drawbacks. Then, recently developed configurations that can appear in commercial systems in the coming years are explored. These configurations are based on vapor compression cascade cycles, including an intermediate heat exchanger, ejector, and three-stage. It is seen that despite the increase in complexity and investment of the advanced configurations, the decrease in coefficient of performance is still notable, causing an increase in the operating cost. Apart from the additional elements or stages, working fluids used in these configurations are critical parameters for increasing the resulting energy performance and cooling capacity, ending with more sustainable ultra-low temperature freezers.

C. M. Udroiu · A. Mota-Babiloni (✉) · J. Navarro-Esbrí
ISTENER Research Group, Department of Mechanical Engineering and Construction, Universitat
Jaume I (UJI), Av. de Vicent Sos Baynat s/n, 12071 Castelló de la Plana, Spain
e-mail: mota@uji.es

C. Espinós-Estévez
Centro Nacional de Investigaciones Cardiovasculares Carlos III (CNIC), Melchor Fernández
Almagro 3, 28029 Madrid, Spain

© The Author(s), under exclusive license to Springer Nature Singapore Pte Ltd. 2022   309
R. J. Howlett et al. (eds.), *Smart and Sustainable Technology for Resilient Cities
and Communities*, Advances in Sustainability Science and Technology,
https://doi.org/10.1007/978-981-16-9101-0_22

1   Impact, Requirements, and Challenges of Low and Ultra-Low Temperature Refrigeration on Biomedical Research

Biomedical research and medical clinics need refrigerated storage to conserve critical samples, medications, vaccines, organs, etc., at very precisely controlled temperatures. Regular refrigerators and freezers are not appropriate as they fail to provide uniform temperature stability. Therefore, they compromise the integrity of the stored samples and hence reproducible results or even life-threatening scenarios. Also, biomedical research can often require explosion-proof refrigerators and freezers for flammable materials. In addition to negligible temperature fluctuations, these refrigerators are highly recommended to have quick temperature recoveries, given that their doors are continually opened and closed. Moreover, room temperature samples are sometimes stored directly at ultra-low refrigerators, avoiding the thawing of neighboring samples at the unit. Also, when this equipment fails to keep the optimal temperature range, audible and visual alarms are needed to ensure the contents are protected. These readouts are often pointed reads, but bulk data recording is highly recommended for monitoring if any abnormality is found in the samples at a certain point during the long-term storage.

Regarding ~4 °C refrigerators, they store most of the daily chemical reagents used in biomedical research. They span from cell culture compounds (culture medium, L-glutamine, non-essential amino acids, gelatin, etc.) to biomolecular reagents such as antibodies, enzyme-linked immunoassays (ELISAs), flow cytometry compounds, protein, and ribonucleic (RNA) extraction reagents.

Regular ~−20 °C freezers host many other different samples such as stable dilutions of deoxyribonucleic acid (DNA), complementary DNA (cDNA), primers, polymerase chain reaction (PCR) reagents, enzymes, some drugs, and vaccines. However, specific samples and more complex compounds, due to their nature, need ultra-low temperatures according to thermostability studies. Examples of these types of compounds are unstable RNA solutions, liquid nitrogen (LN_2) snap-frozen tissues, and protein extracts. All these samples contain a vast amount and variety of destructive enzymes called hydrolases. For instance, RNase (destruction of RNA), proteases (destruction of proteins), or phosphatases (dephosphorylation) can threaten the integrity of the samples. Conversely, at very low temperatures (around ~−80 °C), all these catalyzed degradation reactions happen more slowly; it is the same idea as freezing food in the food industry to keep it from spoiling.

Finally, on the other extreme of the temperature range, cell lines need temperatures under −135 °C for long-term storage to keep their properties and grow and multiply correctly when thawed. Cells are probably some of the most delicate tools used in a biomedicine laboratory. They need a controlled freezing gradient of −1 °C per min until they reach −80 °C to avoid ice formation injury and then transferred to freezers that reach between −196 and −135 °C temperatures. All methods used for this ultra-low temperature storage have advantages and disadvantages that must be assessed by the research center, such as the electric freezer with LN_2 back-up, liquid phase nitrogen freezers, and vapor phase nitrogen freezers. Unfortunately, this type of freezer is not the most common in research centers or hospitals because not many samples require these ultra-low temperatures. Moreover, they encompass many special safety issues such as the risk of asphyxiation, and thus dedicated liquid nitrogen storage areas get dangerous restricted access rooms.

As an example, recently, with the appearance of the Sars-CoV-2 pandemic and vaccines, a problem that has been present in our society for a long time has become evident, deep freezing. Vaccines can be produced by numerous different mechanisms giving rise to the wide offer of effective, safe, and lasting vaccines. They can be inactivated, live-attenuated, viral vectors, toxoids, messenger RNA (mRNA), and subunits/recombinant/polysaccharides/conjugates vaccines. Depending on the development mechanism, their storage requirements and shelf lives may be different. Storage and logistical transport at refrigerators or freezers from −20 to 8 °C are virtually feasible everywhere, except in countries with limited access to refrigerators. This is the case of Johnson & Johnson (2 to 8 °C), Astra Zeneca (2 to 8 °C), and Moderna (from −25 to −15 °C, but it can be stored during one month at 2 to 8 °C) vaccines, which can face better the already established infrastructure of hospitals and vaccination centers and fit better the current logistical cold chains. However, Pfizer-BioNTech vaccine based on mRNA fails to be kept at −20 °C like Moderna, which has some stabilizing technologies such as lipid nanoparticles that allow better conservation and prevents RNA degradation as the temperature rises.

However, these technologies, together with modified stabilized nucleosides, are far from solving the cold-chain problem after the stress testing drugmaker companies have performed and reported. Also, −80 °C freezers (Fig. 1) are not a daily basis

Fig. 1 Biomedical refrigerator

device that can be found in hospitals at huge numbers, hence hindering, even more, the administration of these doses. The so-called pizza boxes have been designed by Pfizer to transport 195 vials in each box in payloads with dry ice, which can store vials in the package at −80 °C for some days. This transport method is widely used to ship frozen items. Nevertheless, dry ice production and transport logistics cannot harbor these massive deliveries of doses. Additionally, dry ice cannot provide a constant and uniform temperature around the parcel, jeopardizing the integrity of the mRNA and hence provoking a life-threatening situation in a global pandemic.

The following chapter will recapitulate the main characteristics, advantages, and drawbacks of different ultra-low temperature configurations. The aim is to shed some light on not widely used but existing technologies that can provide better and cost-effective efficiencies in low temperature freezers, aiming to provide alternatives to gold-standard refrigeration devices.

2   Configurations

There are numerous technologies in refrigeration, and they are classified then according to configurations. The concept of configurations refers to arranging the components into a circuit, obtaining different groups with typical peculiarities.

The most common configurations in refrigeration are the following: simple, indirect expansion, cascade, intermediate exchanger, ejector, multi-stage, absorption, and auto-cascade.

2.1   Single-Stage

This configuration is the basis for all configurations and contains the essential elements of the refrigeration machine for a reliable and safe operation. Figure 2 shows its four main elements.

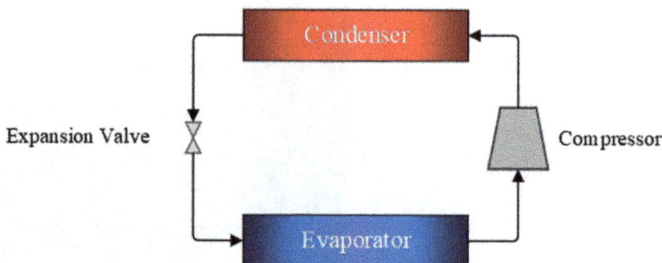

Fig. 2   Scheme of a single-stage cycle

Fig. 3 Scheme of an
indirect expansion cycle

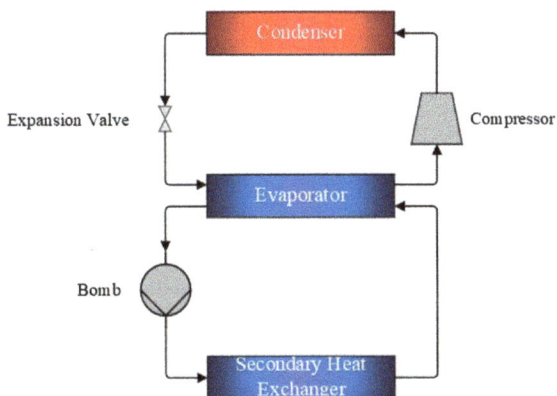

The compressor has the function of compressing the refrigerant, absorbing the energy supplied externally. After the compressor, the refrigerant passes through the condenser, an exchanger that generally expels the heat to the environment, space, or fluid. In contrast, the refrigerant is cooled down by changing the phase from superheated gas to subcooled liquid to pass through the throttling device mechanism, which expands to generate a pressure drop, be it an expansion valve, capillary tube, or another. Finally, the refrigerant passes through the evaporator, which is another exchanger. Still, this process simultaneously transfers heat from a space or fluid to the refrigerant, turning it into a vapor state (with a certain degree of overheating) and returning to the compressor.

2.2 Indirect Expansion

This configuration requires an additional heat transfer fluid (at low temperatures, mixtures based on glycol or CO_2), whose only function is carrying the cooling effect to an evaporator located in another position. When using a flammable, highly toxic or high global warming potential (GWP) refrigerant, the refrigerant charge is reduced and contained in the primary circuit. In its most basic configuration, apart from the evaporator of the primary vapor compression circuit, it has an evaporator that exchanges heat between the two circuits, a pump for the secondary fluid and the external heat exchanger, as shown in Fig. 3.

2.3 Cascade

The cascade system consists of different vapor compression circuits linked by an additional exchanger (cascade heat exchanger in Fig. 3). Different refrigerants may be

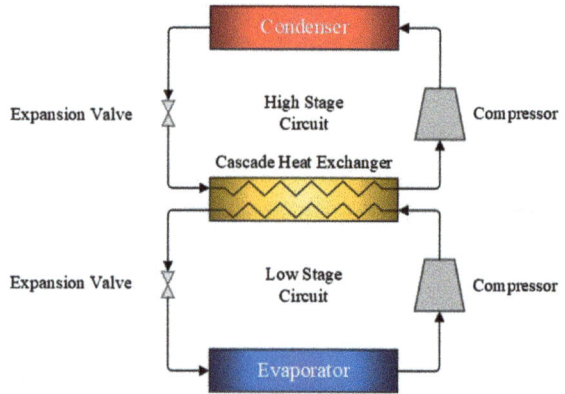

Fig. 4 Scheme of a cascade cycle

used in each circuit without mass exchange according to the most suitable temperature range, as shown in Fig. 4.

According to the optimal temperature range for performance (in operational and energetic terms), one of the most significant benefits of this system is using different refrigerants in each circuit. The higher stage circuit is subjected to higher pressures and temperatures. In contrast, the refrigerant in the low stage circuit covers lower temperatures at acceptable pressures (they are known as the high and low temperature circuits, respectively). The compression pressure ratio and discharge temperatures are significantly reduced for high-temperature lifts, but their construction is more complex than single-stage cycles. Therefore, it offers several possibilities for optimization and increasing energy efficiency.

Usually, the interest is in the low stage circuit's refrigeration, for which an evaporator is used in the cooling process. Contrary to indirect refrigeration systems, both circuits are vapor compression cycles, so the fluid goes through the four main processes typically observed.

2.4 Intermediate Heat Exchanger (IHX)

As shown in Fig. 5, this configuration introduces an additional heat exchanger (different from those mentioned in the indirect expansion and cascade cycles, Sects. 2.2 and 2.3, respectively), achieving two fundamental effects. On the one hand, it superheats the refrigerant before the compressor inlet, ensuring it is in vapor phase. On the other hand, it subcools the refrigerant in the liquid line, increasing the refrigeration effect (evaporator enthalpy difference) and ensuring it is in the liquid phase. However, it increases the compressor suction temperature, decreases the mass flow rate, and increases discharge temperature.

The use of the intermediate heat exchanger (IHX, also known as liquid-to-suction heat exchanger) does not guarantee an increase in COP (coefficient of performance).

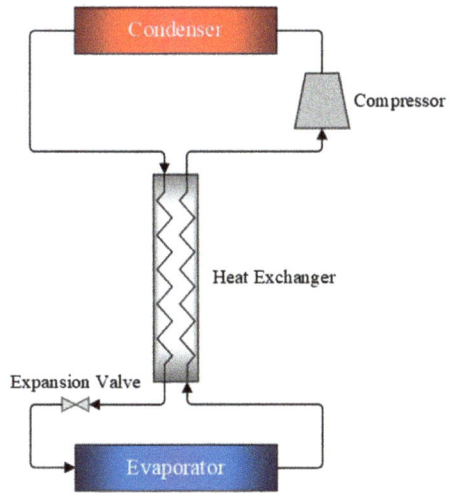

Fig. 5 Scheme of a cycle with intermediate heat exchanger

It is only observed when the increase in the refrigeration effect is more significant than resulting in specific compression work. The final effect depends on the refrigerant and the operating conditions, and it should be studied on a case-by-case basis. It is commonly seen in other applications due to its simple modification. In small capacity systems, it can be observed in the form of a capillary tube-suction line heat exchanger. The capillary tube is rolled with the suction line, so pressure drop and subcooling in one side and superheating are co-occurring.

2.5 Ejector

The ejectors have the role of occupying the function of the expander component of the circuit, as shown in Fig. 6.

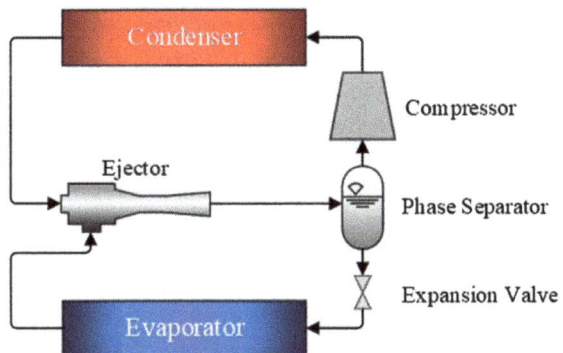

Fig. 6 Scheme of a cycle with ejector

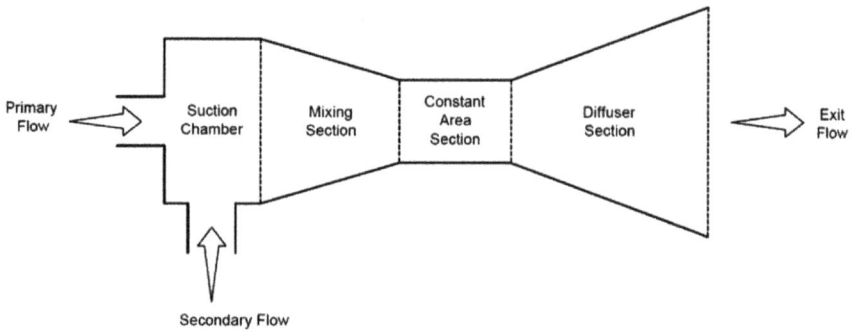

Fig. 7 Schematic representation of an ejector and its main parts

The ejector consists of different parts described in Fig. 7. The first part is the nozzle, which is responsible for converting the potential energy of the primary fluid (the one that comes from the condenser) into kinetic energy, a divergent-convergent transformation zone.

In turn, the rest of the ejector can be divided into three zones: a converging section that works as a mixing chamber, a constant area section, and finally, a diffuser as a diverging section. In the convergent part, the primary and secondary fluids are mixed and directed to the second part, taking advantage of the kinetic energy. The fluid passes through the second part, a zone of a constant area where the pressure increases, and the flow becomes subsonic to pass to the diffuser, an inverted cone. In this zone, the high pressure and temperature flow stream (that of the primary fluid) is transferred to the low pressure and temperature stream (secondary), suctioning the secondary flow so that it mixes in the step described above. The fluid experiences a significant pressure loss in this last stage until it reaches the target outlet pressure.

2.6 Multi-stage

It uses several compressors, dividing the compression process into several stages, allowing a higher temperature lift. The refrigerant discharged by the low stage compressor is desuperheated by mixing it with the refrigerant from another part of the circuit in the liquid or vapor phase using an additional heat exchanger or a phase separator. Subsequently, it goes through the high stage compressor, thus decreasing the partial pressure ratio and final discharge temperature more than if it just comprised a compressor. The total electricity consumption is usually reduced for higher temperature lifts. Figure 8 shows an example of a vapor injection two-stage cycle, in which a part of the refrigerant is injected at an intermediate pressure in vapor phase. The main drawback is that the total mass flow rate compressed by the high stage compressor is higher than single-stage cycles.

Fig. 8 Scheme of a
multi-stage cycle

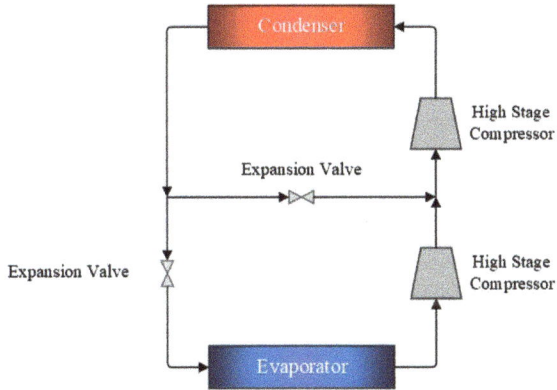

2.7 *Absorption*

In this configuration, like in the ejector, one of the main components of the cycle
is replaced—in this case, the compressor. For this configuration, the compressor is
substituted by a subcircuit called the absorption system.

The absorption system can work based on two processes, as explained by Srikhirin
et al. [1]. In the first process, two vessels are connected, having a binary solution of a
working fluid formed by one part of the refrigerant and the other absorbent, each in
a container. The absorbent in the container takes the refrigerant vapor from the other
container provoking a pressure drop. While the refrigerant vapor is being absorbed,
the temperature of the remaining refrigerant decreases due to its vaporization. Consequently, it cools the refrigerant container. At the same time, the solution within the
container that previously contained only the absorbent becomes more diluted due to
the higher content of absorbed refrigerant.

When the solution becomes saturated, the refrigerant is extracted from the diluted
solution, heat is applied to the vessel, and the refrigerant vapor returns to the other
container. It condenses when transferring heat to the surroundings.

Nevertheless, these two processes do not have to occur in the same vessels. Since
the first process occurs at a higher pressure than the second, an expansion valve and
a pump are necessary, Fig. 9.

As a result of this system, vessels on the right side have the role of a condenser
and evaporator, yielding and absorbing energy, respectively. Thus, it has the same
effects as in the simple cycle described above in this chapter.

This system has the disadvantage of offering very low COPs because of being an
inefficient system, partly due to the need to use thermal energy in the compression
system, which is inconvenient. However, it is not based on synthetic refrigerants
as most vapor compression cycles neither require a compressor, which consumes
electricity and produces noise and vibrations and increases the maintenance cost of
the installation.

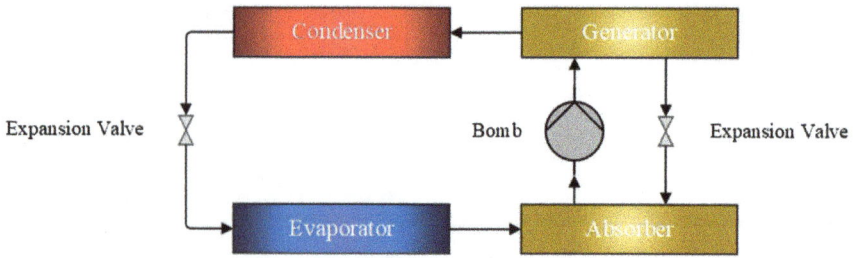

Fig. 9 Absorption cycle

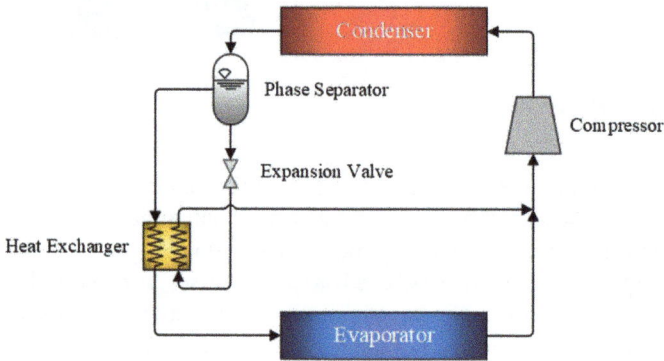

Fig. 10 Scheme of an auto-cascade cycle

2.8 Auto-Cascade

The auto-cascade configuration follows the premise of the cascade of separating the refrigerants for different temperatures-pressures required. A cycle is designed in which the refrigerants flow through separate lines due to their thermodynamic properties and later put them back together. Still, this system is not achieved by separating them in different cycles but by causing their separation in a single physical cycle. Figure 10 shows a basic scheme of a simple auto-cascade cycle configuration.

The benefits of the cascade system are varied, allowing, for example, to have a more significant temperature difference without the need of separating the refrigerants in different cycles. The main drawbacks are the choice of refrigerants and the difficulty when designing the cycles so that refrigerants separate.

2.9 CO_2 Sublimation

Recently, a new type of cycle appeared that allows CO_2 to work at temperatures below its triple point. This point occurs when the CO_2 reaches a temperature of −

57.57 °C and 5.1 bar of pressure, converging the three states simultaneously. Below this point, the refrigerant would theoretically go from the gaseous state directly to the solid state. But, CO_2 would be able to reach up to approximately -78 °C while maintaining a metastable liquid phase.

This can be achieved through cycle modifications such as those already exposed above, for example, double compression. Additionally, the introduction of an intermediate exchanger or the use of cascade systems can also be considered.

3   Ultra-Low Temperature Configurations' Recent Findings

A review of the systems used in ultra-low temperature refrigeration and refrigerants can be read in the article by Mota-Babiloni et al. [2]. Cascade and auto-cascade systems are the most widely used in commercial low or ultra-low temperature freezers. This is because they present considerable advantages when tackling high-temperature lifts, as shown by Mumanachit et al. [3], who compared the cascade system with the double stage. They concluded that the cascade is more efficient below the optimal point of the COP and more cost-effective below -46.2 °C.

After that, all the configurations reviewed come from modifications of the basic cascade refrigeration cycle.

3.1   Two-Stage Cascade

Relative to simple cascade cycles, Di Nicola et al. [4] compared different hydrofluorocarbons (HFCs) to the natural refrigerant R-717 at temperatures of -70 °C, concluding that ammonia is around 5% superior in terms of COP.

Lee et al. [5] investigated the optimum condensation temperature to maximize COP and decrease energetic losses. They observed that COP increases with increasing evaporation but decreases with increasing condensation temperature and temperature variation. Alberto Dopazo et al. [6] confirmed the influence of operating temperature variations in the COP. They quantified a 70% increase in the COP when the evaporator temperature goes from -55 to -30 °C. When the condenser temperature rises from 25 to 50 °C, it causes a 45% drop in the COP. In the same way, they found that an IHX temperature increase from 3 to 6 °C reduces COP by 9%. A proper IHX optimization is essential to have an optimal operating temperature, according to the research of Sun et al. [7]. It also observed that R-41 is a good substitute for R-23.

Experimentation has also been carried out in cascades with not-in-kind refrigerants such as HFE-7000 and HFE-7100. Adebayo et al. [8] concluded that the pair that obtains the highest COP is R-717/R-744, while the opposite is observed with HFE-7100/R-744.

3.2 Cascade with Intermediate Heat Exchanger

The same research mentioned above by Di Nicola et al. proposes using an IHX in the low stage (LS) circuit, concluding that it could benefit energy performance. It is not until years later, when Bhattacharyya et al. [9] studied a cascade system with IHXs, positioning one in the high stage circuit (HS) and the other in LT, trying to optimize it and first observed that the performance of the other system is independent of the exchangers' performance.

Liu et al. [10] carried out a theoretical and experimental study of a cascade system with an IHX in the LS cycle but could turn it into dual. They observed that the COP was lower if only the LS IHX worked, so the dual cycle has a higher potential for energy performance improvement. Also, Dubey et al. [11] proposed a cascade system with two IHXs, observing that the COP with HS IHX was greater than that of the LS. They proved that the higher the temperature difference in the cascade heat exchanger, the lower the COP. Also, the lower the temperature of the LS evaporator, the lower the COP.

3.3 Cascade with Ejector

Aghazadeh Dokandari et al. [12] proposed two ejectors in a two-stage cascade, placing one in each subcircuit. They reported a 7% COP increase in comparison with a standard cascade system. Li et al. [13] compared a cascade circuit with an LS IHX and the same circuit but adding an LS ejector. They concluded that the ejector reduces energy consumption by 4.8%, but they will extend this study in the future.

3.4 Cascade with Three Stages

Another way to cascade is to use even more than two stages. Johnson et al. [14] developed a three-stage cascade with dynamic control, showing effectiveness against flow disturbances in the secondary fluid. A comprehensive comparison of refrigerants is found in the article by Sun et al. [7] in which the following groups of refrigerants are recommended: R-1150/41/717, R-1150/41/152a, R-1150/41/161, R-1150/170/717, R-1150/170/152a, and R-1150/170/161.

4 Conclusions

The slower development of ultra-low temperature refrigeration compared to other refrigeration and heat pump applications makes evident the gap existing in this field yet to be discovered and analyzed. Existing challenges are related to difficulties in proper and long-life system operation and improvement in energy efficiency. The need of preserving vaccines at $-80\ ^\circ\text{C}$ is increasing the attention devoted to refrigeration at extreme or not-in-kind conditions, such as ultra-low temperature.

Most of the commercial ultra-low temperature freezers today are based on two-stage cascades. Ejectors in refrigerators are still developing early, even in more typical refrigeration applications such as commercial supermarket refrigeration or air conditioning. Therefore, it is still far from being a reality in this application. On the other side, the other promising configuration, auto-cascades, is preferred for lower evaporating temperatures because its complexity is still seen as a significant problem. Three-stage cascades are not considered a firm candidate for this temperature lift, in which a two stage is considered a good trade-off between energy efficiency and complexity/cost. Only the internal heat exchanger is considered as an improvement for two-stage cascades.

Several projects have concluded over the last few years and pointed out that refrigeration at these temperatures should be based on cascade cycles. However, the sector has not been developed and studied in-depth beyond other configurations, three-stage cascades, or more complex cycles, optimizing each circuit's working fluids. Many refrigerants have been theoretically considered in research projects, but it seems that the hydrocarbons R-290 in high stage and R-170 are going to phase-down HFCs traditionally used. Given the current opposition of some countries and sectors to the use of 4th generation HFO and HCFOs, it is likely that they will not enter the ultra-low temperature market.

Acknowledgements The authors acknowledge the funding provided by the program "Proyectos de I+D+I 2020" of the Spanish Ministry of Science and Innovation (PID2020-117865RB-I00). Adrián Mota-Babiloni acknowledges the postdoctoral contract "Juan de la Cierva-Incorporación 2019" of the Spanish State Research Agency (IJC2019-038997-I). Carla Espinós-Estévez acknowledges the "la Caixa" Foundation (ID 100010434) fellowship under the code LCF/BQ/DR19/11740012.

References

1. Srikhirin P, Aphornratana S, Chungpaibulpatana S (2000) A review of absorption refrigeration technologies. Renew Sustain Energy Rev 5(4):343–372. https://doi.org/10.1016/S1364-032 1(01)00003-X
2. Mota-Babiloni A et al (2020) Ultralow-temperature refrigeration systems: configurations and refrigerants to reduce the environmental impact. Int J Refrig 111:147–158. https://doi.org/10. 1016/j.ijrefrig.2019.11.016
3. Mumanachit P, Reindl DT, Nellis GF (2012) Comparative analysis of low temperature industrial refrigeration systems. Int J Refrig 35(4):1208–1221. https://doi.org/10.1016/j.ijrefrig.2012. 02.009
4. Di Nicola G, Giuliani G, Polonara F, Stryjek R (2005) Blends of carbon dioxide and HFCs as working fluids for the low-temperature circuit in cascade refrigerating systems. Int J Refrig 28(2):130–140. https://doi.org/10.1016/j.ijrefrig.2004.06.014

5. Lee TS, Liu CH, Chen TW (2006) Thermodynamic analysis of optimal condensing temperature of cascade-condenser in CO_2/NH_3 cascade refrigeration systems. Int J Refrig 29(7):1100–1108. https://doi.org/10.1016/j.ijrefrig.2006.03.003

6. Alberto Dopazo J, Fernández-Seara J, Sieres J, Uhía FJ (2009) Theoretical analysis of a CO_2-NH_3 cascade refrigeration system for cooling applications at low temperatures. Appl Therm Eng. https://doi.org/10.1016/j.applthermaleng.2008.07.006

7. Sun Z, Wang Q, Dai B, Wang M, Xie Z (2019) Options of low global warming potential refrigerant group for a three-stage cascade refrigeration system. Int J Refrig 100:471–483. https://doi.org/10.1016/j.ijrefrig.2018.12.019

8. Adebayo V, Abid M, Adedeji M, Dagbasi M, Bamisile O (2021) Comparative thermodynamic performance analysis of a cascade refrigeration system with new refrigerants paired with CO_2. Appl Therm Eng 184:116286. https://doi.org/10.1016/j.applthermaleng.2020.116286

9. Bhattacharyya S, Garai A, Sarkar J (2009) Thermodynamic analysis and optimization of a novel N_2O-CO_2 cascade system for refrigeration and heating. Int J Refrig 32(5):1077–1084. https://doi.org/10.1016/j.ijrefrig.2008.09.008

10. Liu XF, Liu JH, Zhao HL, Zhang QY, Ma JL (2012) Experimental study on a −60 °C cascade refrigerator with dual running mode. J Zhejiang Univ Sci A 13(5):375–381. https://doi.org/10.1631/jzus.A1100107

11. Dubey AM, Kumar S, Das Agrawal G (2014) Thermodynamic analysis of a transcritical CO_2/propylene (R744-R1270) cascade system for cooling and heating applications. Energy Convers Manag 86:774–783. https://doi.org/10.1016/j.enconman.2014.05.105

12. Aghazadeh Dokandari D, Setayesh Hagh A, Mahmoudi SMS (2014) Thermodynamic investigation and optimization of novel ejector-expansion CO_2/NH_3 cascade refrigeration cycles (novel CO_2/NH_3 cycle). Int J Refrig 46:26–36. https://doi.org/10.1016/j.ijrefrig.2014.07.012

13. Li Y, Yu J, Qin H, Sheng Z, Wang Q (2018) An experimental investigation on a modified cascade refrigeration system with an ejector. Int J Refrig 96:63–69. https://doi.org/10.1016/j.ijrefrig.2018.09.015

14. Johnson N, Baltrusaitis J, Luyben WL (2017) Design and control of a cryogenic multi-stage compression refrigeration process. Chem Eng Res Des 121:360–367. https://doi.org/10.1016/j.cherd.2017.03.018

Lightning Source UK Ltd.
Milton Keynes UK
UKHW020615020323
417913UK00002B/7

9 789811 691034